# Complete Guide to Videocassette Recorder Operation and Servicing

# Complete Guide to Videocassette Recorder Operation and Servicing

Mr. Shiung Y. Lo
P.O. Box 10567
Chicago, Ill. 60610

## JOHN D. LENK

*Consulting Technical Writer*

PRENTICE-HALL, INC., *Englewood Cliffs, N.J. 07632*

*Library of Congress Cataloging in Publication Data*

Lenk, John D.
  Complete guide to videocassette recorder
operation and servicing.

   Includes index.
   1. Video tape recorders and recording.
2. Video type recorders and recording—
Maintenance and repair.   I. Title.
TK6655.V5L46       621.388′33       81-22750
ISBN 0-13-160820-7                  AACR2

Editorial/production supervision by *Ellen Denning*
Interior design by *Ellen Denning*
Manufacturing buyer: *Gordon Osbourne*

© 1983 by Prentice-Hall, Inc., Englewood Cliffs, N.J. 07632

All rights reserved. No part of this book
may be reproduced in any form or
by any means without permission in writing
from the publisher.

Printed in the United States of America

10  9  8  7  6  5

ISBN  0-13-160820-7

PRENTICE-HALL INTERNATIONAL, INC., *London*
PRENTICE-HALL OF AUSTRALIA PTY. LIMITED, *Sydney*
PRENTICE-HALL CANADA INC., *Toronto*
PRENTICE-HALL OF INDIA PRIVATE LIMITED, *New Delhi*
PRENTICE-HALL OF JAPAN, INC., *Tokyo*
PRENTICE-HALL OF SOUTHEAST ASIA PTE. LTD., *Singapore*
WHITEHALL BOOKS LIMITED, *Wellington, New Zealand*

To Irene, Lady of the Lake
and
Lambie, Keeper of the Cream Carmel

# Contents

**PREFACE** xi

# 1 INTRODUCTION TO VIDEOCASSETTE RECORDERS 1

1-1. VCR versus VTR, *2*

1-2. Basic Black-and-White Television Broadcast System, *2*

1-3. Basic Color-television Broadcast System, *13*

1-4. Basic Magnetic Recording Principles, *18*

1-5. Problems in Recording the Video Signal on a VCR, *22*

1-6. Relationship of Rotating Heads to Video Fields and Frames, *26*

1-7. Relationship of Luminance and Chrominance Signals in a VCR, *30*

- 1-8. Introduction to the Beta System, *31*
- 1-9. Introduction to the VHS System, *52*
- 1-10. Basic VCR Operating Procedures and Record Lockout Provisions, *70*

## 2 VCR TEST EQUIPMENT AND TOOLS 73

- 2-1. Safety Precautions in VCR Service, *74*
- 2-2. Signal Generators, *78*
- 2-3. Color Generators, *87*
- 2-4. Oscilloscopes, *90*
- 2-5. Meters, *92*
- 2-6. Frequency Counters, *93*
- 2-7. Miscellaneous Test Equipment, *94*
- 2-8. Tools and Fixtures Required for Service, *96*

## 3 TYPICAL BETA VCR CIRCUITS 100

- 3-1. Introduction to Beta Circuits, *101*
- 3-2. Beta Video Circuits, *101*
- 3-3. Beta Servo Circuit I, *120*
- 3-4. Beta Servo Circuit II, *136*
- 3-5. Beta System Control Circuit, *144*
- 3-6. Beta Audio Circuit, *156*
- 3-7. Beta Timer Circuit, *160*
- 3-8. Beta Tuner Circuit, *165*

## 4 TYPICAL VHS VCR CIRCUITS 171

4-1. VHS Luminance Signal Circuits during Record, *172*

4-2. VHS Luminance Signal Circuits during Playback, *178*

4-3. VHS Color Signal Circuits during Record, *188*

4-4. VHS Color Signal Circuits during Playback, *189*

4-5. VHS Servo System, *192*

4-6. VHS System Control Circuit, *206*

## 5 MECHANICAL OPERATION OF TYPICAL VCRS 222

5-1. Typical Beta Mechanical Operation, *223*

5-2. Typical VHS Mechanical Operation, *242*

## 6 TYPICAL ADJUSTMENT, CLEANING, LUBRICATION, AND MAINTENANCE PROCEDURES 245

6-1. Beta Electrical Adjustments, *245*

6-2. Beta Mechanical Adjustments, *266*

6-3. Cleaners, Lubrication Oils, and Maintenance Timetables, *276*

6-4. Cleaning and Lubrication Procedures, *277*

6-5. VHS Electrical Adjustments, *287*

6-6. VHS Mechanical Adjustments, *307*

## 7 TROUBLESHOOTING AND SERVICE NOTES 316

7-1. The Basic Troubleshooting Functions, *316*

7-2. The Universal Troubleshooting Approach, *318*

x    Contents

7-3.   Trouble Symptoms, *322*

7-4.   Localizing Trouble, *323*

7-5.   Isolating Trouble to a Circuit, *326*

7-6.   Locating a Specific Trouble, *329*

7-7.   Typical VCR Installation, *332*

7-8.   Typical VCR Operating Procedures, *340*

7-9.   Basic VCR Troubleshooting Procedures, *347*

7-10.  VCR Service Notes, *349*

**INDEX    359**

# Preface

The main purpose of this book is to provide a simplified, practical system of operation and service for the many types and models of videocassette recorders (VCRs). It is virtually impossible in one book to cover detailed operation and service for all VCRs. Similarly, it is impractical to attempt such comprehensive coverage, since rapid technological advances soon make such a book's details obsolete.

To overcome the problem, this book concentrates on a *basic approach to VCR operation and service,* an approach that can be applied to any VCR (both those now in use and those to be manufactured in the future). The approach here is based on the techniques found in the author's *Handbook of Practical Solid-State Troubleshooting, Handbook of Basic Troubleshooting, Handbook of Simplified Television Service,* and *Handbook of Practical CB Service.*

Chapter 1 is devoted to the basics of videocassette recorders, including the basics of television and magnetic recording. With basics established, the chapter goes on to describe operation of both Beta and VHS (the two most popular forms of home-entertainment VCRs).

Chapter 2 describes test equipment and tools required for practical VCR operation and service. The descriptions include many examples of the special tools and fixtures required for each model of VCR. The discussion describes how the features and outputs found in present-day test equipment relate to specific problems in VCR test and service.

Chapter 3 describes the theory of operation for a number of Beta VCR circuits. By studying the circuits found in this chapter, you should have no difficulty in understanding the schematic and block diagrams of similar Beta VCRs.

Circuit descriptions are supplemented with partial schematic and block diagrams that show such important areas as signal flow paths, input/output, adjustment controls, test points, and power source connections (the areas most important in service). Chapter 4 provides similar coverage for a cross section of VHS circuits.

Chapter 5 describes operation for the mechanical sections of typical VCRs. Both Beta and VHS systems are covered. By studying this information you should have no problem in understanding the mechanical operation of similar VCRs. This understanding is essential for logical troubleshooting and service, particularly since most VCR faults are the result of failure (or tampering) with the mechanical section.

Chapter 6 describes typical adjustment, cleaning, lubrication, and maintenance procedures for VCRs. Both Beta and VHS systems are described. To show what typical adjustments involve, the chapter describes complete electrical and mechanical adjustment procedures for sample Beta and VHS units, as recommended by the manufacturers. Using these examples, you should be able to relate the procedures to a similar set of adjustment points on most VCRs. Where it is not obvious, the chapter also describes the purpose of the procedure. The waveforms measured at various test points during adjustments are also included here. By studying these waveforms, you should be able to identify typical signals found in most VCRs, even though the signals may appear at different points for your particular unit.

Chapter 7 describes a series of troubleshooting and service notes for a cross section of VCRs, both Beta and VHS. Both electrical and mechanical problems are discussed. Typical operating and installation procedures are also included. These notes represent the combined experience and knowledge of many VCR service specialists and managers.

Many professionals have contributed their talent and knowledge to the preparation of this book. The author acknowledges that the tremendous effort to make this book such a comprehensive work is impossible for one person, and he wishes to thank all who have contributed directly and indirectly. The author wishes to give special thanks to the following: Thomas Roscoe of Hitachi; John Bailey of Burson-Marsteller for JVC; Magnavox; E. I. Sheppard of Mitsubishi; James Knox of Panasonic; Thomas Lauterback of Quasar; J. W. Phipps of RCA; Donald Woolhouse of Sanyo; Larry Benson of Sony; Bob Conroy of Technicolor; and Sara Wiebort of Zenith.

The author extends his gratitude to Dave Boelio, Hank Kennedy, John Davis, Jerry Slawney, Art Rittenberg, and Don Schaefer of Prentice-Hall. Their faith in the author has given him encouragement, and their editorial/marketing expertise has made many of the author's books best sellers. The author also wishes to thank Joseph A. Labok of Los Angeles Valley College for his help and encouragement.

JOHN D. LENK

# Complete Guide
to Videocassette Recorder
Operation and Servicing

# 1

# Introduction to Videocassette Recorders

This chapter is devoted to the basics of videocassette recorders (VCRs). To understand the operation of VCRs, it is essential that you first understand the basics of television and magnetic tape recording. On the assumption that you may not know how television and magnetic tape recorders work, or that you need a refresher, we start with descriptions of TV broadcast and reception for black and white as well as color programs, and then go on to cover magnetic tape recording principles. These descriptions of television and recorders are kept to the block diagram level since you are interested in VCRs. It is not intended that this introduction provide a complete course in TV or tape recording, merely a summary.

With the basics established, we then go on to describe the operation of the two types of home-entertainment VCRs in common use: Beta and VHS. The Beta system, developed and manufactured by Sony Corporation as Betamax, is also found in compatible systems manufactured by Sanyo and Zenith. The VHS system is used by Hitachi, JVC, Magnavox, Mitsubishi, Panasonic, Quasar, RCA, and Technicolor. The two systems, although similar in operation and identical in overall purpose, are not compatible. The videocassettes of the two systems are not physically interchangeable. Even if you wound a VHS-recorded tape on a Beta cassette for playback, or vice versa, there would be no playback, because different frequencies and recording techniques are involved.

Before we go into any descriptions, let us define the type of VCR covered in this book.

## 1-1 VCR VERSUS VTR

A VCR is a form of VTR, or video tape recorder. VTRs have been used in the television industry since about 1955 to record programs. Based on the same electromagnetic principles as an audio tape recorder, a VTR has the ability to record on magnetic tape and play back electronic video signals coming from a video camera or a TV station. This process is called *video recording,* and has many advantages over older TV recording processes and over motion picture film. For example, both picture and sound (video and audio) can be recorded at the same time, there is no need for developing or processing the tape after recording, long-term preservation is possible, many recordings can be made using the same tape over again, continuous recording and playback are possible over relatively long periods of time, and a VTR is relatively compact and easy to use.

Figure 1-1 shows the evolution of magnetic tape used with VTR equipment. At the left is shown the 2-in. reel introduced to professional broadcasters in 1955 and still in use. One-inch tape was developed in 1964, but VTRs were still limited to open reel (reel-to-reel) tape at that time. Videotape in cassettes came into being in the early 1970s in the $\frac{3}{4}$-in. "U" format. In the mid-1970s, $\frac{1}{2}$-in. tape (such as the T-60 VHS tape shown at the right in Fig. 1-1) made its debut, leading to the first true portable equipment and to mass-market consumer VCRs. The $\frac{1}{2}$-in. consumer VCR is the type of equipment we concentrate on in this book. The $\frac{1}{4}$-in. videotape contained in the cassette in the foreground, developed by Technicolor, weighs less than 2 ounces and is used in a portable VCR weighing less than 7 pounds. This VCR is generally for use with portable

**FIGURE 1-1.** Evolution of magnetic tape used with VTR equipment. (Courtesy of Technicolor.)

video cameras, but it can also be used with a home TV receiver. Playing time is limited to $\frac{1}{2}$-hour for the $\frac{1}{4}$-in. cassette shown.

Figures 1-2a to 1-2g show some typical VCRs (Beta and VHS) described in this book. Although the VCRs use the same principles as an audio tape recorder, operation of a VCR is much more complex. In addition to audio information, the VCR must also record and play back video information (both black and white, and color) as well as control or synchronization information.

Figure 1-3 shows the basic functional sections of the VCR which include a tuner (VHF and UHF), an RF section, a timer section, and a mechanical section (including tape transport, stationary audio head, stationary control head, and rotating video heads). Note that the same heads used for record are also used for playback. Also, there are two rotating heads for video record/playback, whereas there is only one audio head and one control head. This configuration is typical for consumer VCRs.

We discuss full circuit details throughout the book. For now, let us consider the functions of the VCR sections shown in Fig. 1-3. The tuner section is similar to tuners found in TV sets, and functions to convert broadcast signals picked up by the antenna to frequencies and formats suitable for use by the VCR. All TV channels, 2 through 83, are covered. Typically, tuner output to the record circuit is 1 V (peak to peak) for video, and 0 dB (0.775 V) for audio.

The record circuits function to convert tuner output into electrical signals used by the heads to record the corresponding information on the magnetic tape along tracks. There are three tracks as shown (audio or sound, video or picture, and control or synchronization). Note that the audio and control tracks are parallel to the tape, whereas the video track is diagonal. As is explained in the following sections, the video track is recorded diagonally to increase tape writing speed, and thus to increase the frequency range necessary to record video signals.

The playback circuits function to convert information recorded on the tracks and picked up by the heads into electrical signals used to modulate the RF section. In the simplest of terms, the RF section is a miniature TV broadcast station operating on an unused TV channel (typically channel 3 or 4, but possibly 5 or 6). The output of the RF section is applied to the TV set.

During normal operation, you select the channel you wish to record using the VCR tuner controls. This need not be the channel being watched on the TV set. Similarly, the TV set need not be on while recording on the VCR. You then turn on the timer and the program or programs are recorded. Typically, the VCR will record up to about 6 hours, depending on the timer setting and tape speed.

When you are ready to play back the recorded program, you select the appropriate unused TV channel (3 or 4) using the TV-set channel controls. Then you turn on the timer and play back the program using the TV set as a display device.

Many present-day VCRs also provide for recording directly from a video camera and microphone. In this case, the video camera and microphone are con-

**FIGURE 1-2a.** Typical Beta-type VCR. (Courtesy of Sanyo Electric, Inc.)

**FIGURE 1-2b.** Typical Beta-type VCR. (Courtesy of Zenith Radio Corporation.)

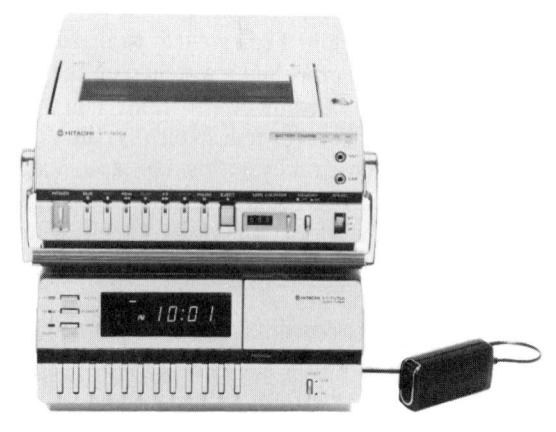

**FIGURE 1-2c.** Typical VHS-type VCR. (Courtesy of Hitachi Sales Corporation of America.)

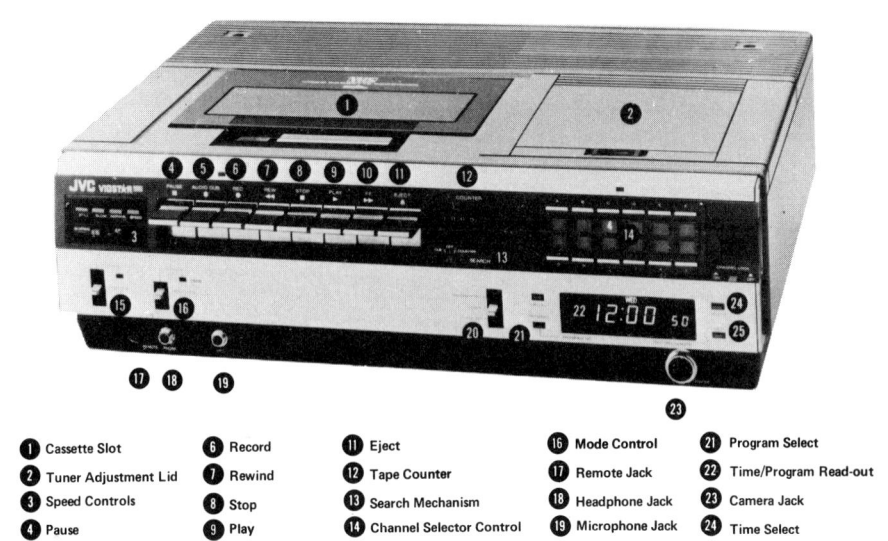

| | | | | | |
|---|---|---|---|---|---|
| ❶ Cassette Slot | ❻ Record | ⓫ Eject | ⓰ Mode Control | ㉑ Program Select |
| ❷ Tuner Adjustment Lid | ❼ Rewind | ⓬ Tape Counter | ⓱ Remote Jack | ㉒ Time/Program Read-out |
| ❸ Speed Controls | ❽ Stop | ⓭ Search Mechanism | ⓲ Headphone Jack | ㉓ Camera Jack |
| ❹ Pause | ❾ Play | ⓮ Channel Selector Control | ⓳ Microphone Jack | ㉔ Time Select |
| ❺ Audio Dub | ❿ Fast Forward | ⓯ Power Switch | ⓴ Clock Set | ㉕ Set |

**FIGURE 1-2d.** Typical VHS-type VCR. (Courtesy of JVC.)

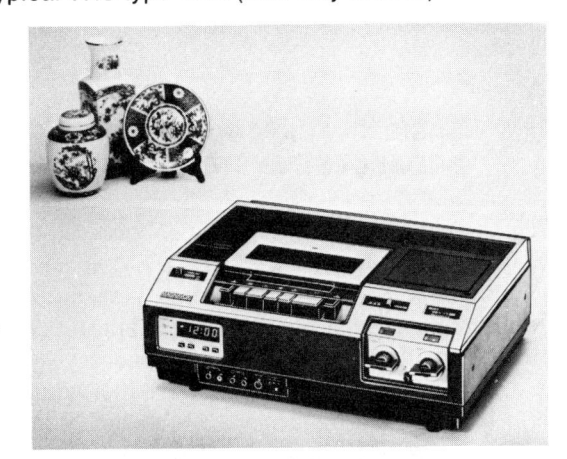

**FIGURE 1-2e.** Typical VHS-type VCR. (Courtesy of N.A.P. Consumer Electronics Corporation-Magnavox Product.)

**FIGURE 1-2f.** Typical VHS-type VCR. (Courtesy of Mitsubishi Electric Sales America, Inc.)

**FIGURE 1-2g.** Typical VHS-type VCR. (Courtesy of Quasar Company.)

nected into the VCR at the same point where the tuner output is applied to the record circuits. Once the material is recorded, operation of the VCR during playback is the same, no matter what the recording source (off-air programming or video camera).

Now that we have established the type of VCR covered in this book, let us review the principles of television broadcast and reception, as well as magnetic recording.

## 1-2 BASIC BLACK-AND-WHITE TELEVISION BROADCAST SYSTEM

As shown in Fig. 1-4, the basic TV broadcast system consists of a transmitter capable of producing both AM and FM signals, a television video camera, and a microphone. TV program sound (or audio) is transmitted through the microphone, which modulates the FM portion of the transmitter. The picture (or video) portion of the program is broadcast by means of the AM transmitter.

The TV broadcast channels are approximately 6 MHz wide, with the sound (FM) carrier at a frequency 4.5 MHz higher than the picture (AM) carrier. For example, on VHF channel 8, the picture (AM) is transmitted at a frequency of 181.25 MHz, whereas the sound (FM) is transmitted at 185.75 MHz. Channel 8 occupies the band of frequencies between 180 and 186 MHz.

The sound portion of the TV signals uses conventional FM broadcast principles, which need not be described here. However, the picture portion of the signal is unique to TV in that both picture information and synchronizing pulses must be transmitted on the AM carrier.

In the simplest of terms, operation of the TV picture channel can be understood by reference to Fig. 1-5, which shows the relationship between the picture tube in the TV receiver and the station camera picture tube. Both tubes have an electron beam that is emitted by the tube cathode and strikes the tube surface. Both tubes have horizontal and vertical sweep systems that deflect the beam so as to produce a rectangular screen (or "raster") on the tube surface. The vertical sweep is at a rate of 60 Hz, whereas the horizontal sweep is at 15,750 Hz.

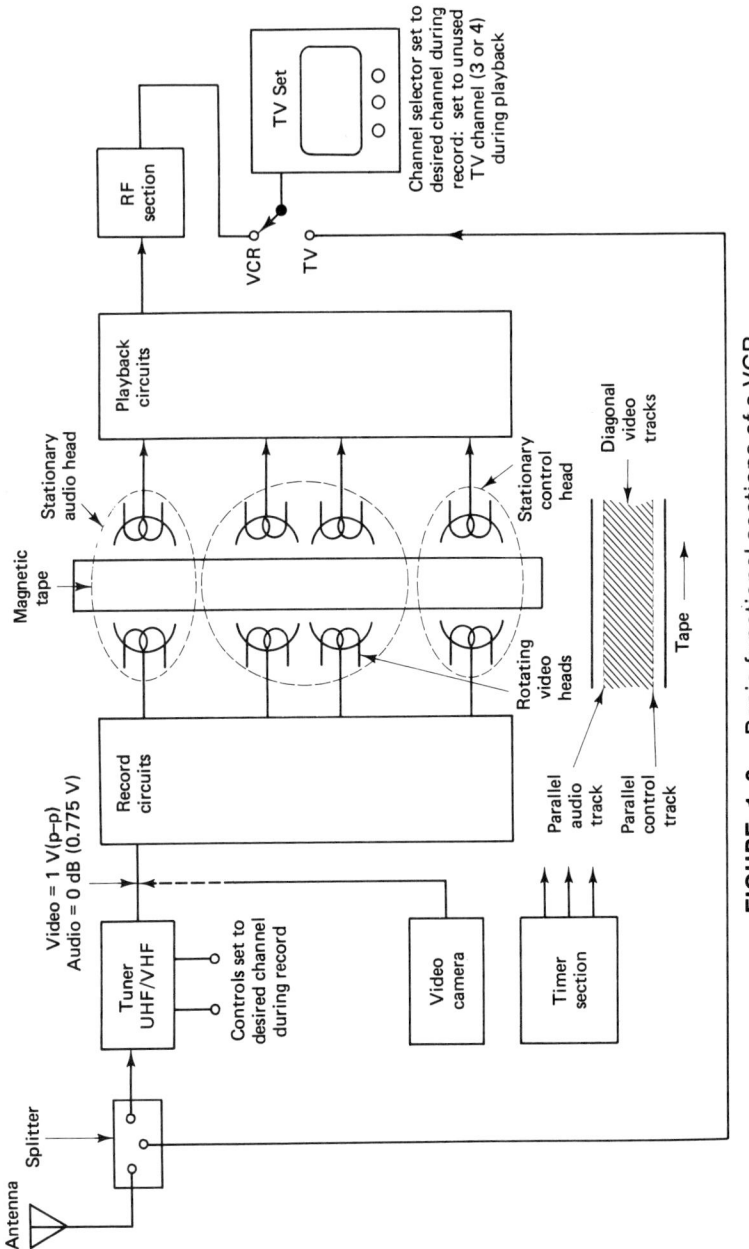

**FIGURE 1-3.** Basic functional sections of a VCR.

*8  Introduction to Videocassette Recorders*

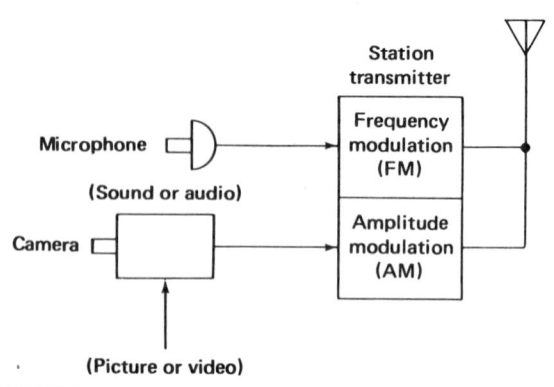

**FIGURE 1-4.** Basic television broadcast system.

The electron beam of the station camera picture tube is modulated by the amount of light that strikes the camera tube screen. In turn, the transmitted AM signal is modulated by the electron beam. Thus, the amplitude of the transmitted AM signal is determined by the intensity of the light at any given instant. The position of the electron beam at a given instant is set by the horizontal and vertical sweep circuits which, in turn, are triggered by pulses at the camera. These pulses are transmitted on the AM carrier and act as synchronizing pulses ("sync pulses") for the TV receiver horizontal and vertical sweep circuits.

The electron beam of the TV receiver is modulated by the transmitted AM signals so as to "paint" a picture on the tube screen. The amplitude of the AM signal determines the intensity of the light produced on the TV screen at any instant. For example, if the camera sees an increase in light at a particular point, the electron beams in both tubes (camera and receiver) are increased, and the TV receiver tube shows an increase in light. The TV receiver tube horizontal and vertical deflection systems are triggered by the horizontal and vertical synchronizing pulses (H-sync and V-sync) transmitted on the AM portion of the TV broadcast signal. Thus, the electron beam of the TV receiver follows the beam in the camera tube. Both beams are at the same corresponding spot, at the same instant.

Assume that the camera is focused on a white card with a black numeral 3 at the center, as shown in Fig. 1-6. As the camera tube electron beam is swept across the surface, light is reflected from the card onto the camera tube surface. When the beam passes the white card background portion of the reflected light, the beam intensity is maximum. Beam intensity drops to a minimum when the beam is at any portion of the light reflected from the numeral 3. This varying electron beam modulates the transmitted AM carrier.

The electron beam starts at the top of the camera tube screen, sweeps across to one side, and is blanked while it returns to the other side (retraces), until the beam finally reaches the bottom of the screen. The beam is then blanked and returned to the top of the screen. Thus, if the camera is focused on the numeral 3 pattern, the camera translates the entire pattern, line by line, into a picture signal (a voltage that varies in amplitude with intensity of the reflected light).

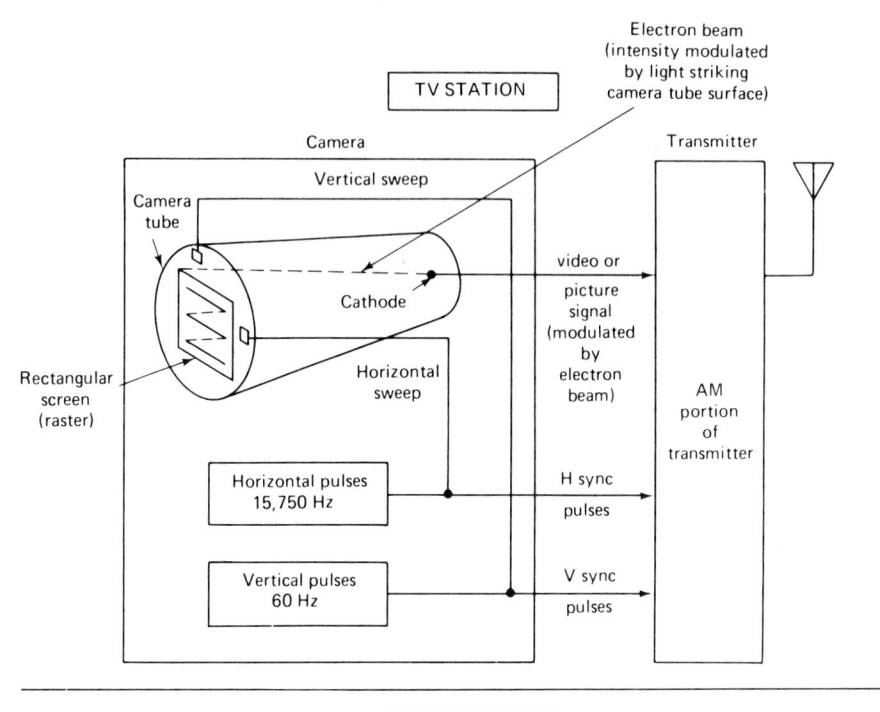

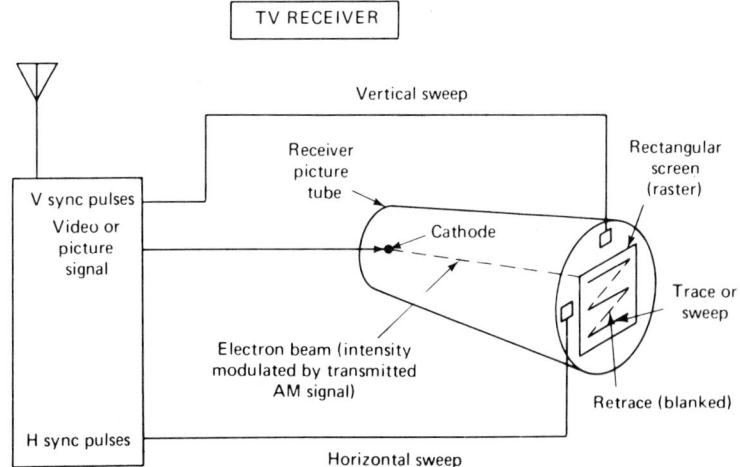

**FIGURE 1-5.** Relationship between camera and receiver picture tubes.

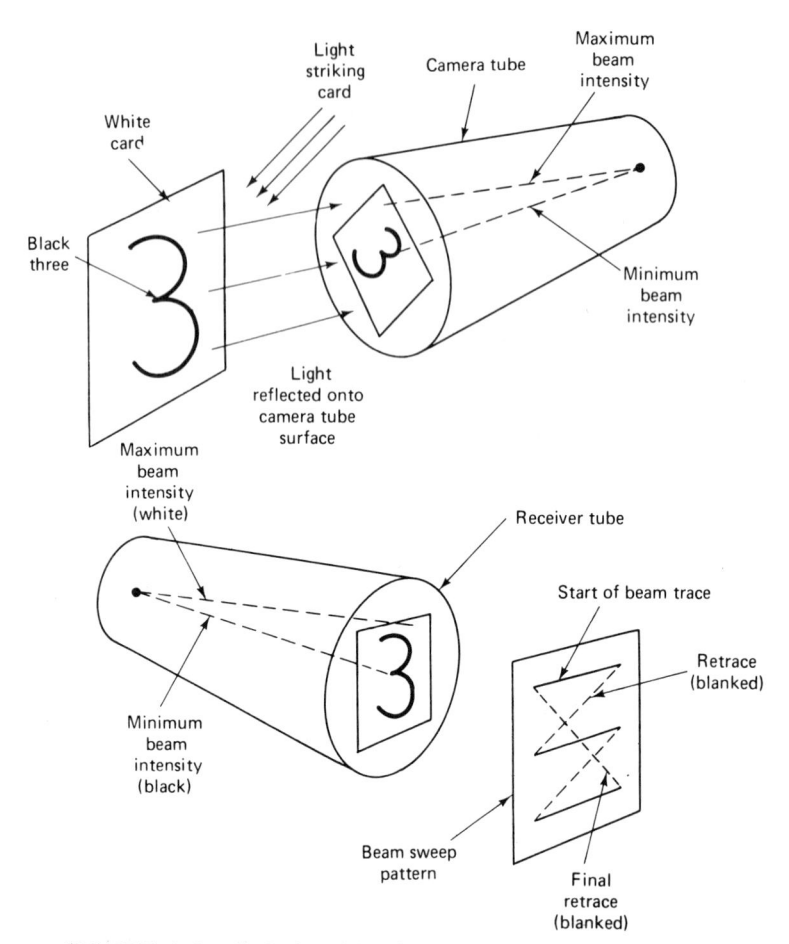

**FIGURE 1-6.** Relationship of electron beams in camera and receiver picture tubes.

At the TV receiver, the picture tube electron beam follows the camera beam in both position and intensity. In the example of Fig. 1-6, both electron beams increase when they are swept across any of the white background. Similarly, both beams decrease when they are swept across any portion of the numeral 3. Thus, the TV receiver picture tube reproduces the numeral 3 in black and white (black, or minimum intensity, on the numeral portion of the sweep; white, or maximum intensity, on the card background portion).

### 1-2.1 Interlaced Scanning

Going further into the scanning system, the image on the TV screen does not change by one complete frame at a time in the same way as a motion picture. Rather, the image is made up of thousands of tiny dots as in a newspaper photo.

## Basic Black-and-White Television Broadcast System    11

In the U.S. and Canadian TV broadcasting systems, these dots number 525 from top to bottom and 700 across (525 x $\frac{4}{3}$), as shown in Fig. 1-7. Beginning at the upper left-hand corner of the screen, the dots are arranged in horizontal lines, called "scanning lines" because they are "scanned" from left to right by the beam. When the end of one line is reached, the beam goes all the way back to the left and begins scanning the next line.

The total number of dots is 525 x 700 = 367,500, and it is this tremendous number of tiny points of light that make up one picture on your TV screen. If the scanning lines were scanned sequentially, at a rate of 60 Hz, there would be a noticeable flicker. Instead, the lines are scanned alternately. First, the odd lines of the 525 scanning lines are scanned from top to bottom, then the even lines (262.5 of them) are scanned to cover the screen a second time. The combination of these two scans produces one complete picture. This method of scanning first odd, then even lines is called "interlaced scanning."

The rough picture produced by scanning half the lines is called a "field." Two fields together form one complete picture called a "frame." Therefore, the picture on the TV set screen consists of 30 frames or 60 fields per second. Figure 1-8 shows the pulses produced by the two fields as they would appear on an oscilloscope screen.

Since a TV picture has 30 frames per second, the picture is capable of changing completely 30 times every second. In terms of dots on the screen, this results in 367,500 x 30 = 11,025,000 per second. If one alternation from light to dark is considered as 1 Hz, the frequency of the video signal is 11,025,000 divided by 2 = 5,500,000 Hz, or 5.5 MHz. When the entire screen is white or black, the frequency is 0. Thus, the video signal bandwidth theoretically uses the range of

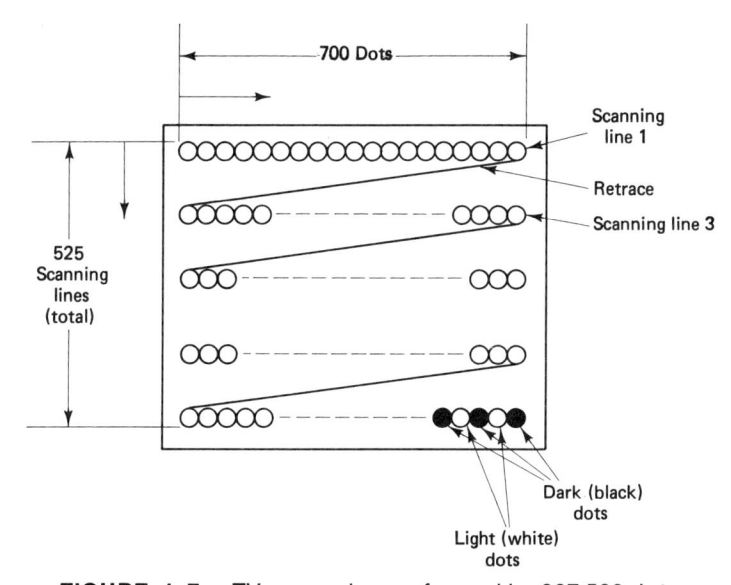

**FIGURE 1-7.** TV screen image formed by 367,500 dots.

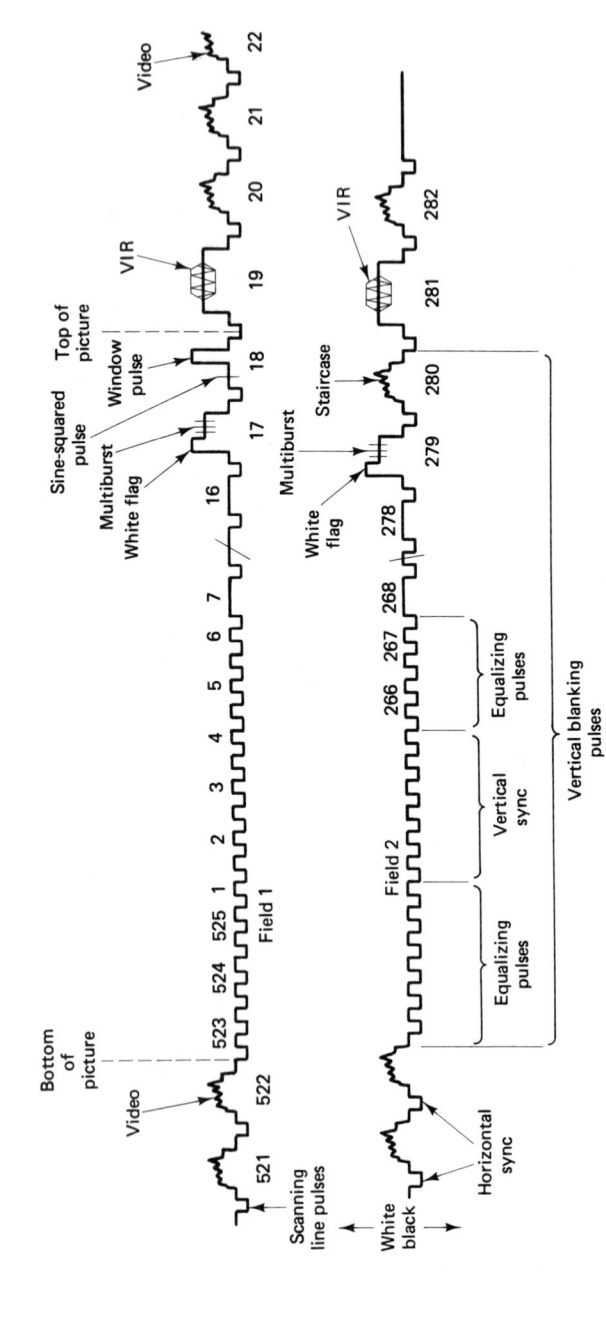

**FIGURE 1-8.** Scanning pulses and other picture information as it would appear on an oscilloscope screen.

frequencies from 0 to 5.5 MHz. In actual practice, the video frequency range is from 0 to about 4 MHz. This is one of the reasons why the video signal is separated in frequency from the sound signal by 4.5 MHz.

## 1-3 BASIC COLOR-TELEVISION BROADCAST SYSTEM

As shown in Fig. 1-9, the basic color-TV broadcast system consists of a transmitter, a color camera, a signal matrix, a brightness amplifier, a color amplifier, and a 3.58-MHz oscillator. As in the case of black and white, the TV program sound is transmitted through an FM transmitter. The picture portion of the program is broadcast by means of amplitude modulation. The same channels and frequencies are used for color and for black and white.

In addition to the horizontal and vertical sweeps, the color signal is made up to two components: brightness and color information. The brightness portion contains all the information pertaining to the details of the picture and is commonly referred to as the "Y" or "luminance" signal. The color portion is called the "chroma" or "chrominance" signal and contains the information pertaining to the hue and saturation of the picture.

The camera in Fig. 1-9 contains three separate image pickup tubes, one tube for each of three colors: red, green, and blue. The camera tubes divide the scene being scanned into three colors (red, yellow, and blue). The relative intensity of the scene is also contained in this signal. The output signal from the camera contains the luminance and chrominance information.

The signal is separated in the signal matrix, and the individual components are amplified by the brightness and color amplifiers. The color information (chrominance) is impressed on a 3.58-MHz subcarrier (in the form of phase shift)

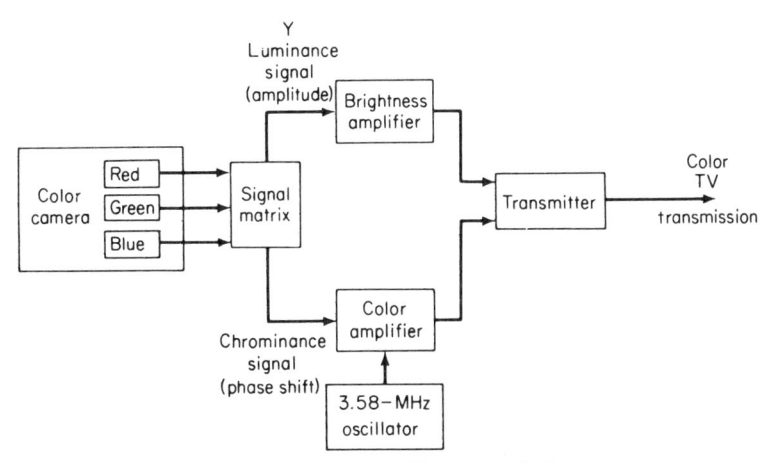

**FIGURE 1-9.** Basic color-TV transmitting system.

## 14  Introduction to Videocassette Recorders

and is then transmitted together with the Y signal (which is in the form of instantaneous amplitude shift or level). This method of color broadcast provides a compatible signal that can be reproduced in both black-and-white as well as color receivers. Black-and-white receivers require only the luminance or Y signal (which is the equivalent of the picture information transmitted on the AM carrier of a black-and-white broadcast).

A compatible color signal must fit within the standard 6-MHz television channel and have horizontal and vertical scanning rates of 15,750 and 60 Hz, plus a video bandwidth not in excess of 4.25 MHz. The color information is interleaved with the video information and transmitted within the 6-MHz television channel on the 3.58-MHz subcarrier, as shown in Fig. 1-10. The color subcarrier is actually 3.579545 MHz above the video carrier, where it does not interfere with reception on black-and-white receivers. (Note that the actual horizontal sync rate is 15,734.26 Hz during color transmission.)

The color signals are synchronized by transmitting approximately eight cycles of the 3.58-MHz subcarrier oscillator signal (of the correct phase) together with the color signal. This signal is called the "burst" sync signal (or may also be known as the "color burst") and is added to the "backporch" of the horizontal sync pulse, as shown in Fig. 1-11. The location of the burst signal does not interfere with horizontal sync on black-and-white receivers, since the burst signal occurs during the blanking or retrace portion of the sync pulse.

### 1-3.1  Black-and-White versus Color Signals

A black-and-white presentation requires only that the amplitude or luminance (or Y) signal be transmitted. A color broadcast also requires that phase-shift information (representing colors) be transmitted. In color television, the hue of the color is determined by the instantaneous amplitude of the signal.

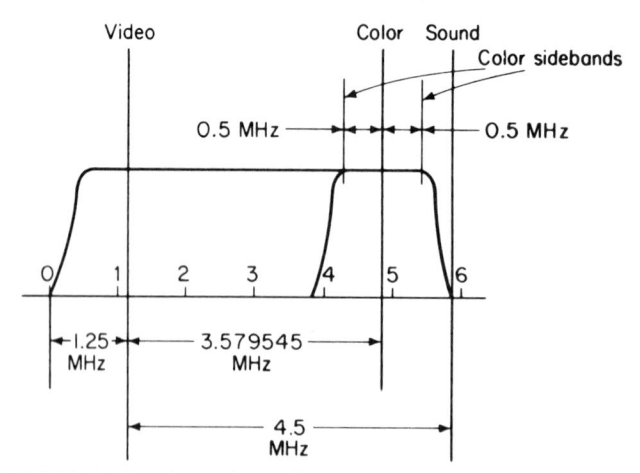

**FIGURE 1-10.** Location of color subcarrier in 6-MHz TV channel.

Basic Color-Television Broadcast System   15

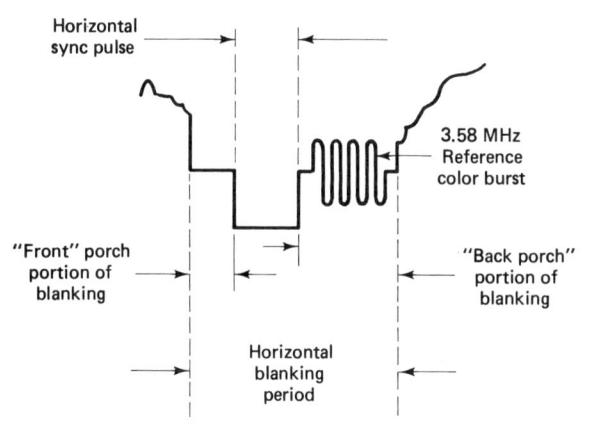

**FIGURE 1-11.** Location of 3.58-MHz reference color burst on horizontal blanking signal.

Figure 1-12 shows the phase and amplitude required to produce each of the six basic colors. For example, at any given instant, if the amplitude is at 0.64 (with an amplitude of 0 for black and 1.0 for white) and the phase of the 3.58-MHz signal is 76.6° (with the reference burst considered to be at 0°), the receiver picture tube produces pure red. If the amplitude is changed to 0.45 and the phase to 192.8°, a pure blue is produced. If the amplitude is changed to 0.59 and the phase to 119.2°, the picture tube produces magenta (which is a combination of red and blue). If the phase of the 3.58-MHz signal is swept from 0 to 360°, a complete rainbow of colors results.

## 1-3.2  Basic Color Receiver Circuits

Figure 1-13 is the block diagram of a color TV receiver. The diagram is actually a composite of several types of color TV receivers, and is presented as a point of reference to understand the operation of a VCR. As shown in Fig. 1-13, the cir-

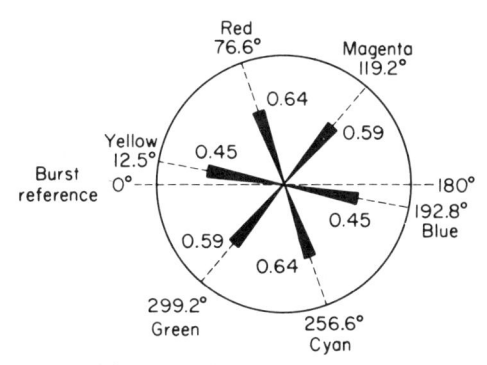

**FIGURE 1-12.** Phase and amplitude relationships of NTSC color signals.

**FIGURE 1-13.** Typical color receiver circuit.

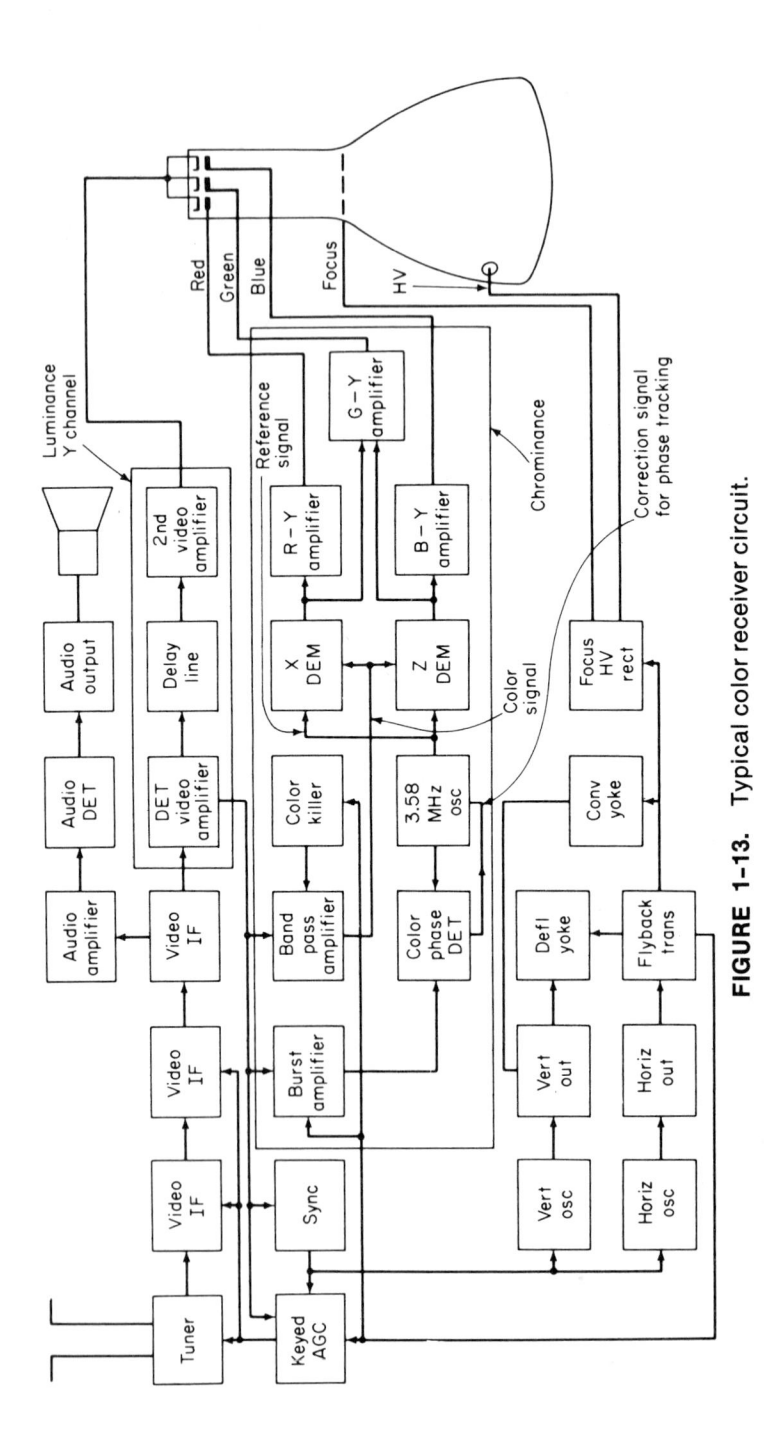

cuits of a color receiver are very similar to those of a black-and-white receiver, with two exceptions: the luminance or Y channel and the chrominance channel. If you are not familiar with either black-and-white or color television receivers, your attention is invited to the author's *Handbook of Simplified Television Service* (Englewood Cliffs, N.J.: Prentice-Hall, Inc., 1977).

### *1-3.3 Luminance Channel Operation*

The luminance or Y channel consists of the detector, first video amplifier, delay line, and second video amplifier. (Note that a separate sound detector is used in the signal path prior to the video detector in most color receivers. This provides good separation of the sound and video signals.)

The composite video signal is detected and amplified in the first stage of the luminance channel. A portion of this signal is fed to the sync and color stages. The rest of the signal continues on to the delay line. This delay is required so that the luminance information (amplitude) arrives at the picture tube at the same time as the color information (phase relationship). The narrow bandwidth of the color circuit causes the chrominance signal to take longer to reach the picture tube. Additional amplification is given to the luminance signal in the second video amplifier before the signal is applied to the cathodes of the picture tube.

### *1-3.4 Chrominance Channel Operation*

The composite video signal is fed to the bandpass amplifier (sometimes called the color IF) from the first video amplifier. The chrominance signal is separated from the composite color signal, is amplified, and is fed to the inputs of the demodulators. The burst amplifier removes the burst signal from the chrominance signal and applies the burst signal to the color phase detector.

A color killer stage is used to prevent signals at frequencies near 3.58 MHz from passing through the color circuit during the reception of black-and-white broadcasts. In the absence of a color broadcast (no color burst signals being transmitted), the color killer biases the bandpass amplifier off, thus disabling all of the chrominance channels. In most circuits, the color killer also receives pulses from the horizontal sweep circuits (usually from the flyback transformer). The horizontal pulses operate the color killer so as to bias the bandpass amplifier off during the horizontal sweep retrace (so that no color information passes during the retrace period).

The 3.58-MHz oscillator provides a locally generated reference signal for demodulation of the composite color signal. The color-phase detector compares the 3.58-MHz oscillator signal with the burst signal and develops a correction voltage to keep the oscillator locked in phase with the burst signal. In most circuits, the burst amplifier (or its control circuit) receives pulses from the horizontal sweep. These pulses gate the burst amplifier on "on" only during the burst signal (immediately after the horizontal sync pulses, during the retrace blanking period, as shown in Fig. 1-11). Thus, the 3.58-MHz oscillator is locked in phase with the burst at the beginning of each horizontal sweep.

As shown in Fig. 1-13, the X and Z demodulators detect the amplitude and phase variations of the chrominance signal to recover color information. At any given instant, the X and Z demodulators receive two signals: a 3.58-MHz signal from the reference oscillator (locked in phase to the reference burst) and a 3.58-MHz signal from the bandpass amplifier (at a phase and amplitude representing the color at that instant).

The R-Y (red luminance), B-Y (blue luminance), and G-Y (green luminance) amplifiers amplify the color signals and apply them to the picture tube inputs (grids in this case).

### 1-3.5 Makeup of Television Colors

Color picture tubes produce most colors by mixing three colors (red, blue, and green). Any color can be created by the proper blending of these three colors; that is, any color can be created if the three colors are present, each at a given level of intensity or brightness. This is due to a characteristic of the human eye. When the eye sees two different colors of the same brightness level, the colors appear to be of different brightness levels. For example, the eye is more sensitive to green and yellow than to red or blue.

The combination of any two primary colors produce a third color. In color television, the process of mixing colors is additive and is dependent on the self-illuminating properties of the picture tube screen (not on the surrounding light source, as found with the *subtractive* color-mixing process familiar to paints and painting).

If two primary colors, such as red and green, are combined, the primary colors produce a secondary color of yellow. The combination of blue and red produces magenta. Also, a secondary color added to its complementary primary produces white. For example, yellow + blue = white, cyan + red = white, and magenta + green = white. Thus, any color can be produced if the three colors (red, blue, and green) are mixed in the proper portions (that is, if the amplitudes of the three signals at the picture tube inputs are at the proper levels).

For the purposes of our discussion concerning operation of VCRs, it is sufficient to understand that the three picture tube inputs (Fig. 1-13) receive signals of correct amplitude proportions (identical to the proportions existing at the color camera for any given instant) provided that (1) the receiver 3.58-MHz oscillator is locked in phase to the burst signal, and (2) the demodulators receive a 3.58-MHz color signal that is of a *phase and amplitude corresponding to the color proportions* existing at the color camera.

## 1-4 BASIC MAGNETIC RECORDING PRINCIPLES

Before going into how magnetic tape recording is used in VCRs to record television programming, let us review the basic principles of magnetic tape recording.

## 1-4.1 Magnetic Recording

As shown in Fig. 1-14, and iron bar placed next to a magnet becomes magnetized in such a way that the end of the bar next to the north pole of the magnet becomes an south pole. The end of the bar next to the magnet's south pole becomes the bar's north pole. This is called "residual magnetism" and is the essential of magnetic tape recording.

Also as shown in Fig. 1-14, if the magnet moves along a larger bar, the bar is magnetized with alternating north-south-north-south poles. These poles are at the same positions that the magnet's poles had at the corresponding points Residual magnetism thus appears as an NSNS pattern along the bar.

## 1-4.2 Magnetic Recording on VCR Tape

In a VCR, the recording heads (video head, audio head, control track head) take the place of the magnet. The cassette tape takes the place of the iron bar. As shown in Fig. 1-15, a recording head is a form of electromagnet made up of a core wrapped with a coil. There is a very narrow space, called the "gap," between the two poles of the electromagnet. The video cassette tape generally consists of a polyester base on which has been painted a binder material containing very small magnetic particles, as shown in Fig. 1-15. The tape surface is polished to a smooth finish.

In the simplified explanation illustrated by Fig. 1-14, the magnet (head) moves, while the iron bar (tape) stands still. The same result (a pattern of residual

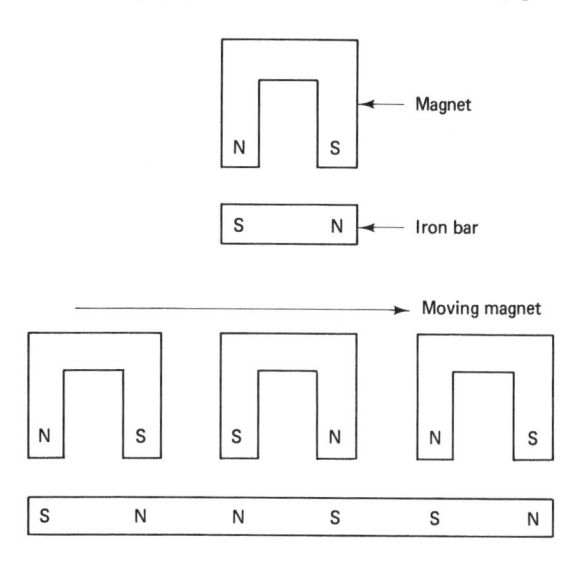

**FIGURE 1-14.** Basic elements of residual magnetism used in magnetic tape recording.

20  Introduction to Videocassette Recorders

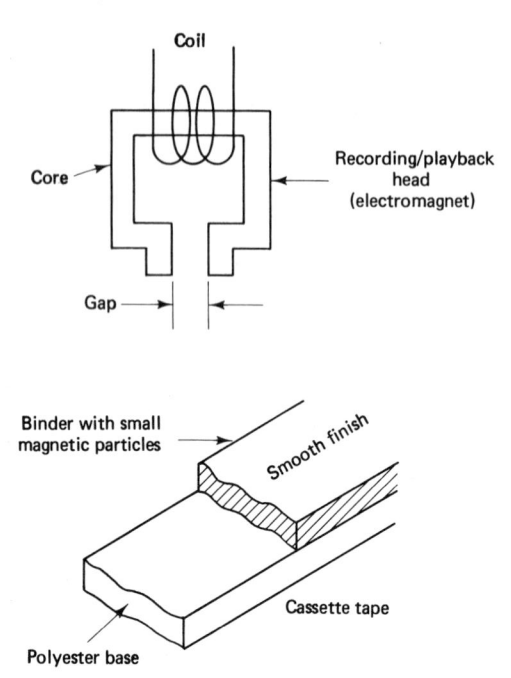

**FIGURE 1-15.** Basic elements of a record/playback head and cassette tape.

magnetism) is produced when the tape moves and the head remains stationary, as shown in Fig. 1-16. Going further, it is not necessary to physically change the position of the head to change the polarity of the poles. Instead, the poles can be changed by applying an alternating current to the electromagnet coil. Since the audio, video, and control signals used in television are a form of alternating cur-

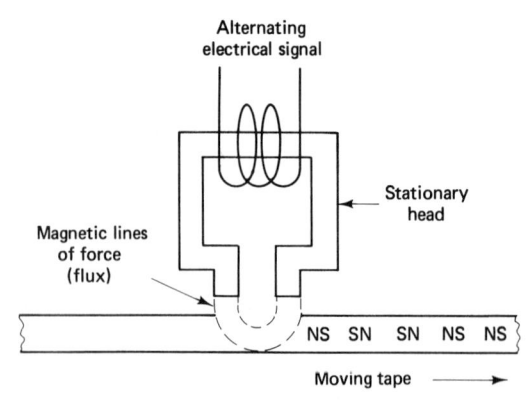

**FIGURE 1-16.** How a pattern of residual magnetism is recorded on a moving tape by a recording head with an alternating electrical signal.

rent, these signals are applied to the corresponding record heads (after much processing), and produce patterns of residual magnetism on the tape that correspond to the audio, video, and control information.

The poles of the recording head change polarity depending on the direction of current flow in the coil. The strength of the magnetism varies with the strength of the current. In this way, a recording is made as patterns of residual magnetism in the magnetic particles on the tape. If the signal being recorded is a sine wave, as shown in Fig. 1-17, the strength of the magnetization on the tape varies along the tape length in a sine-wave pattern. Such a recording takes the form of lines of tiny

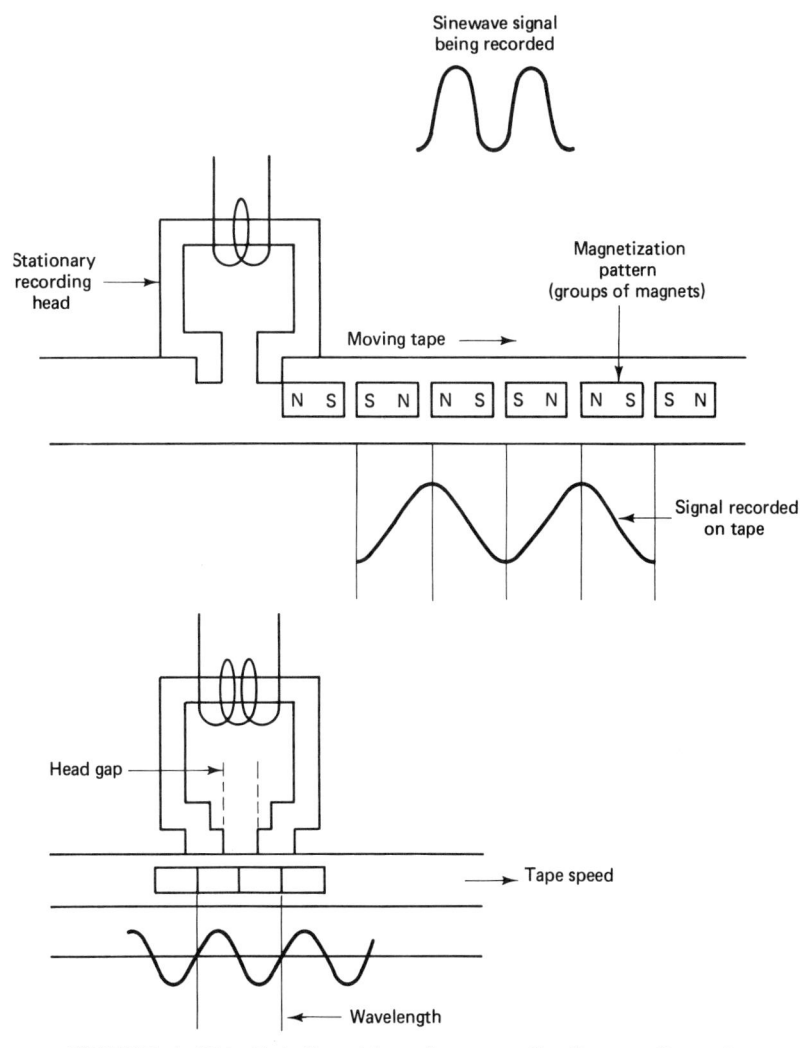

**FIGURE 1-17.** Relationship of magnetization pattern to recorded signal.

magnets on the tape. The length of the groups of these magnets represents changes in the direction of the signal current, while the strength of the magnetism represents the strength of the current.

The wavelength of the recorded signal along the tape corresponds with one cycle of the input signal. However, the magnetized signal pattern on the tape has a wavelength that is proportional to tape speed and inversely proportional to recorded signal frequency, as indicated in the following relationship:

$$\text{wavelength} = \frac{\text{tape speed}}{\text{input signal frequency}}$$

Keep in mind that for the rotating video head of a VCR, tape speed is the *relative speed* of the rotating head and the moving tape. The audio and control heads are stationary, so wavelength depends only on tape speed (and input signal frequency).

### 1-4.3 Magnetic Playback on VCR Tape

During playback, the heads used are the same ones that recorded the corresponding signals on the tape. The residual magnetism on the tape has lines of force that enter the core of the heads. The strength of this flux depends on the amount of magnetism on the tape at the point contacting the head gap. As the tape moves (and, in the case of video, the heads rotate), the amount of flux passing through the core is constantly changing. In this way, a voltage is generated in the coil that corresponds to the changing magnetic pattern on the tape. In theory, the voltage developed at the head during playback is identical in waveshape and frequency to the corresponding voltage during record. To do this, the tape must move at exactly the same speed during record and playback. Similarly, in the case of video, the video heads must rotate at exactly the same speed for both record and playback.

## 1-5 PROBLEMS IN RECORDING THE VIDEO SIGNAL ON A VCR

Recording of the audio and synchronization control signals is a relatively simple matter when compared to recording the video signals. The control signals are typically 60 Hz, whereas the audio signals rarely go below about 20 Hz or above about 20 kHz. (A typical audio range is 50 Hz to 10 kHz.) The methods used in audio tape recorders are adequate for both of these signals. However, typical video signals go from direct current (0 Hz) up to about 4.2 MHz. There are three methods used to increase the frequency range or writing speed to accommodate the video signals used in VCRs: *frequency modulation* (FM), *micro head gaps,* and *rotating heads* to increase *relative speed* between head and tape. Let us review the advantages of all three methods.

## 1-5.1 FM Recording and Playback

There are several advantages in using FM for record and playback in VCRs. First, the output voltage of any tape playback head varies in amplitude with changes in frequency. Figure 1-18 shows the typical output voltage versus frequency relationship of a playback head. At low frequencies, output is near zero. As frequency increases, the output increases at about 6 dB per octave (output doubles as frequency doubles) until a peak is reached. This is followed by one or more voltage nulls where the output drops to zero.

If only a narrow portion of this frequency range were required, as is the case with audio recording, the amplitude variation could be ignored. However, with the video frequency range (where the highest frequency is about 200 times that of the highest audio range), the wide variation in amplitude would produce a playback output that was totally distorted when compared to the recorded signal.

An FM recording has no variation in amplitude (theoretically). Any amplitude variations that do occur in either record or playback can be virtually eliminated by limiters and amplifiers that are driven into saturation (just as they are in the audio portion of TV broadcast and reception circuits). Since the amplitude variations can be ignored, there is no need for a bias signal commonly used in direct (amplitude) audio recording.

For these reasons, the video signal is converted to an FM signal by an FM modulator before recording on tape in a VCR. Upon playback, the FM signals coming from the video heads go to amplifiers and limiters where amplitude variations are removed. Typically, the playback head signal is amplified so that the lowest output is beyond the amplitude limit (at saturation) of a limiter circuit. The limiter then reduces all signals to the same amplitude level. These constant-

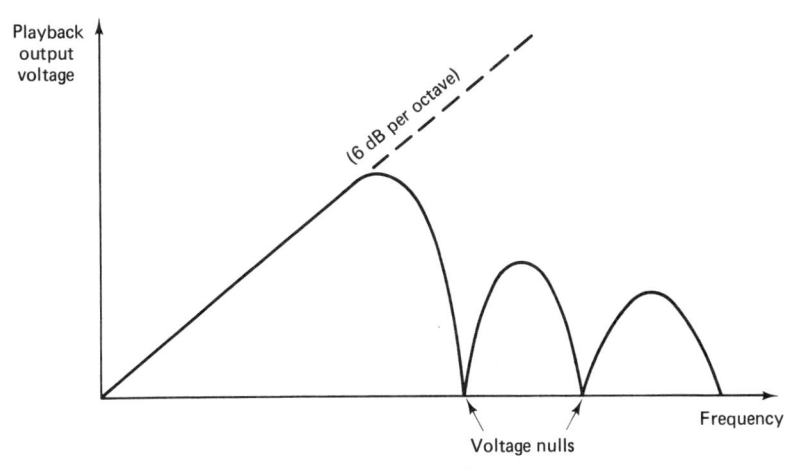

**FIGURE 1-18.** Typical output voltage versus frequency relationship of a playback head.

## 24   Introduction to Videocassette Recorders

amplitude playback signals are then applied to an FM demodulator which converts the FM signal back into (a replica of) the original video signal.

### 1-5.2   Micro Video Head Gaps

By referring back to Fig. 1-17, you can see why the record and playback frequency limits are inversely proportional to head gap (frequency increases as gap decreases). All other factors being equal, a smaller head gap produces a shorter wavelength since the north and south poles are closer together. A shorter wavelength results in a higher frequency. However, there are physical and electrical limits to making the head gap smaller. For example, when the wavelength of the signal recorded on tape is the same as the width of the gap, the playback output is zero. The gap must always be narrower (or shorter) than the wavelength of the highest frequency to be recorded and played back. Although present-day video heads are very small (typically 0.6-micrometer ($\mu$m) gap for Beta and 0.3-$\mu$m gap for VHS), there is an obvious physical limit on how small you can make a practical video head. For this reason it is necessary to increase tape speed to accommodate the video frequency range.

### 1-5.3   Rotating Video Heads

Rotating video heads are used in present-day VCRs to increase the relative speed between tape and head (and thus increase writing speed). If you are familiar with audio tape recording, you know that a slow tape speed is sufficient for recording conversation, but music requires a faster speed (typically 19 cm/s) for good sound quality.

If you assume that the top frequency for recording music is 20 kHz, that the top frequency limit for video signals is 4 MHz (200 times that for audio), and that a 19-cm/s tape speed is sufficient for good quality, the required tape speed is 3800 cm/s (19 × 200 = 3800) or 38 m/s. This works out to about 2280 m/min, and would require a video cassette the size of truck tires (or very thin tape) for 1 hour of playing time!

Instead of using tape at a high speed, the video heads are rotated to produce a high *relative speed* between head and tape. Figure 1-19 shows how the heads and tape move in relation to each other. While the video heads rotate in a horizontal plane (on a drum in Beta and on a cylinder in VHS) the tape passes the heads diagonally. This is known as a *helical scan* system, and produces *slant tracks* or *diagonal tracks* for the video recording.

Note that the audio head and control track head (mounted one above the other) are stationary, and are separate from the video heads, as is the erase head. This head and tape track arrangement is typical for both Beta and VHS. However, because of different drum or cylinder size (diameter and circumference), the relative speed of the Beta system is typically 6.9 m/s (273.2

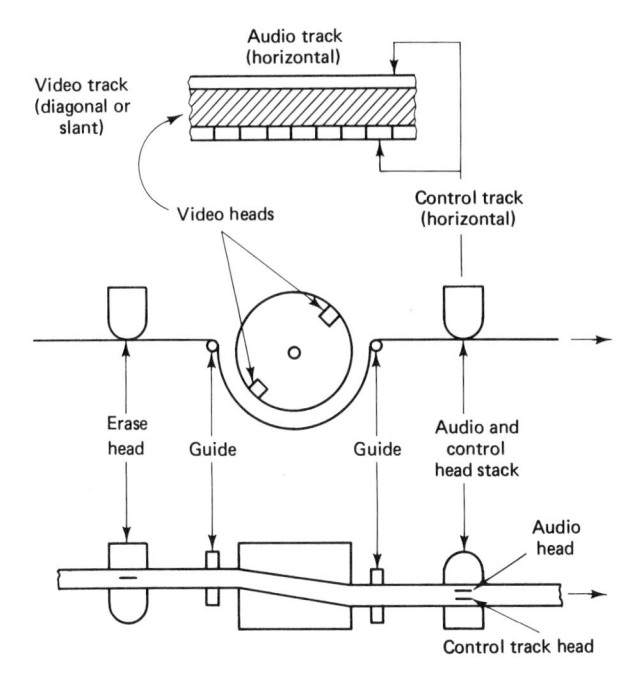

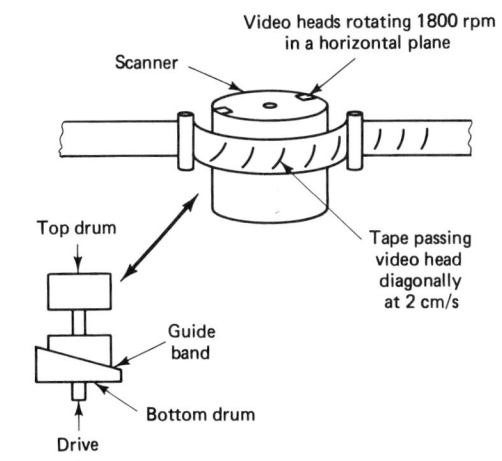

## Problems in Recording the Video Signal on a VCR 25

**FIGURE 1-19.** Relationship of heads and tape movement.

in./s), whereas the typical VHS relative speed is 5.8 m/s (228 in./s), even though actual tape speed is in the range 2 cm/s. Also note that the drum or cylinder (often referred to as the *scanner*) rotates at a speed of 1800 rpm for both Beta and VHS.

## 1-6 RELATIONSHIP OF ROTATING HEADS TO VIDEO FIELDS AND FRAMES

Figure 1-20 shows a simplified diagram of the relationship between the video heads and the video tracks recorded on tape. As shown, video heads A and B are positioned 180° apart on the drum or cylinder, which rotates at a rate of 30 times a second. The tape is wrapped around the drum or cylinder to form an omega ($\Omega$) shape. The tape then passes diagonally across the surface of the drum or cylinder to produce the helical scan. Since there are two heads on the drum or cylinder, which is rotating at 30 rps, each head contacts the tape once each $\frac{1}{60}$ s. Thus, each head completes one rotation in $\frac{1}{30}$ s, and one slant or diagonal track is recorded on the tape during half a rotation ($\frac{1}{60}$).

Since the tape is moving, after the first head has completed one track on the tape, the second head records another track immediately behind the first track, as shown in Fig. 1-20. If head A records during the first $\frac{1}{60}$ s, head B records during the second $\frac{1}{60}$ s. The recording continues in a pattern A-B-A and so on. During playback, the same sequence occurs (the heads trace the tracks recorded on the tape and pickup the signal, producing an FM signal that corresponds to the recorded video signal).

Figure 1-21 shows the theoretical relationship among tracks, fields, frames, and the television vertical sync pulses. Since there are two heads, 60 diagonal tracks are recorded every second. One field of the video signal is recorded as one track on the tape, and two fields (adjacent tracks A and B) make up one frame. In actual practice, there is some overlap between the two tracks. As an example, the video signal recorded by head A (just leaving the tape) is

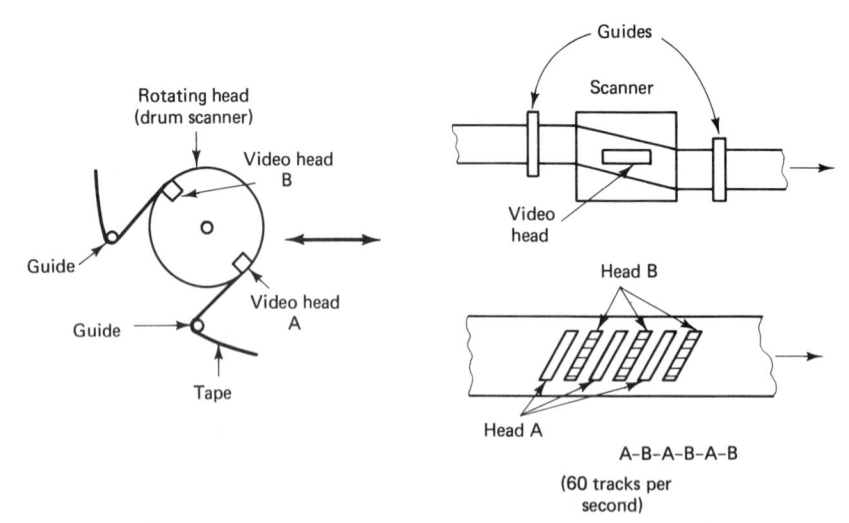

**FIGURE 1-20.** Simplified diagram of the relationship between the video heads and the video tracks recorded on tape.

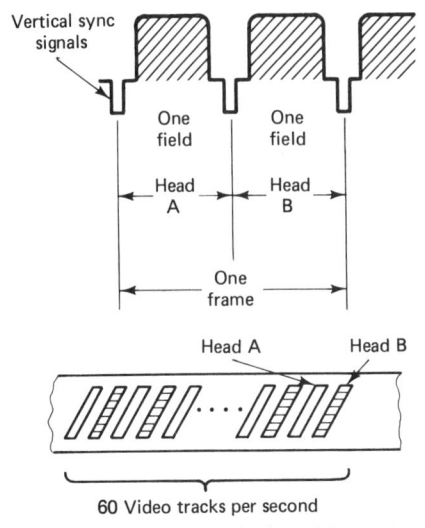

**FIGURE 1-21.** Theoretical relationship among tracks, fields, frames, and the television vertical sync pulses.

simultaneously applied to head B (just starting its track). During playback, this overlap is eliminated by electronic switching so that the output from the two heads appears as a continuous signal.

### 1-6.1 Picture Stability Using a Servo System

It is obvious that no matter how precisely the tracks are recorded, the picture cannot be reproduced if these tracks are not accurately traced by the rotating heads during playback. One step toward making playback trace as accurate as possible is to raise the mechanical precision of all parts in the tape path, including tape guides, head position, and so on. However, this mechanical precision is not enough, especially since the video signal tracks are very narrow. (In the Beta system, the video signal tracks are 0.03 mm wide when tape speed is 2 cm/s.)

In addition to mechanical precision, both Beta and VHS systems use an automatic self-governing arrangement, generally known as the *servo system*. Since there are many different types of servo systems, we discuss the basic or typical servo system here.

Figure 1-22 shows operation of a basic VCR servo system during record. As shown, the vertical sync pulses of the TV broadcast signal are used to synchronize the rotating heads with the tape movement. The heads are driven (by the same motor used to drive the tape capstan) so that the heads turn just a little faster than 1800 rpm (30 Hz). The vertical sync pulses of 60 Hz are applied to a 2:1 divider circuit, producing 30-Hz control signals (often referred to as the CTL signal). This CTL signal is recorded on the tape by the separate stationary control track head (which is on the same stack as the audio head).

*28  Introduction to Videocassette Recorders*

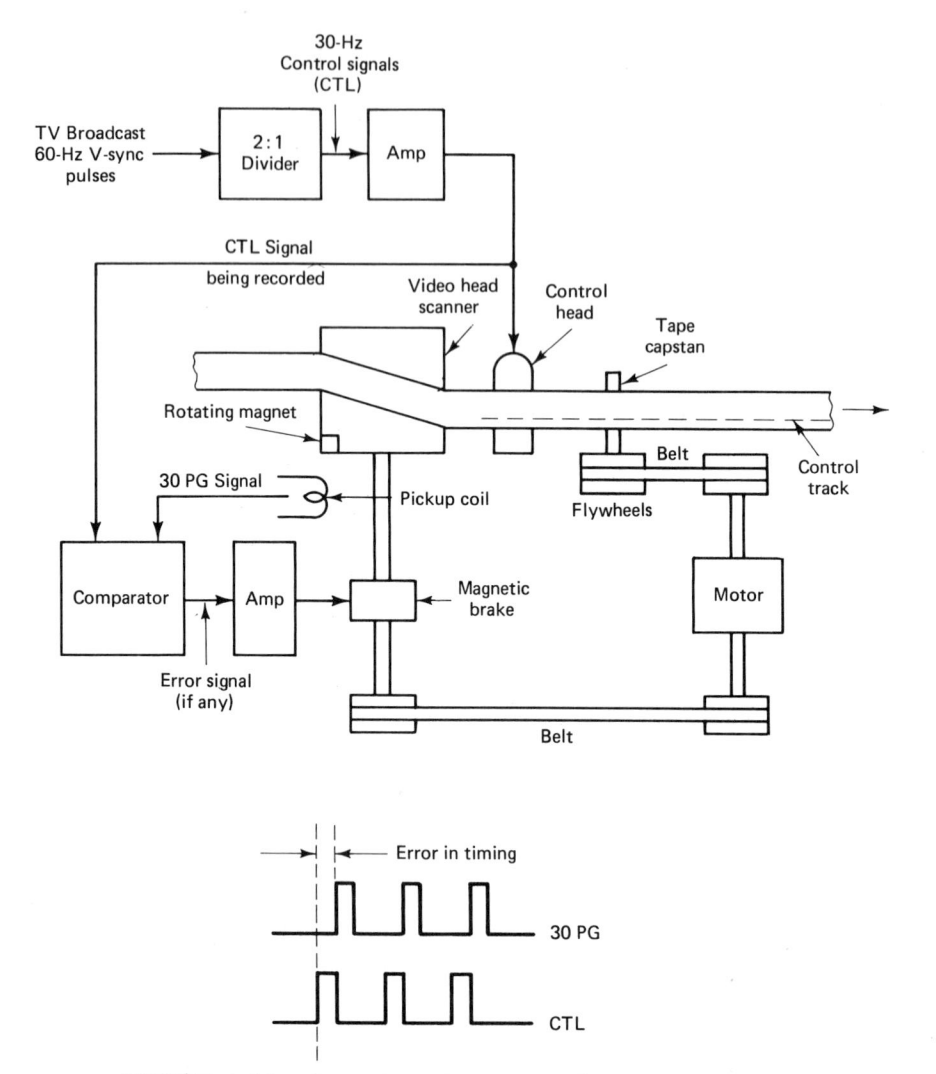

**FIGURE 1-22.** Operation of a basic VCR servo system during record.

A pulse signal (often referred to as the 30PG signal) is generated by detecting the actual rotational speed of the heads. One way to generate the 30PG signal is to use a stationary pickup coil and a rotating magnet. The magnet is rotated together with the heads, and produces a pulse in the pickup coil each time the magnet passes the coil. The 30PG signal is compared with the 30-Hz CTL signal, and any difference (or *error signal*) is amplified. The output of the amplifier is used to control the current of a magnetic brake that controls speed of the rotating heads. The current applied to the brake coil is increased if rotational speed ex-

## Relationship of Rotating Heads to Video Fields and Frames

ceeds 1800 rpm, and is decreased if the speed drops below 1800 rpm. In this way, rotational speed is maintained precisely at 1800 rpm.

Figure 1-23 shows operation of the basic VCR servo system during playback. As shown, the CTL control signal recorded on the tape becomes the standard reference signal during playback. The CTL signal is compared with the 30PG signal of the rotating heads, and any difference or error signal is amplified. Again, the output of the amplifier is used to control operation of the brake (and thus control speed of the rotating heads). In this way, the heads trace the appropriate tracks, and playback is synchronized with record.

Note that the system shown in Figs. 1-22 and 1-23 is a form of *drum servo*. As discussed throughout this book, there is also a *capstan servo* used in many VCRs. Similarly, there are many forms of drum servos and capstan servos in common use. Most of the advanced servo systems used in present-day VCRs not only control speed, but also control the phase relationship between drum and capstan servos.

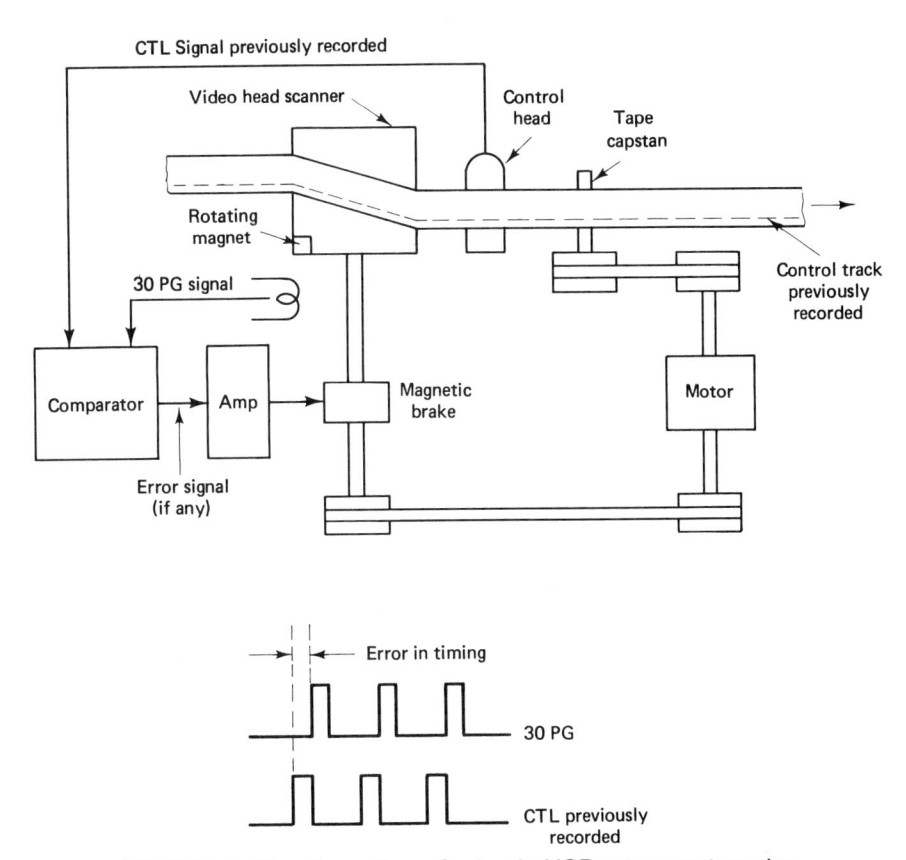

**FIGURE 1-23.** Operation of a basic VCR servo system during playback.

## 1-7 RELATIONSHIP OF LUMINANCE AND CHROMINANCE SIGNALS IN A VCR

Figure 1-24 shows the typical sequence in recording and playback of the luminance signal on a VCR. During record, the entire luminance signal (from sync tips to white peaks) is amplified and converted to an FM signal that varies in frequency from about 3.5 to 4.8 MHz (for Beta systems) or 3.4 to 4.4 MHz (for VHS). During playback, the FM signal is demodulated back to a replica of the original luminance signal. Note that this provides an FM luminance bandwidth on tape of about 1.3 MHz for Beta and 1 MHz for VHS.

As discussed in Sec. 1-3, TV color information is transmitted on the 3.58-MHz chrominance subcarrier. Color at any point on the TV screen depends on the instantaneous amplitude and phase of the 3.58-MHz signal. In VCRs, the 3.58-MHz subcarrier is *hetrodyned* or *down-converted* to a frequency of 688 kHz (for Beta) or 629 kHz for VHS, and recorded directly (AM, not FM) on tape. Figure 1-25 shows the typical sequence for such down-conversion (known as a *color-under* system, since the color signal frequency is always well below the luminance signal frequency).

There are several advantages for the color-under system. For example, no

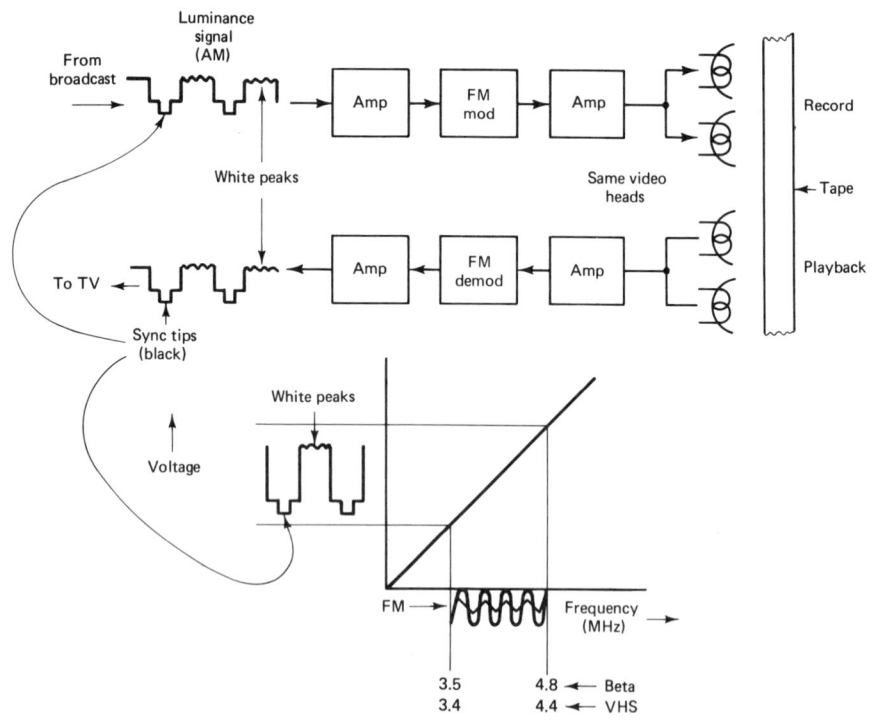

**FIGURE 1-24.** Typical sequence in recording and playback of the luminance signal on a VCR.

30

*Introduction to the Beta System* 31

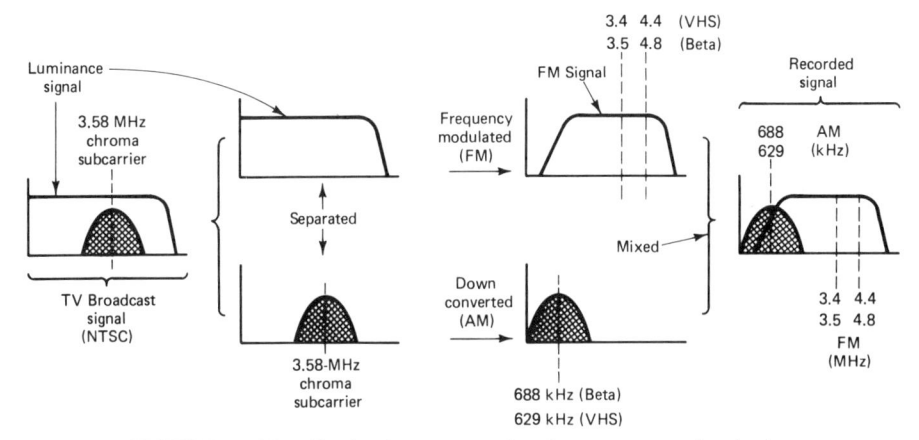

**FIGURE 1-25.** Typical sequence for down-conversion (color-under) of a 3.58-MHz chroma subcarrier.

bias is needed to record the chrominance signal. Since the FM luminance signal is recorded together with the chrominance signal (both signals on the same video heads), the FM luminance signal acts as a bias. Electronic stability is good since the color signal (including sidebands of about ± 500 kHz) is far removed below the lowest luminance signal frequency.

## 1-8 INTRODUCTION TO THE BETA SYSTEM

Now that we have reviewed the operation of VCRs in general, let us go into a bit more detail on the Beta system. An introduction to VHS is provided in Sec. 1-9. Note that the discussion in this section applies to Betamax, developed and manufactured by Sony. However, the principles described here also apply generally to other Beta systems. Also, note that the circuit descriptions here are general in nature. We go into much more detailed Beta functions in Chapter 3.

### 1-8.1 High-Density Recording

Present-day consumer VCRs are high-density recording devices. Because of this, consumer VCRs can handle more information on a given amount of tape (longer playing time on narrower tape) than older studio-type VTRs. One way to accomplish this high-density recording is to make the head drum or scanner smaller to shorten the track length. Another method is to make the track width narrower. Although both of these methods increase recording density, the real improvement of consumer VCRs over studio VTRs is elimination of *guard bands*. This is known as *guard bandless* recording or *zero guard band* recording.

On broadcast VTRs, the tape is driven against the video heads so that there is unrecorded vacant space between video tracks. This blank area, or guard band, is necessary to eliminate *crosstalk* between tracks. Crosstalk occurs when the heads mistrack and play back a portion of two adjacent tracks. Figure 1-26 shows a comparison of tapes with and without guard bands. Note that the tape without guard bands contains much more information. Also note that even though the high-density Beta system is called zero guard band, there are guard bands that separate the audio and control tracks from the video tracks (but no guard bands between adjacent video tracks).

Without special precautions, crosstalk results if the guard bands are eliminated. In the Beta system, the problem of crosstalk is eliminated by two techniques: *azimuth recording* and *phase inversion* (or PI).

### 1-8.2 Azimuth Recording

As discussed, the luminance signal is recorded as a high-frequency (3.5 to 4.8 MHz) FM signal on the tape. If you are familiar with audio tape recorders, you know that considerable high-frequency loss (known as *azimuth loss*) occurs when there is a difference in the angle of the head in relation to the tape between record and playback. This angle (known as azimuth angle) is adjusted so that the head gap is exactly at right angles to the line of tape motion or recorded track. Typically, in audio, the track is horizontal and the head gap is vertical. Any deviation from correct alignment is known as *azimuth error*. One symptom of an audio recorder with azimuth error is that the recorder will play back its own recordings without high-frequency signal loss (since the head angle is the same for both record and playback), but produces considerable loss when playing

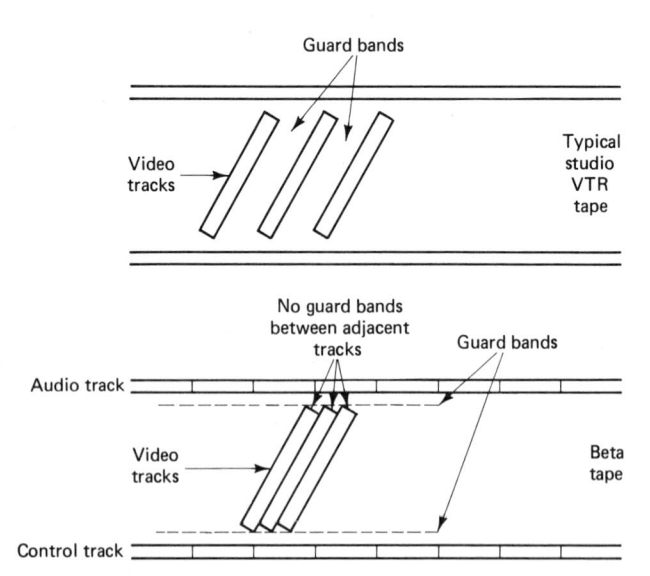

**FIGURE 1-26.** Comparison of tapes with and without guard bands between adjacent video tracks.

back tapes recorded on other machines (where the tapes were recorded with different azimuth adjustment).

In VCRs, this azimuth-loss principle is used to eliminate crosstalk between adjacent tracks. As discussed, video information is recorded on two heads which produce two adjacent tracks. The two video heads are mounted so that one head is at a different angle from the other head. As shown in Fig. 1-27, the angle for one head (arbitrarily called head A) is +7° from the reference point, whereas head B has an azimuth angle of −7° from the reference. Thus, there is a difference of 14° between the heads. During playback, a strong signal is picked up only when head A traces track A. If head A runs over to track B for any reason, the track B high-frequency signal is weak and does not produce interference or crosstalk.

### 1-8.3 Phase-Inversion (PI) Color Recording

Although azimuth-loss recording can be used to eliminate crosstalk for the high-frequency luminance signal, this method is not sufficient when it comes to the chrominance signal (which is recorded directly on tape after conversion to the relatively low frequency of 688 kHz). Phase inversion is used to eliminate

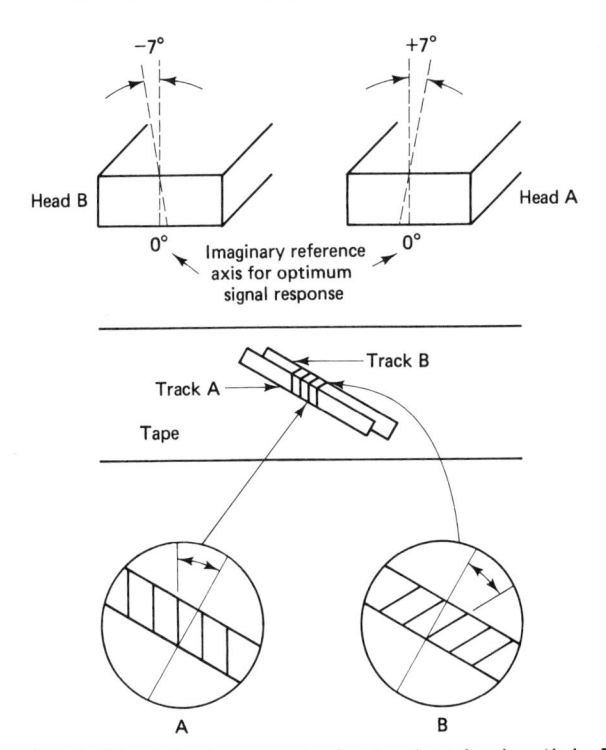

**FIGURE 1-27.** Displacement of video head azimuth in Beta systems.

34   Introduction to Videocassette Recorders

crosstalk of the chrominance signal. In the simplest of terms, the chrominance signal to be recorded on track A is phase inverted by 180° with every *line period*, while the chrominance signal recorded on track B remains continuously in the same phase. (The term "line period" refers to the period of time required to pro-

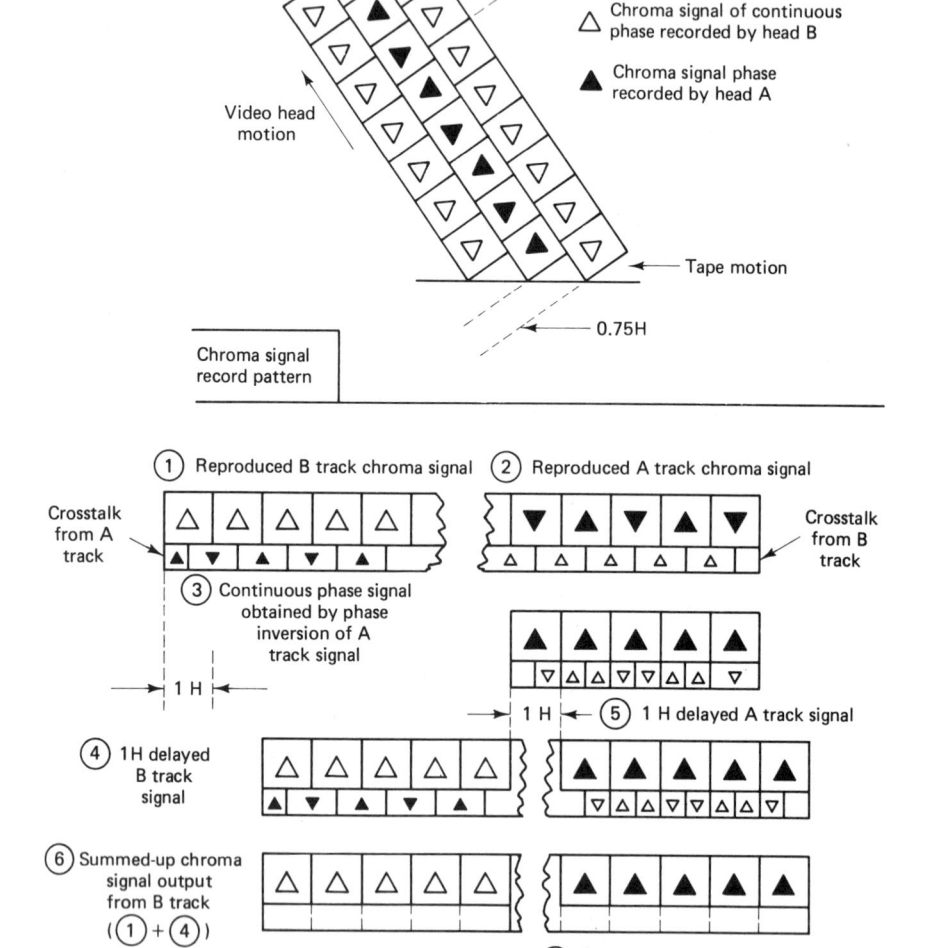

**FIGURE 1-28.** Chroma signal record pattern and summary of how crosstalk can be eliminated from playback chroma signal.

duce one horizontal line on the TV screen, or approximately 63.5 $\mu$sec and is referred to as "1H.") Upon playback, both track A and track B signals are restored to the same phase relationship.

Figure 1-28 shows how the chrominance or chroma signal is recorded using phase inversion. Figure 1-28 also illustrates the phase relationship between the fundamental chroma signal and undesired crosstalk signal contained in the playback signal, that results from phase-inverted recording. The line-by-line phase inversion of the chroma signal is done during the conversion process from 3.58 MHz to 688 kHz and applied to head A. The 3.58 chroma signal is mixed with a 4.27-MHz continuous-wave fixed reference signal which is phase inverted during the repeated time that head A is in contact with the tape, in order to effect the frequency conversion.

The resultant signal frequency is 688 kHz, and is amplified for application to the video heads. When the recorded signal is played back, the 688-kHz signal is hetrodyned back to 3.58 MHz. In an exact counterpart of the recording process, the signal reproduced through head A is again subject to phase inversion using the same 4.27-MHz signal which was used for recording. Adding the recovered line signal (from head A) to the adjacent line signal (from head B) restores the signal to normal. However, the phase of the crosstalk component contained in the 3.58-MHz playback chroma signal remains phase inverted at every other line.

The playback chroma signal is then passed through a *comb filter* using a 1H *delay line* and a resistive matrix, as shown in Fig. 1-29. Both the delayed and nondelayed signals are added together in the resistive circuit with the result that the crosstalk component is canceled out and the normal chroma signal component is double in amplitude.

In addition to canceling crosstalk, the PI color recording system also minimizes the effect of mechanical jitter or flutter in the scanner drum or servo. Such jitter can cause a phase shift and result in poor picture quality. Jitter effects are eliminated by locking the frequency and phase of the 4.27-MHz reference signal to the TV horizontal synchronizing (H-sync) signal (known as fH) during record. At playback, the 4.27-MHz reference signal is locked to the recorded H-sync signal (known as fH*). Thus, if there is any jitter component from any cause (mechanical or electrical), the 4.27-MHz reference is also locked to the jitter, thus eliminating the jitter effects. This feature is similar to locking the scanner speed to the TV vertical synchronizing signal during record, and to the recorded V-sync signal during playback as described in Sec. 1-6.1. However, operation of the PI color recording circuits is far more complex.

### 1-8.4 Basic PI Color Recording Circuit

Figure 1-30 shows the basic circuits involved in PI color recording. As shown, there are two *phase-locked loops*. One is known as the AFC (automatic frequency control) loop, and produces a signal at a frequency 44 times fH, or about 693 kHz. The AFC loop receives its input of fH from the TV video signal during

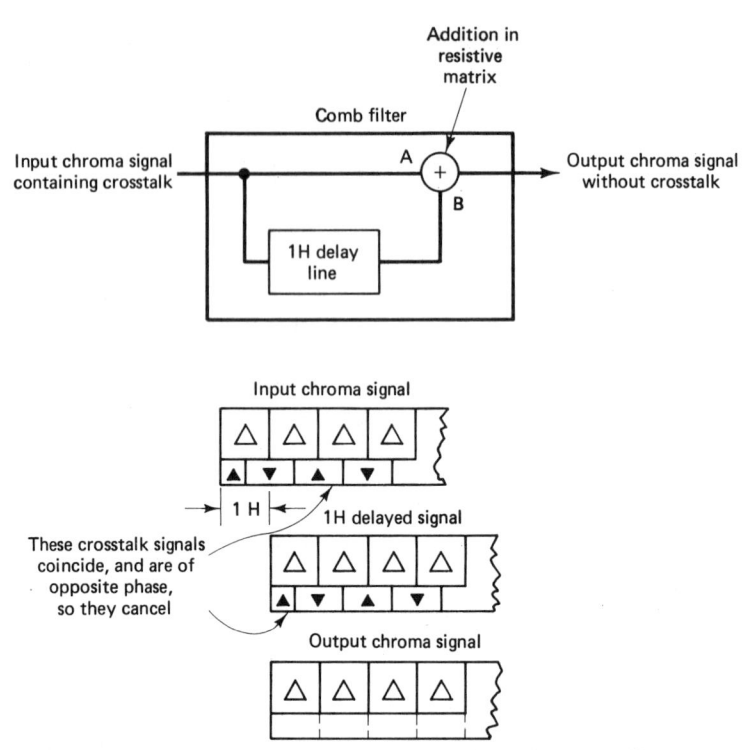

**FIGURE 1-29.** Function of the comb filter in canceling out crosstalk.

record, and an input fH* from the recorded video signal during playback. The output of the AFC loop (either 44fH or 44fH*) is combined with the output of a 3.57-MHz crystal oscillator in frequency converter 2.

The 3.57-MHz oscillator is free running during record, but is locked in phase to the chroma 3.58-MHz signal during playback. (Note that 3.57 MHz is equal to 3.58 MHz, less $\frac{1}{4}$ fH.) Phase lock of the 3.57-MHz oscillator during playback is accomplished by the APC (automatic phase control) loop. In either playback or record, the 3.57-MHz signal is combined with the 44fH signal to produce a 4.27-MHz signal. This 4.27-MHz signal is applied to frequency converters 1 and 3 through a *carrier phase inverter*.

During both playback and record, the 4.27-MHz signal is passed by the *carrier-phase inverter* (usually a center-tapped transformer). The phase inverter is operated by a flip-flop and OR-gate circuit that receives both 30PG and H-sync pulses. Both signals are required since the carrier-phase inverter serves to phase invert the 4.27-MHz with H-sync only when track A is being made. The 30PG pulse overrides the H-sync pulses when track B is being traced by head B.

During record, the 3.58-MHz chroma signal to be recorded is applied to frequency converter 1 where the 4.27-MHz signal is added, resulting in a difference frequency of 688 kHz. Since the 4.27-MHz signal is locked to the H-sync signal, the 688-kHz signal to be recorded is also locked to H-sync.

*Introduction to the Beta System* 37

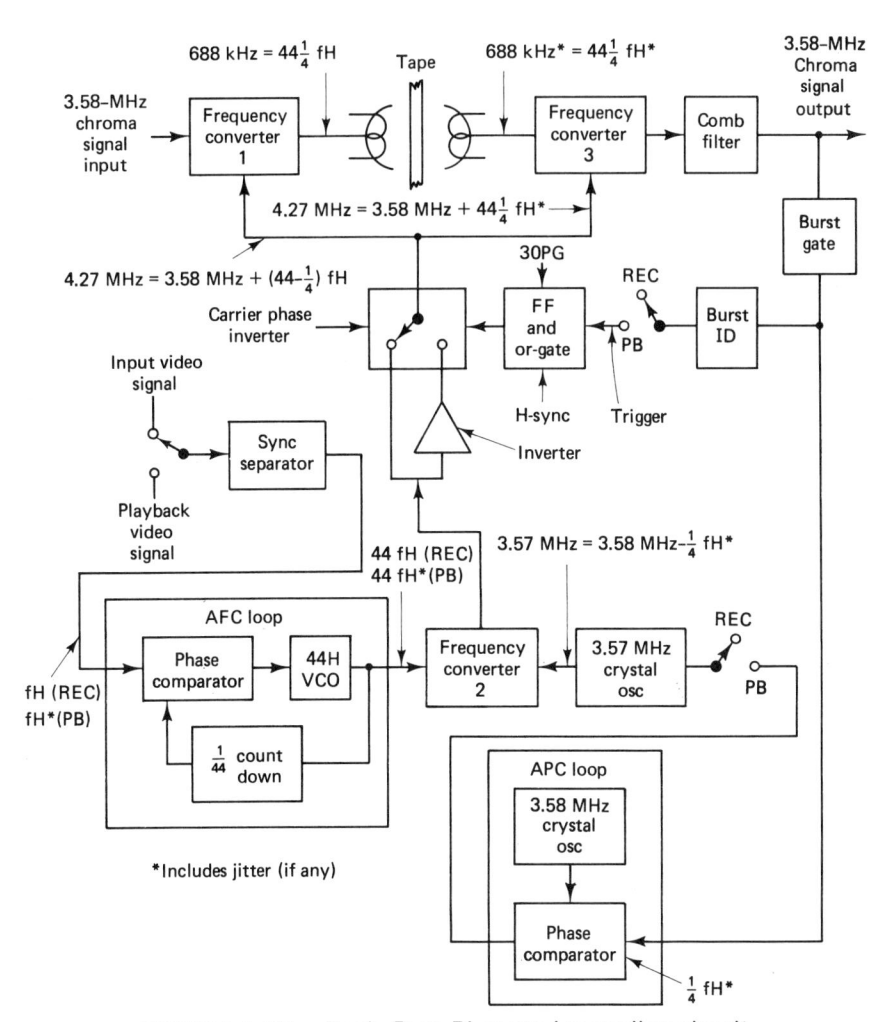

**FIGURE 1-30.** Basic Beta PI control recording circuit.

During playback, the 688-kHz chroma signal from the head is applied to frequency converter 3 where the 4.27-MHz signal is again added, resulting in a difference frequency of 3.58 MHz. The 3.58-MHz chroma output signal is compared with the APC loop oscillator (3.58 MHz). Any phase variations due to jitter are used to shift the 3.57-MHz oscillator signal (free-running during record). Since the phase of the 4.27-MHz signal is controlled by the 3.57-MHz oscillator, any phase shift in the 3.58-MHz chroma output signal is eliminated.

Even with the AFC circuit, there is still a possibility that the 3.58-MHz chroma playback signal burst will lock up on the wrong phase of the 4.27-MHz signal (locked in, but 180° out-of-phase). This condition is prevented by the burst ID (burst identification) circuit that compares the phase of the 4.27-MHz reference signal with the 3.58-MHz chroma signal during playback. If the APC

*38   Introduction to Videocassette Recorders*

system has locked up on the wrong phase, the carrier-phase inverter flip-flop circuit is switched by a trigger pulse developed in the burst ID circuit. The burst ID compares the phase of the 3.58-MHz chroma playback signal for each horizontal line, and produces the corrective pulse whenever the phase-invert flip-flop switch has locked on the incorrect phase.

## 1-8.5  Automatic Tape Loading for the Beta System

The performance of any VCR is determined to a large degree by the mechanical parts. Although a VCR does not differ mechanically from a comparable audio cassette recorder (except for the rotating heads), VCRs incorporate a device that automatically loads the tape around the capstan and head drum. This is generally known as an "auto-loading mechanism" or similar term.

Figure 1-31 shows the auto-loading system for a Betamax VCR. The Beta

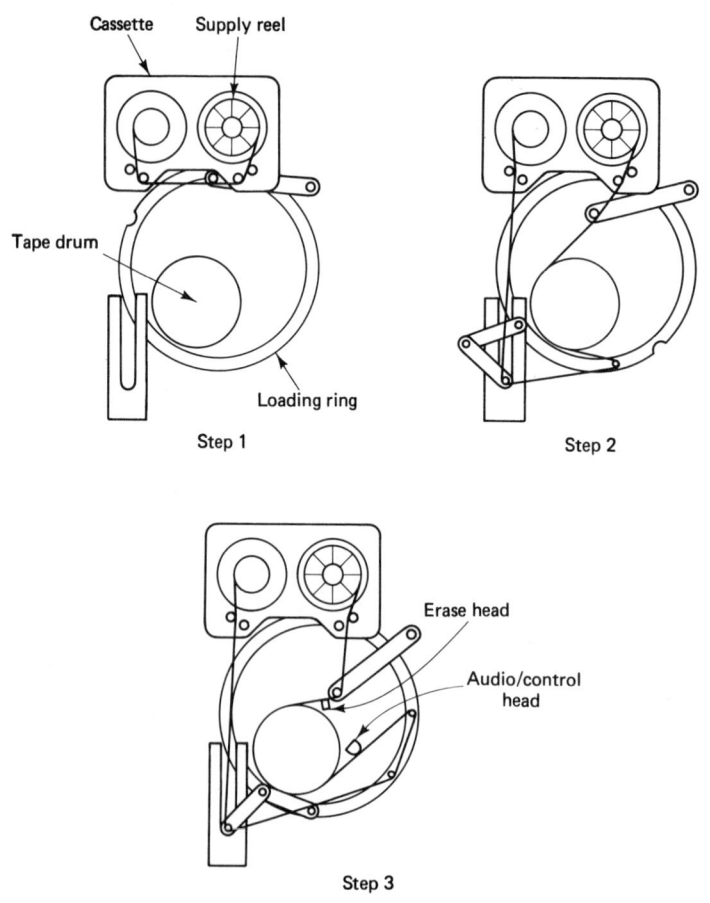

**FIGURE 1-31.**  Basic auto-loading system for a Betamax VCR.

system uses a so-called U loading or threading system, since the tape appears to form the letter U when fully threaded. When the cassette is inserted into the cassette box at the top of the VCR, a loading ring picks up the tape as shown, and then threads the tape around the tape drum in about 3 seconds. When the eject button is pressed, the loading ring turns in the reverse direction, and the excess tape is taken up by the take-up reel inside the cassette. When the loading ring returns to its original position and the tape is all back inside the cassette, the cassette automatically rises and is ejected.

Since video cassettes form a parallel two-reel system, the cassette can be easily removed, even if stopped in the middle during record or playback. Since loading takes place automatically as soon as the cassette is inserted, VCRs are no more complicated to operate than an audio tape recorder. Also, the tape cannot be damaged since no excess pressure is applied and it is not touched by hand.

### 1-8.6 Typical Overall Functions of a Beta VCR during Record

Figure 1-32 is a typical overall block diagram of a Beta VCR during record. The following is a brief explanation of overall operation. In Chapter 3 we describe operation for the circuits of a similar Beta VCR in much greater detail. In this section we also expand on the functions of the video channel during record.

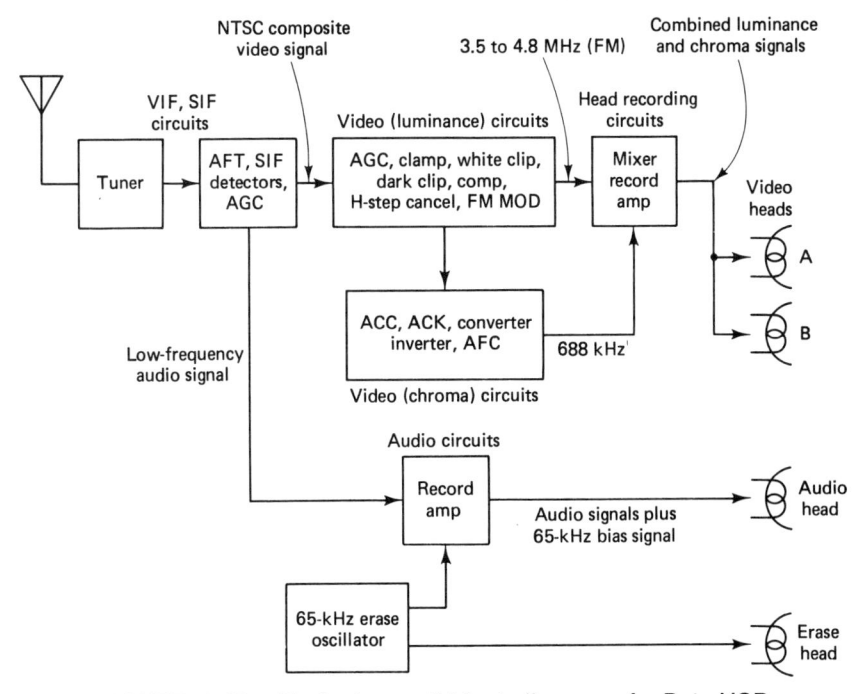

**FIGURE 1-32.** Typical overall block diagram of a Beta VCR during record.

**Tuner (VIF/SIF Detection).** The TV broadcast channel signal is converted to a 45.75-MHz video IF (VIF) and 41.25-MHz sound IF (SIF). The NTSC composite video signal is extracted by a video detector. The audio signal is converted to 4.5 MHz by an audio IF second detector. The low-frequency audio signal is obtained by means of FM detection.

**Luminance Signal (FM).** With AGC maintaining constant NTSC composite video signal levels, the luminance signal and chrominance (chroma) signals are separated. Then, in order to convert the wide-frequency band (0 to 4.2 MHz) luminance signal into an easily recordable FM signal, a clamp circuit (among many other circuits) matches up the level of the luminance signal, which is then processed to reduce crosstalk and improve the signal-to-noise ratio. An FM modulator converts the luminance signal to a 3.5- to 4.8-MHz FM signal.

**Chrominance (Chroma) Signal.** The separated chrominance signal (the 3.58-MHz chrominance subcarrier, or so-called "color burst") is mixed with a 4.27-MHz reference signal and converted to a frequency of 688 kHz. Azimuth recording and phase inversion are used to eliminate the problem of crosstalk between tracks.

**Head Recording Amplifier.** The FM-modulated luminance signal (3.5 to 4.8 MHz) and the chrominance signal (688 kHz) are combined and amplified for optimum recording performance before application to the video heads A and B.

**Audio Recording.** Because of the nonlinear response of magnetic tape, preemphasis and deemphasis are applied respectively during record and playback of the relatively low-frequency audio signal. A 65-kHz bias signal is superimposed on the audio signal during record to raise efficiency and reduce distortion (as is typical for most audio tape recorders). The same 65-kHz oscillator output is also used to erase previously recorded signals on the magnetic tape.

Figure 1-33 shows a typical video signal flow path during record. The following summarizes the functions of the various sections in the signal flow path.

The AGC (automatic gain control) section maintains input signal amplitude fluctuations at a constant level. In the Betamax (Sony), a sync AGC system (similar to that found in many TV receivers) is used. This sync AGC system maintains a constant output level regardless of picture brightness.

The LPF (low-pass filter) removes the sound IF signal (4.5 MHz) and other unnecessary high-frequency components. The ATT (attenuator) reduces the amplitude of the signal level to a suitable value. The Y/C separator separates the luminance (Y) and chrominance (C) components of the color input. The sync tip clamp lines up the sync signal level. The d-c component of the video signal is normally removed by capacitive coupling in the signal path. Clamping is used to

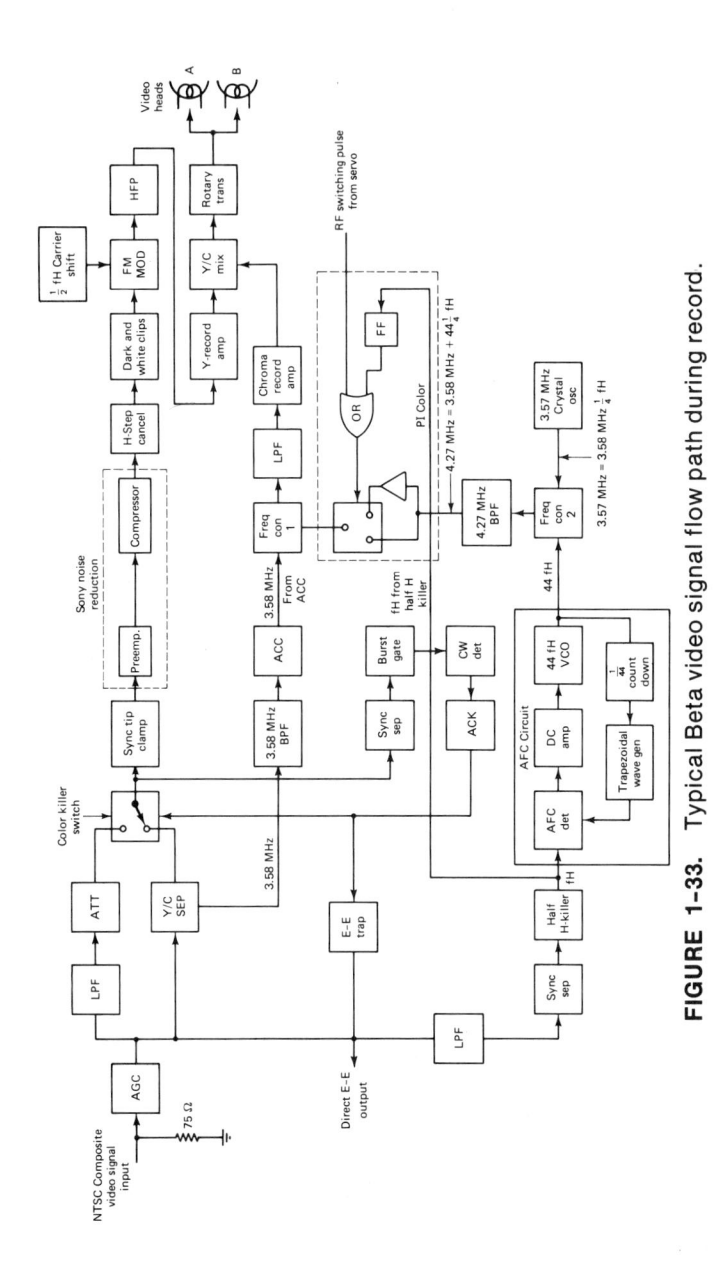

**FIGURE 1-33.** Typical Beta video signal flow path during record.

match up the sync signal since the sync signal tip becomes the low-frequency (3.5-MHz) reference for the FM signal (as shown in Fig. 1-24 and discussed in Sec. 1-7).

The E-E trap ensures that the direct E-E output (for a monitor TV) switches to the black-and-white mode when signal conditions are bad. Under such bad conditions, the chrominance signal level of the input video signal is low, and the VCR color killer (ACK, or automatic color killer) circuit operates, switching the recording mode to black and white. The term "E-E," or electric-to-electric, can be explained as follows. When the VCR is in the record mode, the record output circuit is connected to the playback input circuit so that the video signal to be recorded can be monitored on a TV set (if desired). Since the magnetic components (heads, tape, etc.) have nothing to do with this signal (the signal is passed directly from one electrical circuit to another), the function is called the E-E mode. When the heads and tape are involved in the normal record/playback cycle, the term "V-V" or video-to-video is sometimes used.

The Sony noise reduction system includes both preemphasis and compressor functions to produce a nonlinear emphasis of the video signal during record. In any VCR, emphasis is applied to raise the higher-frequency part of the video signal during record (as is done in most audio recorders). In the Sony noise reduction system, the amount of emphasis applied varies with the fluctuation in the input signal level (a low level receives greater emphasis, and a high-level signal receives less emphasis). This feature produces a constant maximum degree of FM modulation. During playback, deemphasis (having the opposite characteristics of the emphasis applied during record) is used to reproduce the original signal. The use of this noise-reduction system makes possible further improvement in the high-frequency range signal-to-noise (S/N) ratio.

The H difference or step canceler circuit corrects nonuniformity in H-sync spacing. In areas where reception conditions are bad, the H-sync signal may not be uniform, resulting in possible crosstalk problems during playback. The H difference or step canceler circuit maintains constant H-sync spacing.

The dark and white clipping circuit clips unwanted high-level pulse components arising from preemphasis to help stabilize the picture. Dark clipping cuts off excess sync signal excursions, while white clipping prevents whiter-than-white level (overmodulation).

The HPF (high-pass filter) differentiates the output of the FM modulator, and passes the differentiated peaks. Differential recording is used because of the undesirable phenomenon of self-demagnetization by high-frequency components when the luminance signal (FM) is recorded.

The $\frac{1}{2}$fH carrier shift circuit produces a $\frac{1}{2}$fH difference between the carrier frequencies of the field A period (during the A head period) and field B period. Without such a circuit, there is a possibility of hetrodyning (or "beating") due crosstalk between adjacent tracks during playback. The $\frac{1}{2}$fH carrier shift circuit shifts the beat frequency by a $\frac{1}{2}$-offset relationship, making the beat unnoticeable as far as the picture is concerned.

## 1-8.7 Typical Overall Functions of a Beta VCR during Playback

Figure 1-34 is a typical overall block diagram of a Beta VCR during playback. The following is a brief explanation of overall operation. In Chapter 3 we describe operation for the circuits of a similar Beta VCR in much greater detail. In this section we also expand on the functions of the video channel during playback.

**FM Playback Amplifier.** The FM luminance signal and the 688-kHz chrominance signal picked up by the two video heads are amplified by a high-S/N-ratio amplifier, and frequency response is corrected. Then the two playback outputs are mixed to form a single continuous signal.

**Luminance Signal Playback.** A high-pass filter is used to separate the FM component of the playback signal. A dropout compensation circuit (DOC) operates if dropout is present. When there are dirt particles or scratches on the tape, the video signal may not be picked up by the heads, and noise can appear on the screen. These variations in signal level during playback result in a data-reduction error called "dropout." After passing the DOC circuit, the luminance signal is frequency corrected, and passed through a noise canceler circuit before emerging as a proper video signal.

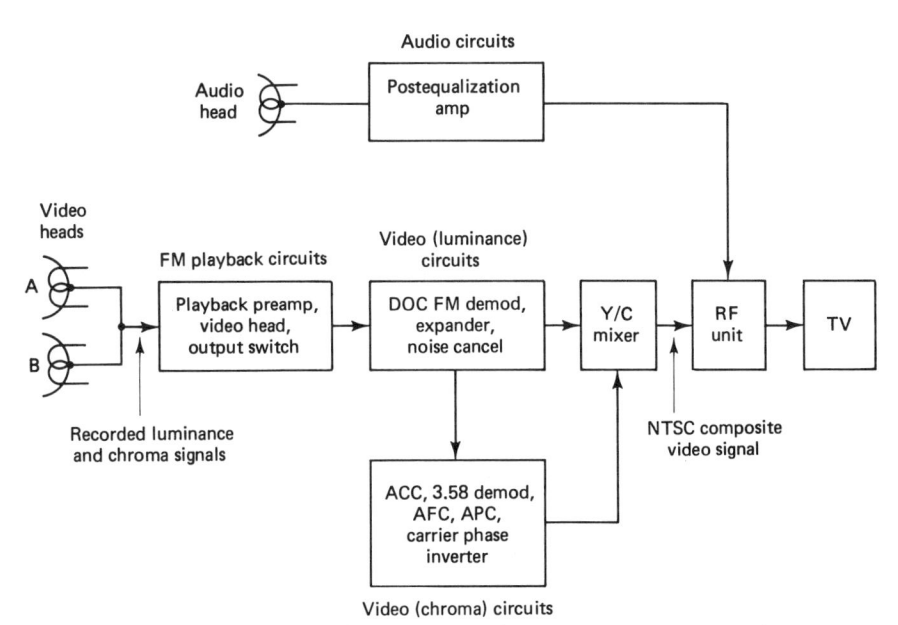

**FIGURE 1-34.** Typical overall block diagram of Beta VCR during playback.

44   Introduction to Videocassette Recorders

**Chrominance Signal Playback.** The 688-kHz low-frequency range chrominance subcarrier is separated from the playback signal by a low-pass filter, undergoes frequency mixing with an internally generated frequency of 4.27 MHz, and is demodulated as a 3.58-MHz chrominance subcarrier signal. At the same time, the comb filter is used to eliminate crosstalk between the two video tracks. As discussed in Sec. 1-8.4, the internally generated 4.27-MHz signal is processed by the APC and AFC circuitry to remove time-base errors (jitter) that may appear during record and playback.

**Y/C Mixing.** The luminance (Y) and chrominance or chroma (C) signals are mixed to form an NTSC composite color video signal.

**Audio Playback.** The low-frequency signal picked up by the stationary audio head undergoes postequalization to correct frequency response and poor S/N ratio, and is then amplified to a suitable level.

**RF Unit.** The NTSC composite color video signal and the audio signal are converted to a television broadcast frequency signal (channel 3 or 4, typically) so that the VCR output signal can be viewed on a conventional color-TV set.

Figure 1-35 shows a typical video signal flow path during playback. The following summarizes the functions of the various sections in the signal flow path.

The head preamplifiers amplify the very low amplitude video signals at the heads. The equalizer corrects the frequency response of the two head signals and adjusts channel balance so that both head A and head B signal levels are equal.

The switcher/mixer combines the signals of two channels and removes any overlap to provide a composite signal output. Since the video tape wraps the drum just a bit more than 180° (as shown in Fig. 1-20), the RF switching pulse from the servo circuit is used to remove the overlap between channels A and B, which are then formed into a composite signal in the mixer circuit.

The ATT (attenuator) matches the level of black-and-white reproduction with that of color reproduction. The HPF (high-pass filter) passes only the luminance (FM) component, and rejects the 688-kHz chroma signal. In the absence of a color signal during playback, the ACK circuit switches the video signal through the HFP so as to remove any chroma signal or chroma noise that might interfere with black-and-white reproduction.

The DOC senses any dropout and compensates by using the preceding horizontal line (1H) signal. There are several DOC circuits in common use. Most DOC circuits use a limiter, detector, adjustable Schmitt trigger, and 1H delay line, as shown in Fig. 1-36. Typically, all of the circuit elements, except for the Schmitt trigger adjustment control, are contained within an IC. The limiter suppresses any waveform distortion and passes the video signal to the detector and adjustable Schmitt trigger. The adjustable control sets the firing point of the

Schmitt trigger and thus sets the point at which dropout compensation occurs. If dropout is present, the Schmitt trigger switches the video signal so that the preceding 1H delayed signal (passed through the 1H delay line) appears at the output. Note that only a few lines of dropout can be handled if the dropout is continuous.

The output of the LPF is applied to the chroma playback circuit, consisting of the frequency converter, comb filter, AFC, APC, PI, and burst ID circuits. Operation of these circuits is as discussed in Sec. 1-8.4.

Output of the DOC is applied to the limiter, which limits amplitude of the FM signal and removes amplitude fluctuation components from the playback signal. The FM demodulator/LPF combination changes the FM signal back into the original video signal. Most video FM playback demodulators use a multivibrator/multiplier circuit rather than the familiar balanced FM discriminators. No matter what form of video FM demodulator circuit is used, the circuit is part of an IC and is neither accessible nor adjustable.

The demodulated video signal is applied through a noise reduction system similar to that described for record operation. Since nonlinear emphasis is applied during record, the opposite process must take place during playback.

The $\frac{1}{2}$fH carrier shift return circuit restores the $\frac{1}{2}$fH carrier shift produced during record. If this carrier shift is not restored, the d-c component will fluctuate when the FM signal is demodulated.

In the noise canceler, the noise component (high-frequency component) of the video signal is removed by a high-pass filter, the phase is inverted and, with reverse phase and the same amplitude, the processed video signal is added back to the original video signal, thus canceling the noise. This sequence is shown in Fig. 1-37.

During rewind, fast forward, and playback servo startup time, a muting signal from the system control circuitry prevents the video signal from being applied to the RF modulator (or to the video output connector).

## 1-8.8 Typical Overall Functions of a Beta VCR Servo System

Figure 1-38 is a typical overall block diagram of a Beta VCR servo control system. The following is a brief explanation of overall operation. In Chapter 3 we describe operation for the circuits of a similar Beta VCR in much greater detail. In this section we also expand on the functions of the servo system during both playback and record.

The servo (a drum servo in this case) has the job of constantly governing rotational speed (1800 rpm) and position of the video heads during record and playback. During record, the servo maintains constant phase and rotational control of the drum by comparing the phase of the 30PG signal with the phase of the vertical sync (V-sync) signal derived from the video input. The servo system also

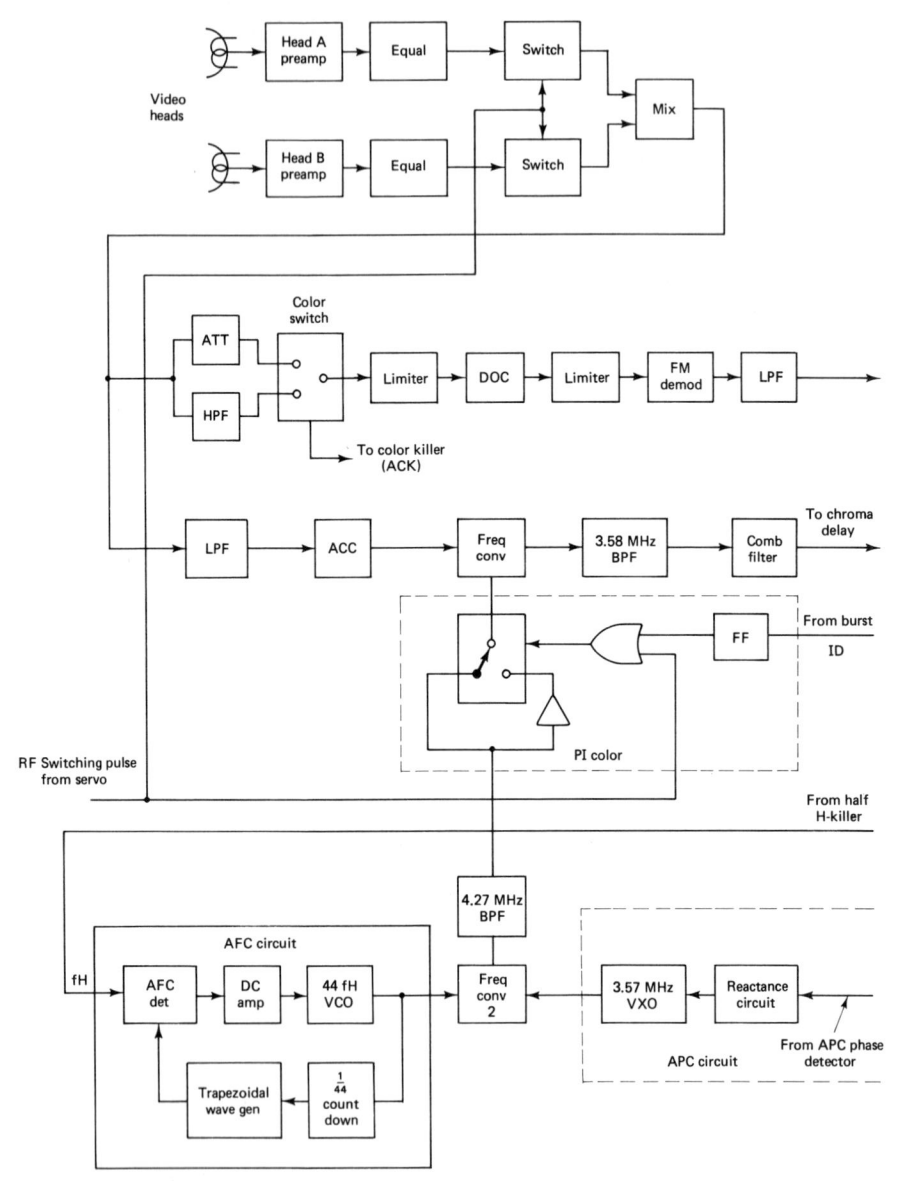

**FIGURE 1-35.** Typical Beta video signal flow path during playback.

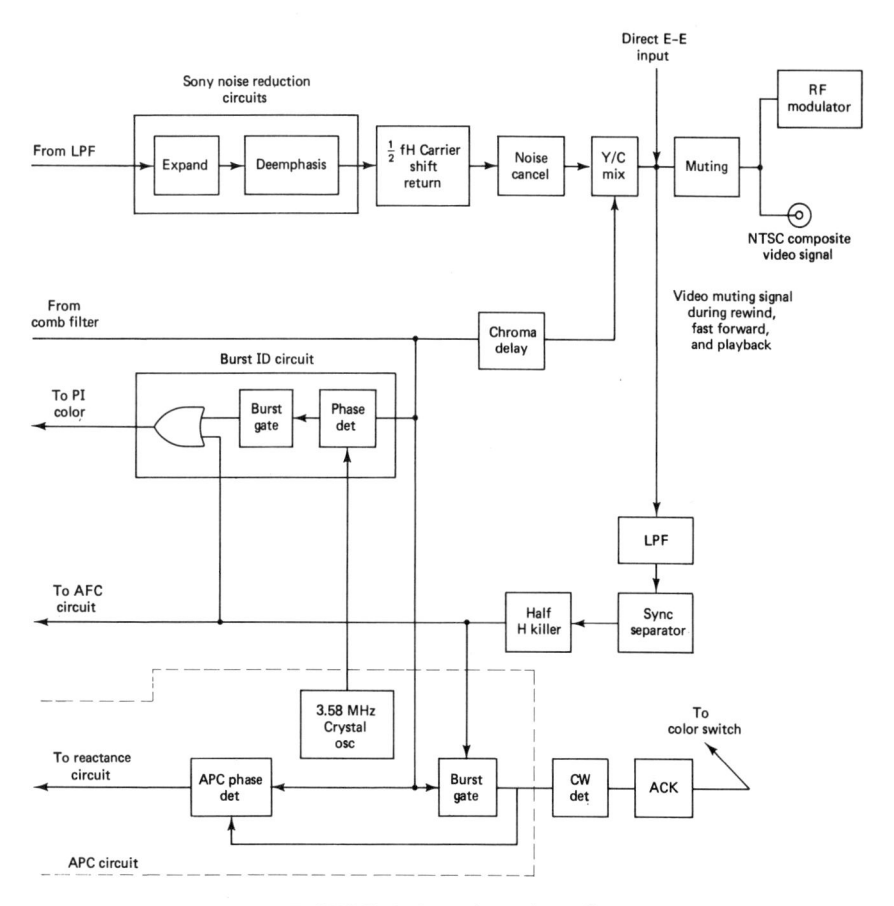

**FIGURE 1-35.** (*continued*)

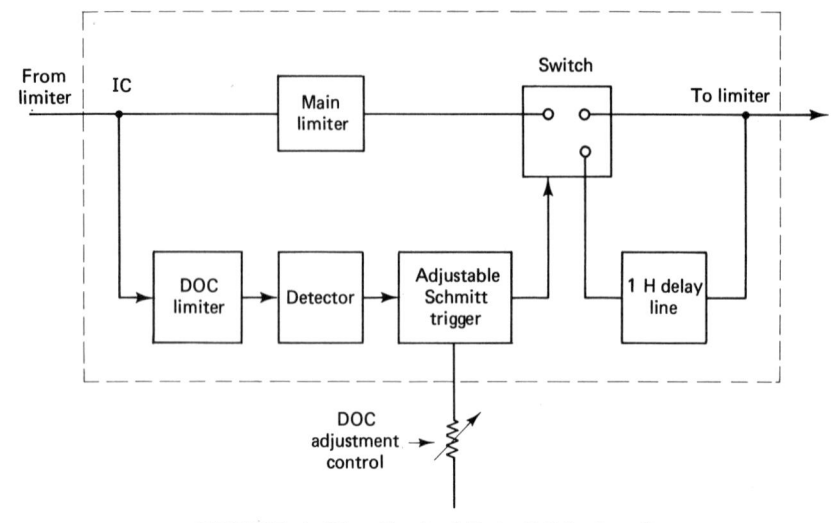

**FIGURE 1-36.** Typical Beta DOC circuit

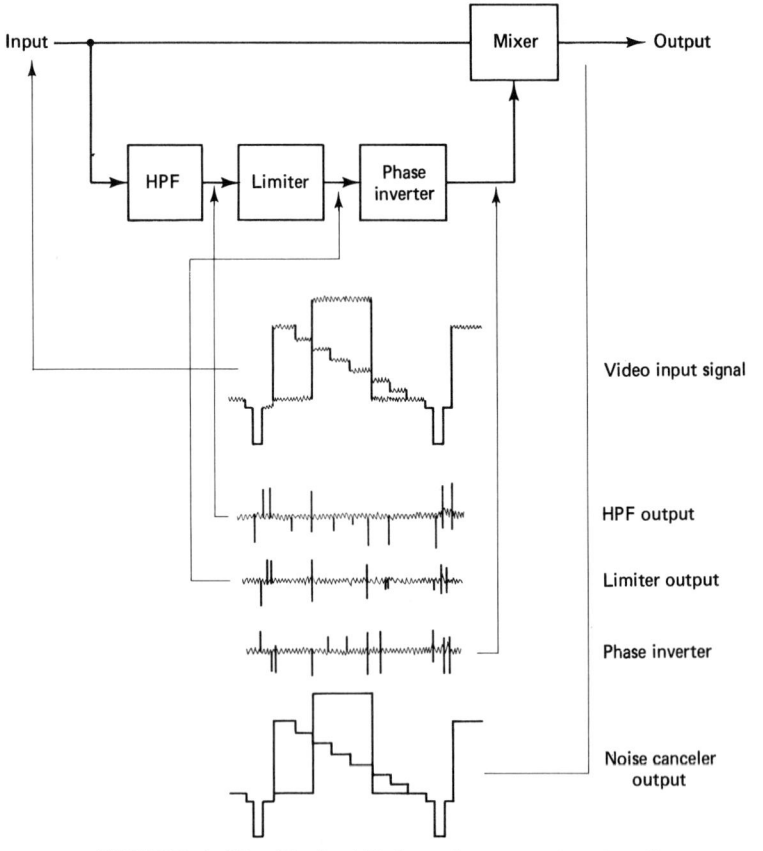

**FIGURE 1-37.** Typical Beta noise canceler circuit.

Introduction to the Beta System    49

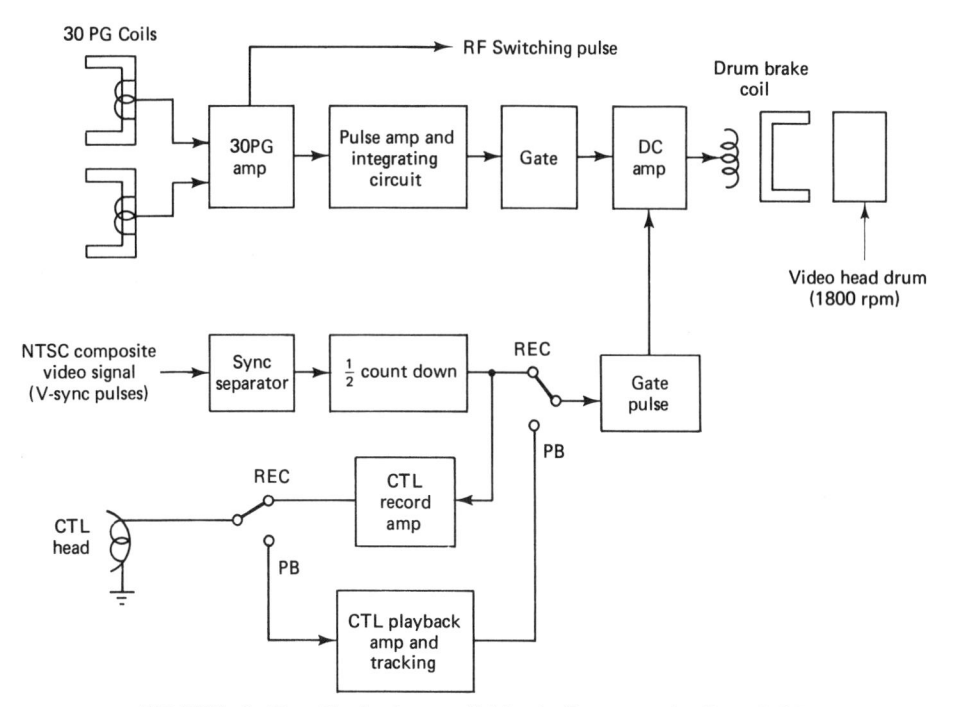

**FIGURE 1-38.** Typical overall block diagram of a Beta VCR servo control system.

supplies the CTL (control) signal to the control head. This control signal is recorded on the control track and is used as the reference signal during playback.

During playback, the phase of the 30PG signal and the reproduced CTL signal (being picked up by the CTL head) are compared to obtain servo control of the video heads for accurate tracing of the magnetic patterns on the tape. In effect, the system operates as a tracking servo.

Figures 1-39 and 1-40 show both the electrical and physical relationships of a typical servo system control circuit. The following summarizes the functions of the various sections in the signal flow path.

The drum is driven by an a-c hysteresis motor via a belt, and drum rotational speed is controlled by a brake coil. The a-c motor is also used to drive the capstan for tape transport. The rotating drum configuration is made up of an upper drum (stationary), a video head disk (rotating), and a lower drum (stationary). Two video heads and two magnets to generate the 30PG signals are attached to the rotating head disk. The video heads are connected to rotary coils as shown in Fig. 1-40. Corresponding stationary coils are mounted on the lower (stationary) drum. The two sets of coil windings (rotating and stationary) form a *rotary transformer* to couple the video heads to the remaining circuits.

Two PG coils (30PG coils A and B) are also attached to the lower (stationary) drum. These PG coils produce pulses when the rotating magnets pass

50   Introduction to Videocassette Recorders

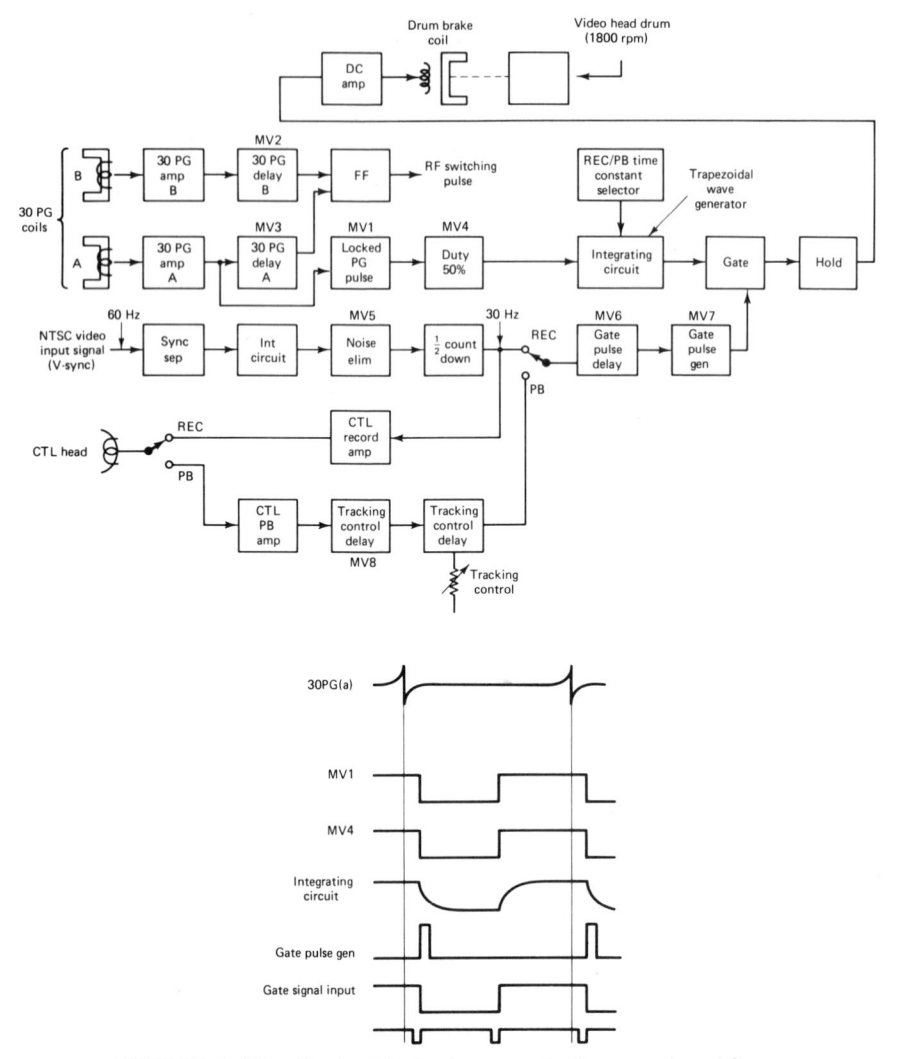

**FIGURE 1-39.** Typical Beta drum control servo signal flow.

by. The 30PG coil A pulse is used for the drum servo. The combination of 30PG coil A and 30PG coil B pulses form the switching pulse used for switching the two video heads, and as a control signal (known as the RF *switching pulse)* for other functions.

30PG pulse A triggers the locked PG pulse multivibrator MV 1 to produce a 30-Hz rectangular wave. After the MV 1 output is formed into a 50% duty cycle waveform by means of MV 4, the output is converted into a trapezoidal wave by an integrating circuit.

The 60-Hz VD signal (V-sync separated from the video input signal) passes

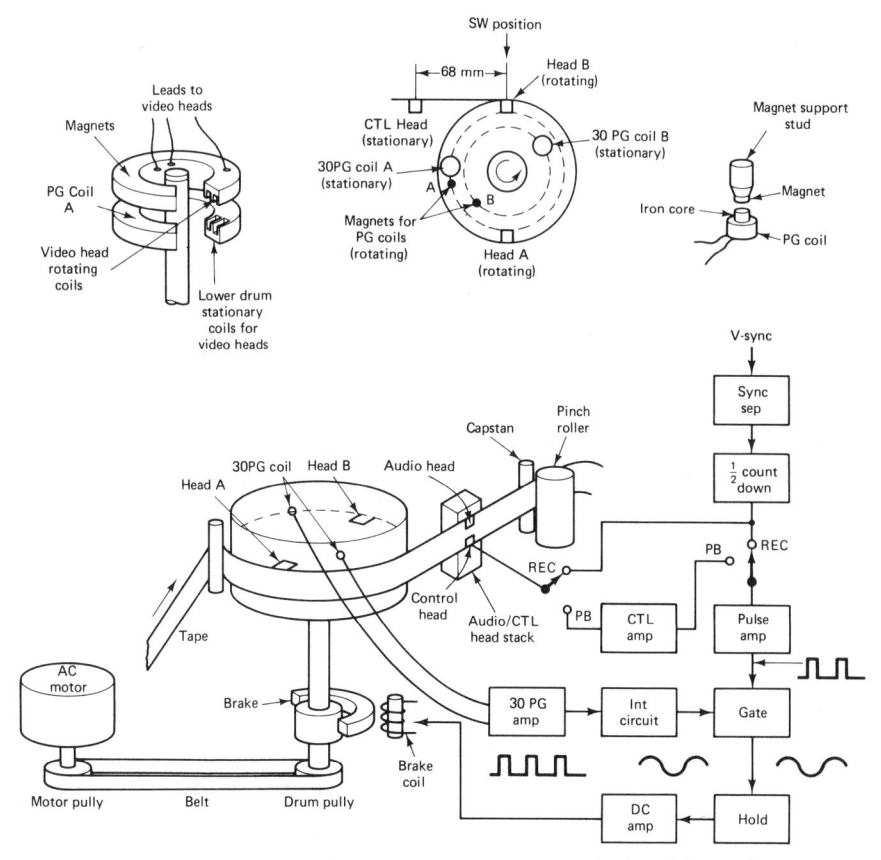

**FIGURE 1-40.** Electrical and physical relationships of a typical Beta drum servo system control circuit.

through the noise eliminator MV 5, and is converted into a 30-Hz rectangular wave by the $\frac{1}{2}$ countdown FF. While the countdown FF output is recorded as the CTL signal, it is also delayed by MV6, and the trailing edge of the trapezoidal wave is sampled by the gate circuit. The gate output voltage thus obtained depends upon the time period from PG pulse generation to the VD signal or, in effect, the phase of the drum in terms of the VD signal. The sampled voltage is held until the next sampling, and is then applied to a combination correction circuit and d-c amplifier. This output is used to drive the drum brake coil, and thus control speed and phase relationships of the scanner drum.

During playback, the CTL signal triggers tracking control multivibrators MV 8 and MV 9. The output of MV 9 is applied to the gate pulse delay MV 6 during playback. (The relationships of the various servo pulses are shown by the timing chart on Fig. 1-39.) Note that the delay produced by MV 9 can be varied by the *tracking control*. This control provides for minor variations between tapes recorded on one machine and played back on another machine. If the physical distance between the control head and video heads is different for the two machines, the playback signals will not be synchronized, even though the servo is locked to the CTL signal. This condition can be corrected by physically moving the control/audio head stack in relation to the scanner. (That is one of the recommended service adjustment procedures for some VCRs.) However, it is far more practical for the user to operate a front-panel tracking control.

### 1-8.9 Typical Overall Functions of Beta VCR System Control Circuits

The system control circuits are not identical for all Beta VCRs. In fact, this is an area where one model of VCR can be quite different from other models. However, most system control circuits have some basic functions in common.

First, the system control circuits coordinate operation of all other VCR circuits during the various operating modes. For example, the system control circuits provide the necessary voltages and signals to keep both the tape and scanner moving during normal record and playback operations, but stop tape movement when the pause operating mode is selected. Since all VCRs do not have the same operating modes, it is not practical to generalize on the functions of the system control circuits.

Another major function of the system control circuits is to provide a failsafe function that stops operation of the VCR in case of failure, as well as at both ends of the cassette tape. Typical failures that stop operation of the VCR include slack tape, excessive moisture on the tape, prolonged operation in the pause mode, and failure of the drum to rotate.

Most system control circuits also provide the control signals for muting audio and video circuits during fast-forward and fast-rewind operating modes. In Beta VCRs, muting also occurs when there is no CTL signal during playback. This produces a blank picture instead of snow when a blank cassette is played back or at the end of a recording. In Chapter 3 we describe operation for the system control circuits of Beta VCRs in much greater detail.

## 1-9 INTRODUCTION TO THE VHS SYSTEM

Now that we have reviewed Beta VCRs, let us go into similar detail on the VHS system. There are many similarities between the two systems. In this section we concentrate on the differences between the two. Again, note that the circuit descriptions here are general in nature. We go into much more detailed VHS functions in Chapter 4.

The VHS system also uses high-density recording to get the maximum amount of program information on a given amount of tape. This involves *zero guard band recording* and results in the crosstalk problem described in Sec. 1-8.3. The VHS system also uses *azimuth recording* and *phase inversion* to

Introduction to the VHS System    53

minimize the effects of crosstalk. The azimuth recording used for VHS is similar to that for Beta. However, VHS uses a ± 6° azimuth difference (resulting in a 12° difference between head A and head B) rather than the ± 7° for Beta. Also, VHS records the chroma or color information at 629 kHz rather than the 688 kHz for Beta. The 629 kHz is obtained by mixing the incoming 3.58-MHz chroma signal with a 4.2-MHz reference signal, which is phase-inverted and locked to the incoming H-sync signal. Note that 629 kHz is 40 times the H-sync frequency of 15,750 Hz (actually 15,734.26 Hz during a color broadcast).

The phase-inversion system used in VHS is entirely different from that in Beta. In the simplest of terms, the phase of the 629-kHz color signal being recorded on head A is advanced in phase in increments of 90° at each successive horizontal line. At the end of four lines, the 629-kHz signal is back to original phase. For example, lines 1, 2, 3, and 4 are shifted 0°, +90°, +180°, and +270° in succession. When head B is recording, the 629-kHz color signal is shifted in phase (retarded) in the opposite direction (0°, 270°, 180°, 90°). This results in the following pattern:

| Line   | 1  | 2    | 3    | 4    | 5  | 6    |
|--------|----|------|------|------|----|------|
| Head A | 0° | 90°  | 180° | 270° | 0° | 90°  |
| Head B | 0° | 270° | 180° | 90°  | 0° | 270° |

Thus, recorded phase shifts for odd-number lines (1, 3, 5) are the same, but are opposite for even-number lines (2, 4, 6).

When the 629-kHz color signal is played back, the 4.2-MHz signal is again phase inverted, and mixed with the 629-kHz signal to restore the 3.58-MHz chroma signal. When both the playback 629-kHz and reference 4.2-kHz signals are phase shifted in the same direction, the effect in the mixer is to restore the 3.58-MHz signal to its normal phase. When the playback 629 kHz and reference 4.2 MHz are phase shifted in opposite directions, the phase of the 3.58-MHz chroma signal is reversed from normal. Thus, the phase of the restored 3.58-MHz signal is shifted on every other line. As discussed in Sec. 1-8.3, when such a signal is passed through a 1H delay line (Fig. 1-29), the crosstalk component is canceled out, and the normal chroma signal component is double in amplitude.

### 1-9.1  Typical Luminance Circuit Operation of a VHS VCR during Record

Figure 1-41 is a block diagram showing luminance (Y) signal flow in a VHS system during record. This illustration is also referenced in Chapter 4.

The video signal from the VIDEO IN terminal is fed to LPF (low-pass filter) 2F4, where the 3.58-MHz color signal is attenuated and the video fed to the AGC circuit Q201 and Q202. The AGC circuit serves to keep the output level constant at all times regardless of input-level variations. The video signal subject

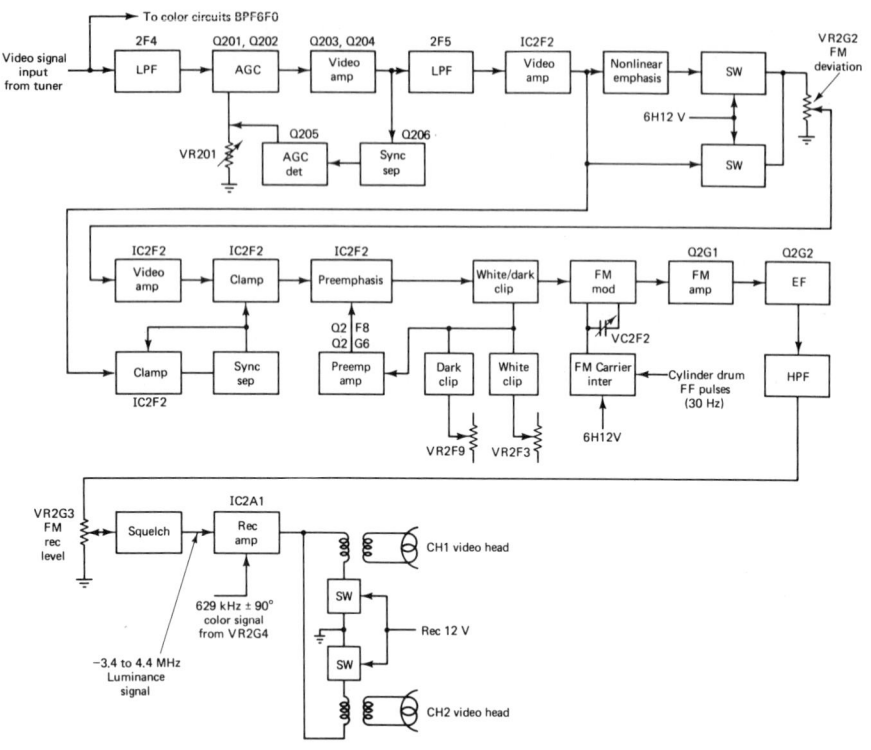

**FIGURE 1-41.** Luminance (Y) signal flow in a VHS system during record.

to AGC is amplified by the video amplifier Q203 and Q204, and fed to LPF2F5. Low-pass filter 2F5 serves as a 3.58-MHz trap which removes the color signal completely. A pure video signal is amplified by video amplifier IC2F2 and is fed to the nonlinear emphasis circuit. This circuit (Q2H1–Q2H5 and D2F6–D2F7) emphasizes the luminance signal frequencies by different amounts, depending on playing time (2 hours or 6 hours). The selection of frequency emphasis is made by a pulse 6H12V from Q402 applied to the switches.

The output of the nonlinear emphasis circuit is applied through VR2G2, which sets the level of FM modulation. Note that the nonlinear emphasis network is completely bypassed on the 2-hour playing mode. In any mode, the signal from the nonlinear emphasis circuit is amplified by the video amplifier (part of IC2F2) and is fed to a clamp circuit where the d-c voltage of the video sync tip remains constant regardless of the fluction in the video signal. This keeps the sync tip at 3.4 MHz (as shown in Fig. 1-24 and described in Sec. 1-7).

The clamped signal is fed to the preemphasis network, where the high-frequency spectrum is emphasized to improve the signal-to-noise ratio in FM modulation. This preemphasis of high frequencies is necessary to reduce noise.

In FM, the higher the modulated frequency, the more liable it is to be influenced by noise.

The preemphasized video signal causes a sharp overshoot at the rise and fall of the video signal. If such a signal were fed directly to the FM modulator, the frequency deviation would be excessive at the rise and fall, and overmodulation would result. Overmodulation could cause a *reverse phenomenon* (or *negative picture*), and will cause a poor S/N ratio. The white and dark clip circuits are included to prevent overshoot above a specified level.

The white and dark clip circuits are set to the correct level by VR2F3 and VR2F9, respectively. Output from the white and dark clip circuits is applied to the FM modulator, which is designed to operate at 3.4 MHz for the sync tips and 4.4 MHz for the white peaks. The FM modulator also receives 30-Hz pulses from the FM carrier interleaving circuits. These circuits are similar to the $\frac{1}{2}$fH carrier shift circuits described for the Beta system in Sec. 1-8.6. As discussed, there is a possibility of hetrodyning due to crosstalk between adjacent tracks during playback. The carrier interleaving circuits advance the video signal phase by $\frac{1}{2}$fH for the channel 2 track. (Note that in some VCR service literature, the heads and tracks are referred to as A and B, whereas the terms "channel 1" and "channel 2" are used to identify the heads and tracks in other literature.)

The FM luminance signal is amplified by the FM amplifier and passed through an emitter follower and high-pass filter. The HPF attenuates the lower end of the FM signal so as not to interfere with the 629-kHz chroma signal that is added later in the signal path. The output of the HPF is applied to squelch circuit through VR2G3, which sets the level of FM modulation.

The squelch circuit serves to prevent the signal from being fed to the record amplifier for about 1.5 s after completion of cassette loading. This prevents the recorded signal from being erased if the tape runs near the drum in a transient tape running condition (in the middle of loading). The signal passing through the squelch circuit is amplified by the record amplifier to the optimum recording level and is supplied to the video heads through a rotary transformer, to be recorded on the tape. Note that the record amplifier also receives the 629-kHz chroma signal from the color recording system, as described in Sec. 1-9.3.

### *1-9.2 Typical Luminance Circuit Operation of a VHS VCR during Playback*

Figure 1-42 is a block diagram showing luminance (Y) signal flow in a VHS system during playback. This illustration is also referenced in Chapter 4.

The reproduced signal from the video heads for channel 1 and channel 2 is supplied to the preamplifier IC2A0 separated for each channel through the rotary transformer. Switch circuit IC2A0 processes the signals from the two channels and removes any overlap to provide a composite signal output. This signal is amplified by mixer amplifier IC2A0 and applied to video amplifier cir-

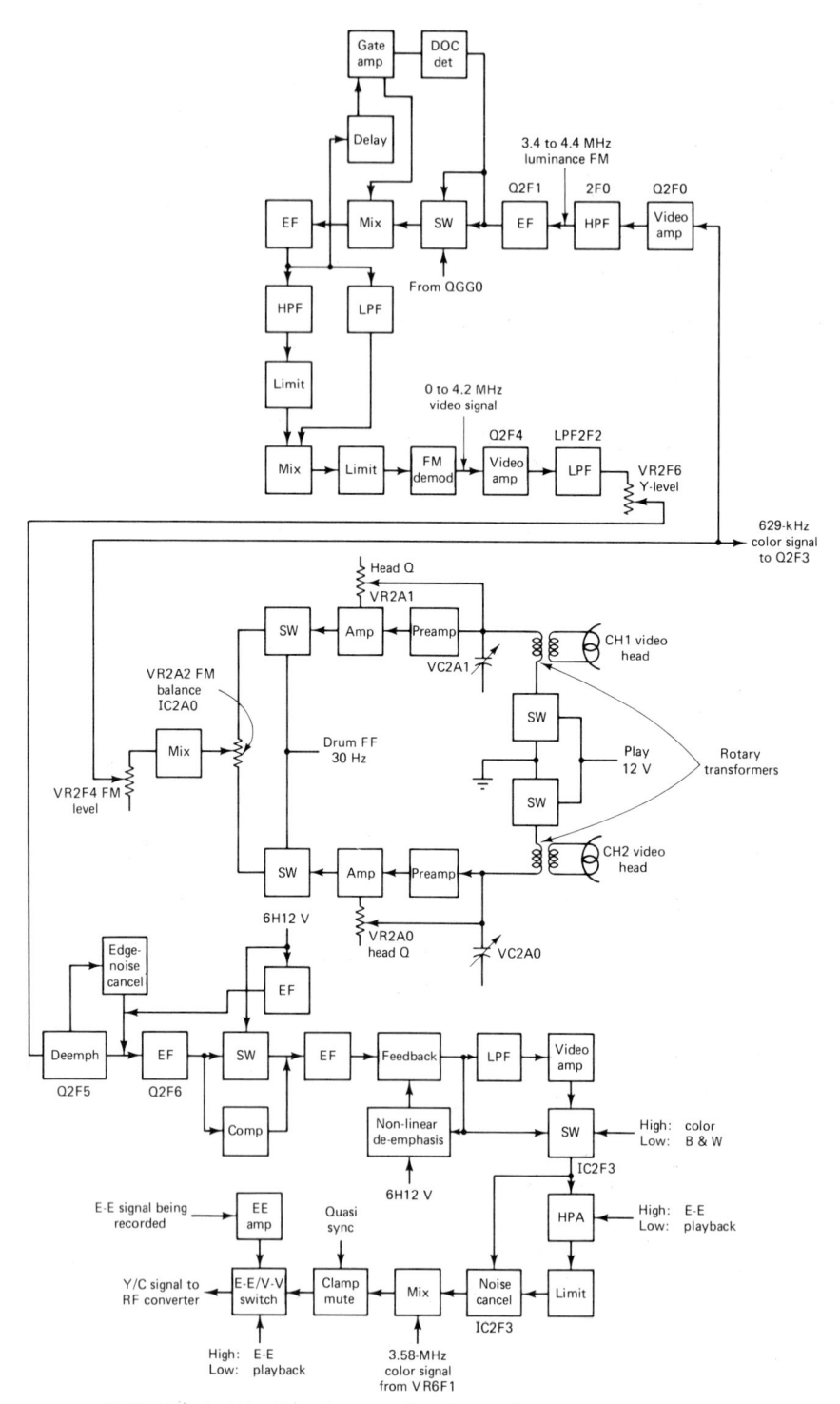

**FIGURE 1-42.** Luminance (Y) signal flow in a VHS system during playback.

cuit Q2F0 through the FM level adjust VR2F4. The reproduced signal is also made available to the color circuits (Sec. 1-9.4) from VR2F4. After being amplified, the signal is passed through HPF2F0 to extract only the luminance (Y) FM signal and applied to a dropout compensation circuit consisting of a mixer amplifier, 1H delay line, a gate amplifier, a DOC, and a detector. This circuit prevents deterioration of picture quality by supplying a 1H preceding signal through the 1H delay line if the FM signal is partially missed (drop out) due to a flaw in the magnetic tape, excessive dirt, and so on. The dropout circuit also provides a pulse to the AFC circuits (Sec. 1-9.6).

The FM signal is then fed to a double limiter circuit consisting of a highpass filter, first limiter, low-pass filter, mixer amplifier, and second limiter. This circuit removes the AM components in the FM signal. The signal is then fed to the FM demodulation circuit, which uses a delay-line type of phase detection. The demodulated signal is amplified and impedance matched by video amplifier Q2F4, and only the video signal is derived from the low-pass filter LPF2F2. The video signal is compensated in frequency response (reverse to the response of preemphasis at recording as described in Sec. 1-9.1) by deemphasis circuit Q2F5. An edge-noise canceler then removes noise from the signal. The video signal is then applied to a compensator circuit through emitter follower Q2F6.

During the 6-hour playing mode, the compensator together with the nonlinear deemphasis and feedback amplifier return the nonlinear emphasis which was supplied during record by the nonlinear emphasis circuit (Sec. 1-9.1). The output of the feedback amplifier is applied to a video amplifier through a low-pass filter (during a color broadcast). For black and white, the low-pass filter is bypassed by action of the color/black and white switch in IC2F3. The low-pass filter is used to remove noise which may arise where the video signal overlaps the demodulated chroma signal.

The signal transmitted through the color/black and white switch circuit is sent to the noise cancel circuit of IC2F3. The noise cancel circuit suppresses pulse noise contained in the video signal. The luminance (Y) and chroma (C) signals are combined in the Y/C mixer circuit, and then applied to E-E/V-V switch circuit through the clamp mute circuit. The output of the E-E/V-V switch is applied to the RF unit, where the NTSC signal is converted to a television broadcast frequency signal (channel 3 or 4). The E-E amplifier and E-E/V-V switch combination permits a signal being recorded to be monitored on a TV set if desired (as discussed in Sec. 1-8.6).

### 1-9.3 Typical Color Circuit Operation of a VHS VCR during Record

Figure 1-43 is a block diagram showing chroma (C) signal flow in a VHS system during record. This illustration is also referenced in Chapter 4.

The video signal from the VIDEO IN terminal is passed through bandpass filter Q6F0 to remove only the chroma signal (3.58 MHz ± 500 Hz), which is

*58 Introduction to Videocassette Recorders*

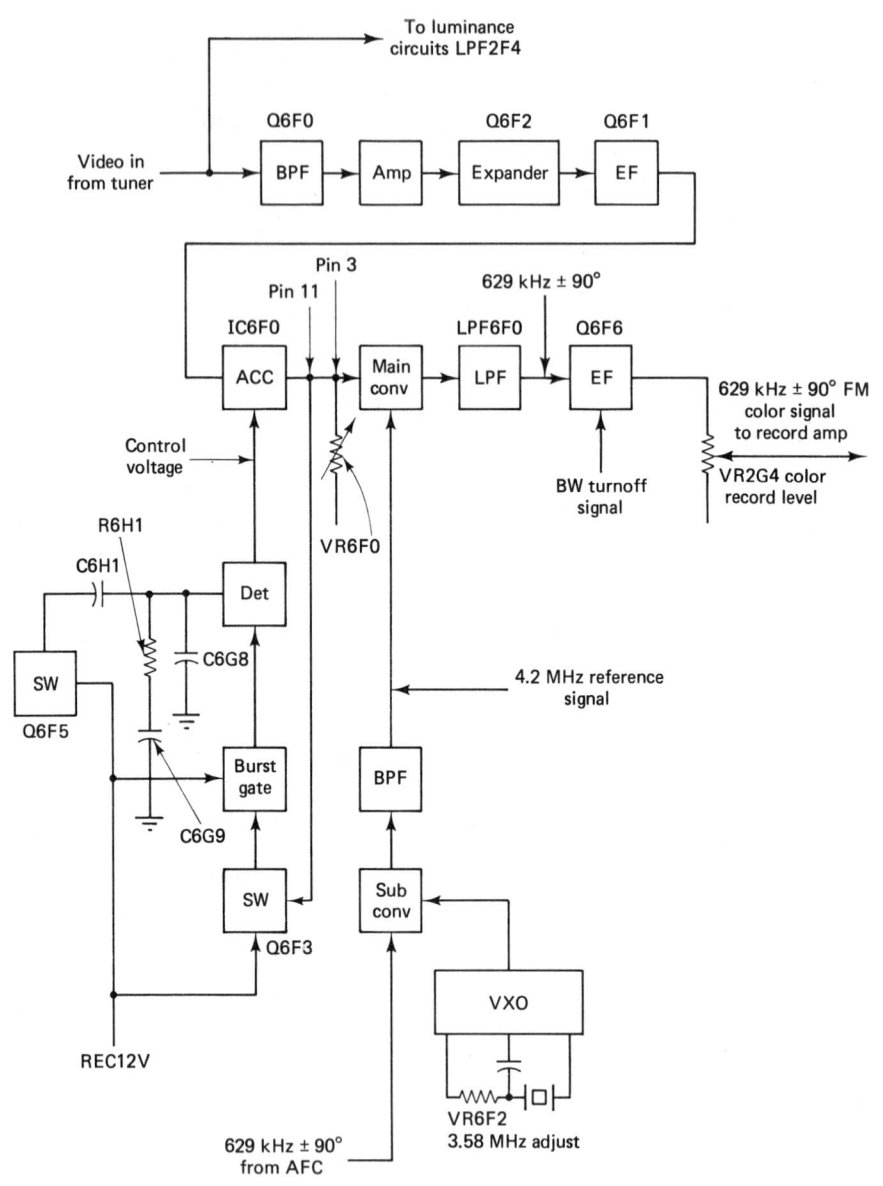

**FIGURE 1-43.** Chroma (C) signal flow in a VHS system during record.

amplified by Q6F0. The color signal is then fed to the expander Q6F2 to boost the burst signal by 6 dB. The burst is fed to the automatic color control (ACC) circuit through impedance-matching emitter follower Q6F1.

The signal at pin 11 of IC6F0 is applied to the color control detector circuit through switch Q6F3 and a burst gate circuit. The peak of the color signal is detected in the detector circuit and produces a control voltage which is applied to the ACC circuit. This detected voltage controls the ACC so as to maintain the color signal at a certain voltage level. The color signal is then applied to the main converter where it is mixed with a reference signal to produce the desired 629 kHz for recording on tape (together with the luminance signal as described in Sec. 1-9.1).

The reference signal applied to the main converter is developed by mixing a 3.58-MHz signal from the VXO (variable crystal oscillator) and a 629-kHz ± 90° signal from the AFC circuit (described in Sec. 1-9.6). Note that the term "±90°" applied to a signal means that the signal has been rotated or shifted in phase every 1H period as described in the introduction to Sec. 1-9. The 3.58-MHz VXO signal and 629-kHz signals are combined in the subconverter to produce a 4.2-MHz signal. This 4.2-MHz signal is passed through a bandpass filter to the main converter, where the signal is combined with the 3.58-MHz chroma signal to produce a 629-kHz ± 90° signal. The resultant signal is then fed to the record amplifier (together with the luminance signal, Sec. 1-9.1) through a low-pass filter, emitter follower, and color record level control VR2G4.

### 1-9.4 Typical Color Circuit Operation of a VHS VCR during Playback

Figure 1-44 is a block diagram showing chroma (C) signal flow in a VHS system during playback. This illustration is also referenced in Chapter 4.

The reproduced signal from IC2F0, applied through the FM level adjust VR2F4 (Sec. 1-9.2), is amplified by Q2F3. At this point, the signal contains both luminance and chroma. Only the 629-kHz ± 90° chroma signal is passed by low-pass filter LPF2F0. This chroma signal is amplified by Q6F0 and applied to the ACC circuit through impedance-matching emitter follower Q6F1.

The chroma signal is maintained at a constant level by the ACC circuit, and is mixed with a 4.2-MHz ± 90° signal from the APC circuit (Sec. 1-9.5) in the main converter. The resultant 3.58-MHz signal is amplified by Q6F7 after being passed by the bandpass filter BPF6F1.

The compressor circuit operates whenever there is a burst gate pulse to reduce gain in amplifier Q6F7. This is necessary to restore the burst signal to a normal level at playback. As discussed in Sec. 1-9.3, the burst signal is increased in amplitude by about 6 dB during record.

The restored chroma signal is passed through a 1H delay line to remove crosstalk from the neighboring video track, as discussed in the introduction to Sec. 1-9. The 1H delay line output is amplified by IC6F1 and is fed to the killer amplifier of IC6F0. This killer amplifier is a switch circuit which allows the signal to pass only when there is a color signal carrier present in the video signal. When the playback is black and white (no color carrier), the color killer prevents a color signal from being applied to color out control VR6F1. This function eliminates noise components from the chroma circuit being applied during a black-and-

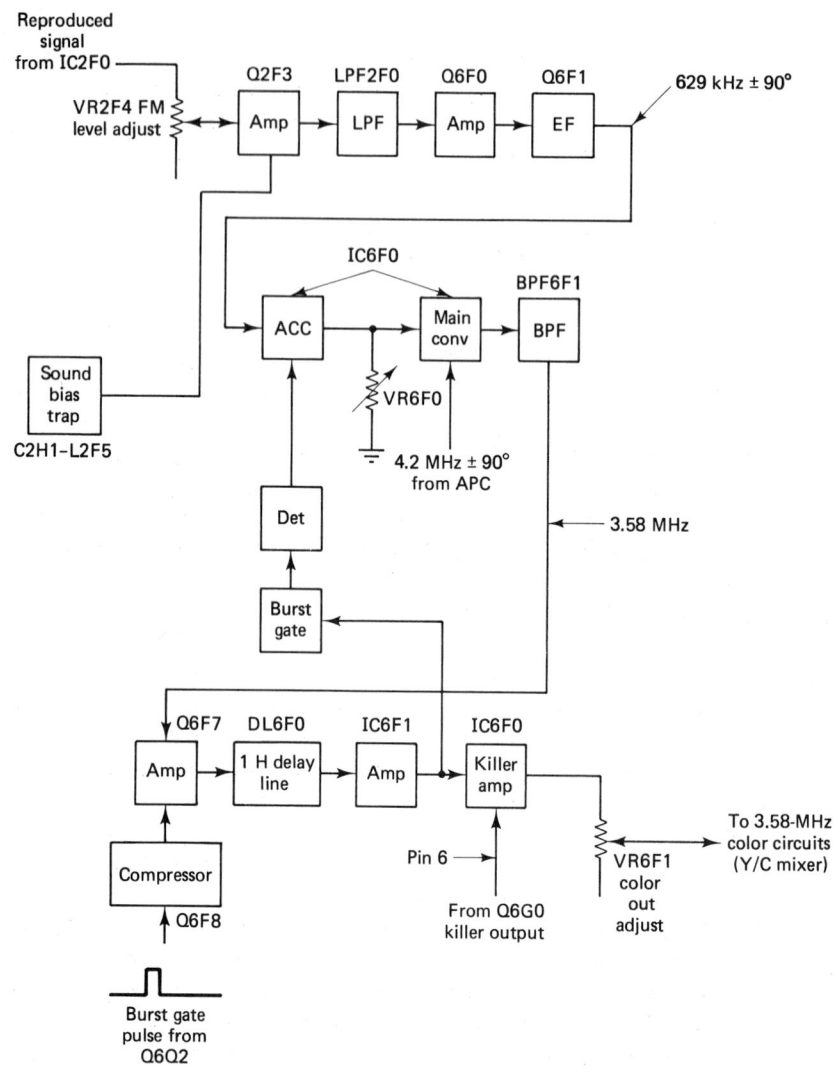

**FIGURE 1-44.** Chroma (C) signal flow in a VHS system during playback.

white playback. The color killer circuit is operated by signals from Q6G0. When pin 6 of IC6F0 is high, the color signal passes through VR6F1 and is superimposed on the luminance (Y) signal (as discussed in Sec. 1-9.2).

## 1-9.5 Typical APC Circuit Operation of a VHS VCR

Figure 1-45 is a block diagram showing operation of the APC (automatic phase control) circuit during record and playback. This illustration is also referenced in Chapter 4.

Introduction to the VHS System    61

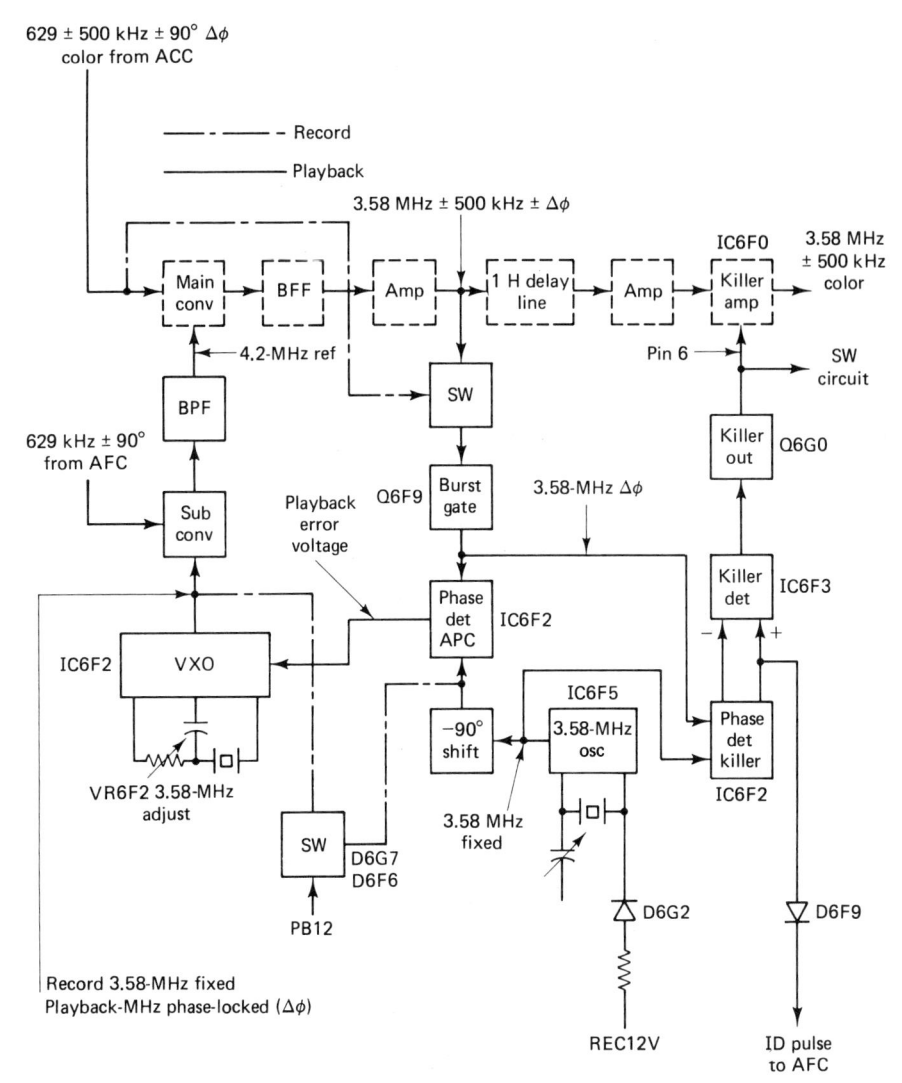

**FIGURE 1-45.**    Typical APC circuit operation.

*During record,* the VXO IC6F2 operates as a fixed 3.58-MHz oscillator. The output of the VXO is mixed with the 629-kHz ± 90° signal from the AFC circuit (Sec. 1-9.6) in the subconverter to form a 4.2-MHz reference voltage, as discussed in Sec. 1-9.3. The VXO output is also applied to the phase detector killer IC6F2 through diode switches D6G6 and D6G7.

*During playback,* the VXO operates as a phase-locked 3.58-MHz oscillator. The phase of the VXO is controlled by an error voltage from the phase detector APC IC6F2. This phase detector receives and compares two 3.58-MHz inputs. One input is the playback color burst (which includes any phase shifts due to jitter), while the other input is from a fixed 3.58-MHz oscillator IC6F5. If

## 62   Introduction to Videocassette Recorders

there are any phase differences between the two signals, the error voltage produced by the phase detector shifts the phase of the VXO to correct the condition. Since the phase of the playback color burst is controlled by the VXO, any phase shift in the playback 3.58-MHz color signal is eliminated.

Note that the term "$\Delta\phi$" or "delta phi" applied to a signal means that the signal has been shifted in phase or is of differing phase. In the case of the 3.58-MHz signal applied to the phase detector, the term means that the signal contains any possible jitter effect which could shift the phase.

The killer circuits shown in Fig. 1-45 have two functions. First, they prevent a color signal from being passed when the signal is black and white only (to eliminate color circuit noise from being mixed with the black-and-white signal). Second, the killer circuits provide an identification pulse (ID) which is used by the AFC circuit to prevent 180° out-of-phase lockup. This is similar to the burst ID pulse used for Beta as described in Sec. 1-8.4.

*During color operation,* the 3.58-MHz color burst signal is passed through the burst gating circuit Q6F9 to the phase detector killer IC6F2, which also receives a 3.58-MHz signal from either the 3.58-MHz oscillator IC5F5 (during playback) or the VXO (during record). The two signals are compared in phase by the phase detector killer. If both signals are of the same phase, the output of the killer detector IC6F3 becomes low, and the output of the killer output circuit Q6G0 becomes high. This high output is applied to pin 6 of killer amplifier IC6F0. As discussed in Sec. 1-9.4, with a high signal at pin 6 of IC6F0, the color signal is passed.

*During black-and-white operation,* there is no 3.58-MHz color burst signal. Therefore, the phase detector killer IC6F2 sees only one signal. The output of the killer detector IC6F3 then becomes high, and the output of the killer output circuit Q6G0 goes low. This low output at pin 6 of IC6F0 cuts off the killer amplifier and prevents passage of color signals (or color noise).

*Also during color operation,* if the 3.58-MHz color burst is exactly 180° out of phase with the IC6F5 3.58-MHz oscillator (locked in phase, but 180° out), the phase detector killer IC6F2 develops a burst identification pulse. This pulse is applied through D6F9 to the AFC circuits (Sec. 1-9.6).

### 1-9.6   Typical AFC Circuit Operation of a VHS VCR

Figure 1-46 is a block diagram showing operation of the AFC (automatic frequency control) circuit. Note that the AFC system operates in much the same way for both record and playback. However, during record, the AFC uses H-sync pulses contained in the video signal from the tuner. During playback, the AFC uses the H-sync signals recorded on tape.

The AFC system has five inputs and one output. The five inputs include the video H-sync pulses, a dropout pulse from the dropout detection circuit (Sec. 1-9.2), a 30-Hz cylinder flip-flop from the servo (Sec. 1-9.7), a color burst ID pulse from the APC circuits (Sec. 1-9.5), and a 3.58-MHz fixed or phase-

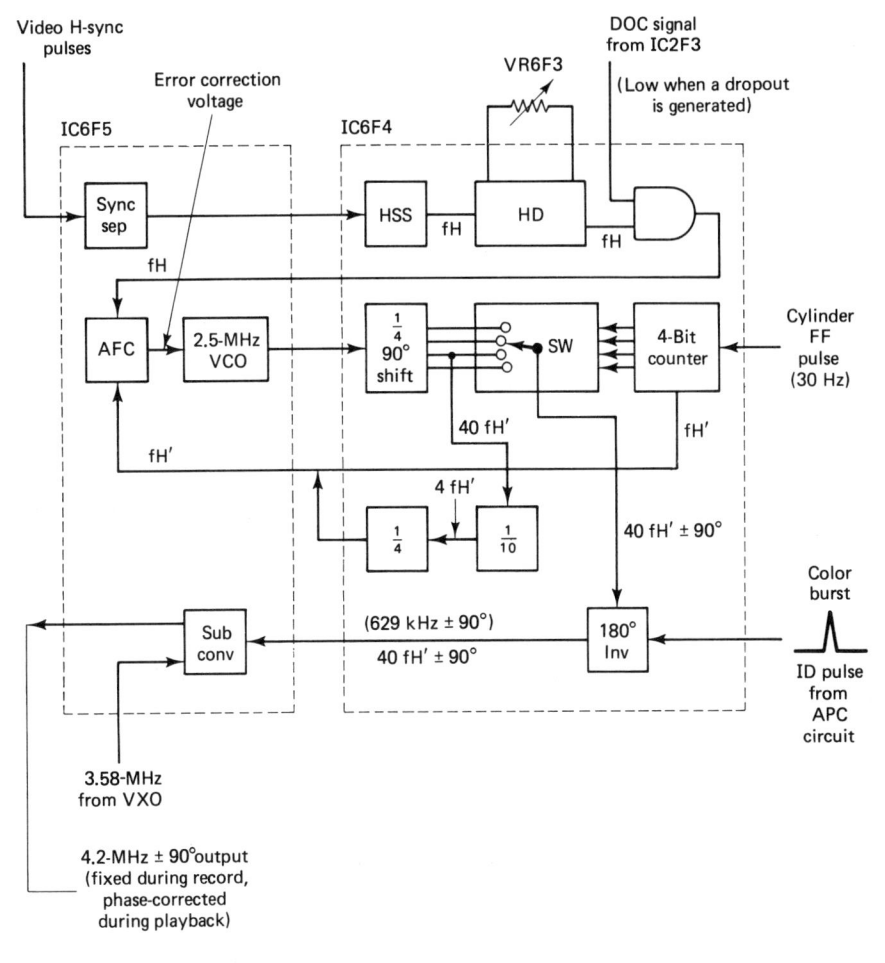

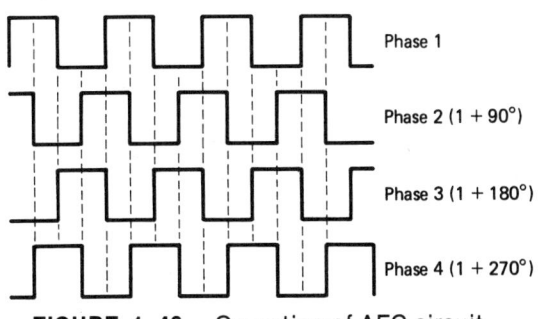

**FIGURE 1-46.** Operation of AFC circuit.

64    Introduction to Videocassette Recorders

corrected signal from the VXO in the APC circuits. The output of the AFC circuit is a 4.2-MHz ± 90° signal (fixed reference during record, or phase corrected during playback).

The video signal (from tuner or playback) is applied to a sync separator where only the vertical and horizontal sync signals are passed. The resultant signal is then applied to the HSS (H-sync separator) where only the H-sync signal is passed. The H-sync signals (or fH as they are referred to in most VCR literature) are shaped into a 2-$\mu$sec pulse by a HD (horizontal drive) circuit. The output from the HD circuit is adjusted to exactly 2 $\mu$sec by VR6F3 and is applied to an AND gate. The other input to the AND gate is normally high so that the 2-$\mu$sec fH pulses can pass. However, if there is a dropout (Sec. 1-9.2) the other AND gate input goes low, preventing the fH pulses from passing.

The output of the AND gate is applied to an AFC circuit within IC6F5. This AFC circuit also receives a fH' (fH prime) signal developed by a 2.5-MHz VCO (voltage-controlled oscillator). Note that the actual frequency of the 2.5-MHz oscillator is 160 times the H-sync frequency of 15,750 Hz (for black and white), or 160 times 15,734.26 Hz (for color). Note that the term "prime" applied to the signal here means that the signal has been locked in frequency to some other signal (to the H-sync signals in this case).

The 160fH' from the VCO is divided by four into 40fH' through operation of a $\frac{1}{4}$ 90° shift and switch circuit which is operated by a 4-bit counter. The 40fH' output of this circuit is further divided by 10 to produce 4fH', and by one-fourth to produce 1fH' (or simply fH'). This fH' is fed back to the AFC circuit. If there is any difference in frequency between the fH signal coming from the AND gate, and the fH' signal originating at the VCO, the VCO is shifted in frequency by an error correction voltage developed in the AFC circuit. Thus, the VCO is precisely locked onto the H-sync frequency.

The 4-bit counter (operated by the fH' and cylinder pulses) produces switch signals which select each of four signals from the $\frac{1}{4}$ 90° shift circuit, in sequence. Each of the four signals is shifted by 90° from the previous signal as shown. In effect, the switch can be thought of as a rotary switch where the rotational direction and speed of the rotor are determined by the 4-bit counter. The counter supplies the pulses to the switch each time an H-sync pulse is applied. As shown in Fig. 1-47, the channel 1 signals are advanced in phase by 90°, whereas the channel 2 signals are retarded or delayed in phase by 90°. These signals (40 fH' ± 90°) are applied through the 180° inverter circuit to be mixed in the subconverter with the 3.58-MHz signal from the VXO, and result in a 4.2-MHz ± 90° that is precisely locked to the VCO.

The 180° inverter is operated by the burst identification (ID) pulse from the APC circuit. As discussed in Sec. 1-9.5, the ID pulse occurs only when the 3.58-MHz color burst is 180° out of phase with the 3.58-MHz oscillator (locked in phase, but 180° out). The inverter normally passes the 40fH' ± 90° signal without change. However, if the burst ID pulse is present (indicating an undesired 180° lockup) the inverter reverses the phase of the 40fH' ± 90° signal to correct the condition.

## 1-9.7 Typical Overall Functions of a VHS Servo System

Figure 1-48 is the block diagram of a typical VHS servo system. The following is a brief explanation of overall operation. In Chapter 4 we describe operation for the circuit of a similar VHS VCR in much greater detail. As in the case of Beta, a VHS servo must keep the speed of the video heads constant relative to the input signal to properly record the video signal on tape. Both the *speed* and *phase* of direct-drive (DD) motors of the cylinder (containing the video heads) and the capstan (that moves the tape) must be servo controlled.

One major purpose of the servo control during record operation is to rotate

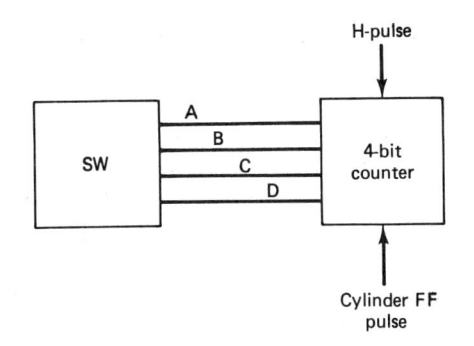

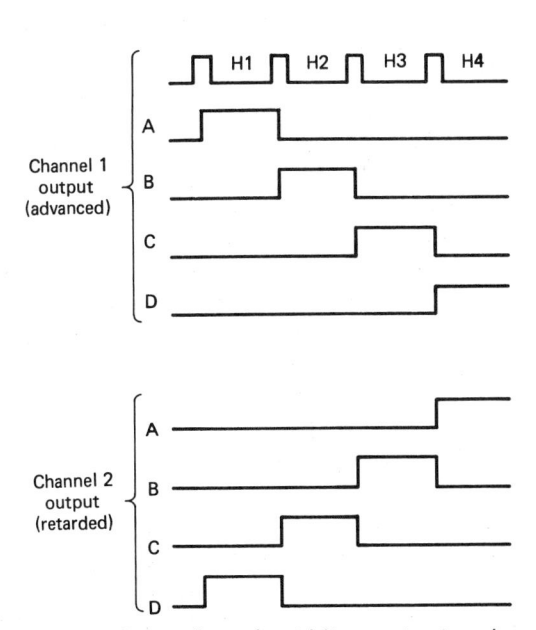

**FIGURE 1-47.** Operation of a 4-bit counter to advance and retard signals by 90°.

65

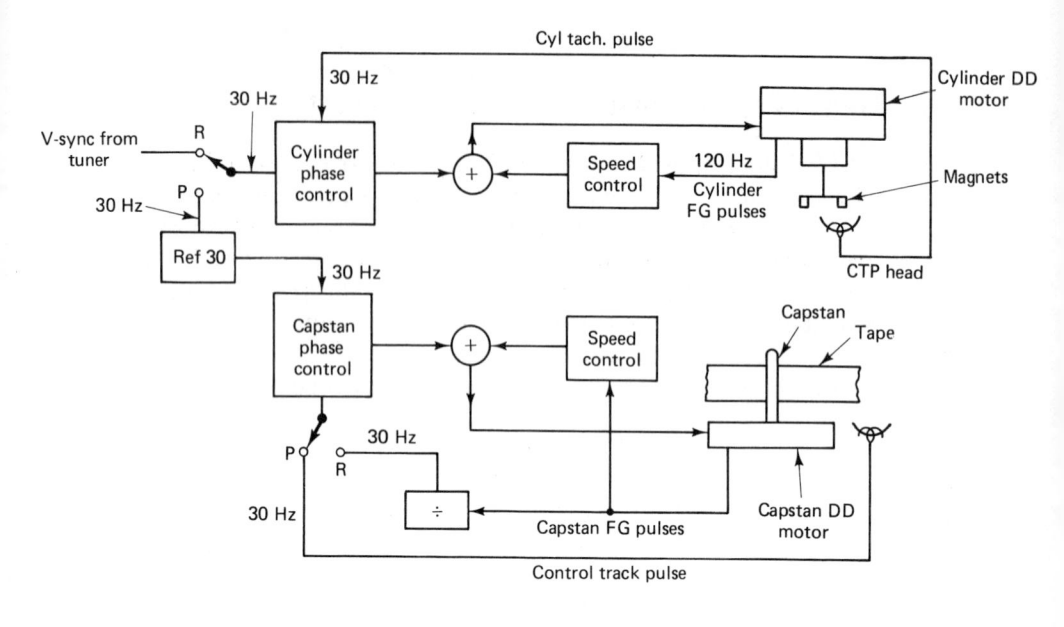

| Pulse | Ferquency (Hz) | |
|---|---|---|
| Cylinder FG | 120 | |
| Capstan FG | SP (2 hour) | 720 |
| | LP ( 4 hour) | 360 |
| | EP (6 hour) | 240 |
| | Slow (slow motion) | 120 |
| | Quick (fast motion) | 720 |
| | Search | 2160 |
| Cylinder tach | 30 | |
| REF 30 | 30 | |
| Control track | 30 | |

| System | | Mode | Reference signal | Waveform | Comparison signal | Waveform |
|---|---|---|---|---|---|---|
| Phase | Cylinder | REC | $\frac{1}{2}$ V-sync | Sample pulse | Cylinder tach pulse | Trapezoid |
| | | PB | REF 30 | | | |
| | Capstan | REC | REF 30 | Trapezoid | 1/24 Capstan FG | Sample pulse |
| | | PB | REF 30 | | Control track pulse | |
| Speed | Cylinder | PB/REC | Cylinder FG | Sample pulse | Cylinder FG | Sawtooth |
| | Capstan | | Capstan FG | | Capstan FG | |

**FIGURE 1-48.** Typical VHS servo system.

the cylinder at precisely 30 Hz, which is one-half the vertical sync frequency (60 Hz) of the input video signal. With this speed, the vertical blanking period can be recorded at any desired point on each video track. In television, the vertical blanking occurs at the bottom of the screen, where it will not interfere with the picture. For this reason, the vertical sync signal is recorded at the bottom (or start) of each video track. This is shown in Fig. 1-49, which is the typical magnetic tape pattern used in VHS.

As in the case of Beta, there are two heads (channel 1 and channel 2) and each head traces one track for each field. Two adjacent tracks or fields make up one complete frame. To ensure that there are no nonimage or blanks in the picture, the information recorded on tape overlaps at the changeover point (from one head to another). This changeover point must also occur at the bottom of the screen, where it will not interfere with the picture. For that reason the vertical sync signal is recorded precisely in the position 6.5H from the changeover time of the channel 1 and channel 2 tracks.

This precise timing requires that the speed and phase of both the cylinder motor and capstan motor be controlled (since the cylinder motor determines the position of the heads at any given instant, while the capstan motor determines the position of the tape). In the servo system of Fig. 1-48, five separate signals are used to achieve the precise timing. The following paragraphs describe each of these signals and how (in general terms) they are used. The tables of Fig. 1-48 summarize the signal functions.

**Cylinder FG Pulses.** As shown in Fig. 1-48, the cylinder FG pulses are developed by a generator in the video head cylinder. The generator consists of an eight-pole magnet installed in the cylinder rotor, and a detection coil in the stator. When the cylinder rotates at 30 rps, the stator coil detects the moving magnetic fields and produces the cylinder FG pulses at a frequency of 120 Hz.

**Capstan FG Pulses.** The capstan FG pulses are developed by a generator in the tape capstan and are applied to the capstan speed control circuits, as well as the capstan phase control circuits (through a divider) during record. The

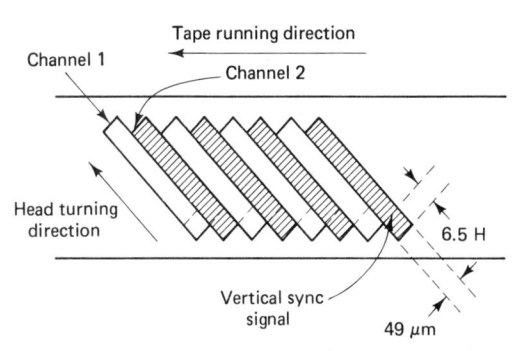

**FIGURE 1-49.** Typical magnetic tape pattern used in VHS.

generator consists of a 240-pole magnet installed in the lower part of the capstan shaft, and a detection coil in the stator. When the capstan rotates, the stator coil detects the moving magnetic fields and produces the capstan FG pulses. The frequency of the capstan FG pulses depends on the speed of the capstan (which also controls tape speed). The tables of Fig. 1-48 show some typical capstan FG pulse frequencies for various playing times and tape speeds. Note that not all VHS machines are capable of the six play modes or tape speeds shown in Fig. 1-48.

**Cylinder Tach Pulse.** The cylinder tach pulses (CTP) are developed by another generator in the cylinder and are applied to the cylinder phase control circuits. The generator consists of a pair of magnets installed symmetrically in a disk in the lower part of the cylinder shaft, and a stationary pickup head. When the cylinder rotates, the CTP pickup head detects the moving magnetic fields. The pulse frequency is a constant 30 Hz. In effect, the tach pulse indicates video head channel switching, and is used as a comparison signal in the cylinder phase control circuits during both record and playback.

**REF 30 Pulse.** The reference signal for the phase control system of both the capstan motor and the cylinder motor is obtained from a crystal oscillator with a frequency of 32.765 kHz. A frequency of 30 Hz is obtained when the crystal oscillator signal is divided. The REF 30 pulse is used for the cylinder phase control circuit only during playback. During record, the cylinder phase control circuit receives broadcast V-sync pulses from the tuner.

**Control Track Pulse.** The 30-Hz control track pulses are the broadcast V-sync pulses recorded on tape during record. At playback, the pulses are picked up by the control head (as described for Beta, Sec. 1-8.8) and applied to the capstan phase control circuit.

### 1-9.8 Tape Loading for the VHS System

The VHS tape loading system is entirely different from that used for Beta (Sec. 1-8.5). Unlike Beta, the VHS tape is not loaded automatically when the cassette compartment lid is closed. Instead, VHS loading starts when the PLAY button is pressed. Basic operation of a typical VHS tape loading system is illustrated in Fig. 1-50. As shown, the VHS system uses a so-called M loading or threading system, since the tape appears to form the letter M when fully threaded. When the cassette is dropped into place, two guide rollers, a tape tension arm, and the capstan are positioned behind the tape (between the tape and the reels). The capstan is stationary, and the two guide rollers are attached to movable arms which are pivoted below the video head scanner. The tape tension arm is spring loaded and pivots near the supply reel.

As shown in Fig. 1-50a, when the PLAY button is pressed, the two guide rollers are moved by the arms to pull the tape against the capstan and the various

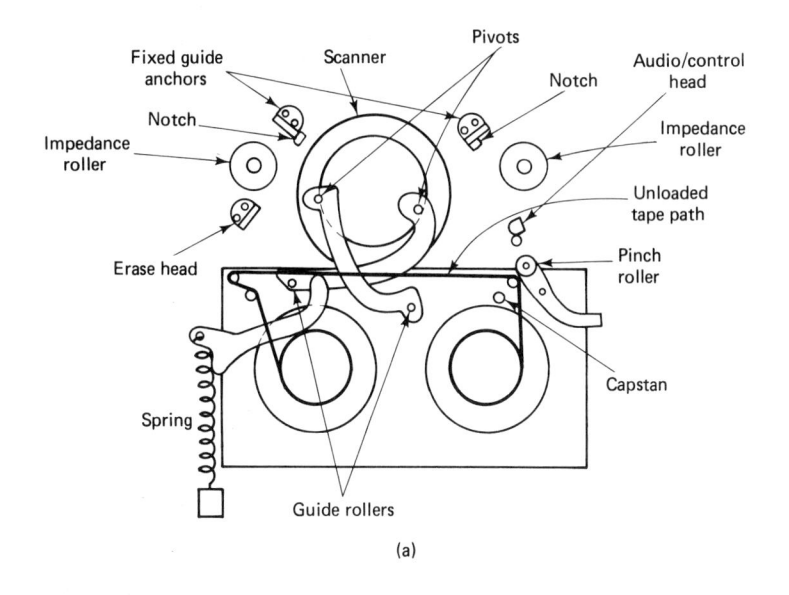

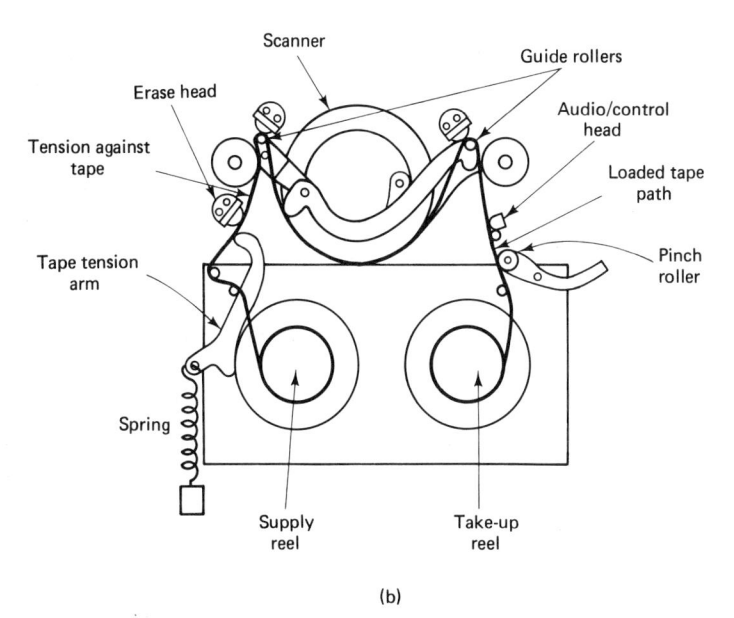

**FIGURE 1-50.** Typical VHS tape loading system.

heads (video, audio/control, erase). The pinch roller is also pulled against the tape. Usually, the arms and pinch rollers are operated by a gear train driven by the capstan motor. The movable arms have guide pins which press into notches on fixed guide anchors when fully extended. These anchors are positioned with considerable accuracy since the guide pins and rollers form entrance and exit tape guides for the scanner.

When tape threading is complete (usually in less than 1 second), the tape tension arm is released to apply spring-loaded tension against the tape. With the tension arm in place, the tape begins to move, the arms are locked in place, and the arm gear train is disengaged from the capstan motor.

Unthreading of a VHS is initiated when the stop button is pressed. The STOP button interrupts both PLAY or RECORD functions. When the STOP mode has been selected, the arms, pinch roller, and tape tension are returned to the position shown in Fig. 1-50a. With the tape back in the cassette, FAST FORWARD and REWIND functions can be performed by pressing the appropriate buttons.

## 1-10 BASIC VCR OPERATING PROCEDURES AND RECORD LOCKOUT PROVISIONS

The operating procedures for VCRs are not standard. Each VCR has its own set of controls and procedures, so we will make no attempt to describe the operating procedure for any particular VCR. However, there are two operating functions in common for most VCRs. The procedures involve inserting and removing the cassette, and a *record lockout* or *malerase* provision for both Beta and VHS.

### 1-10.1 Inserting and Removing a Video Cassette

As shown in Fig. 1-2, both Beta and VHS cassettes are installed in the VCR from the top. To insert a cassette in a typical Beta VCR such as shown in Fig. 1-51, you must apply power to the VCR and set the POWER switch to ON. You then press the EJECT button, which automatically opens and raises the cassette compart-

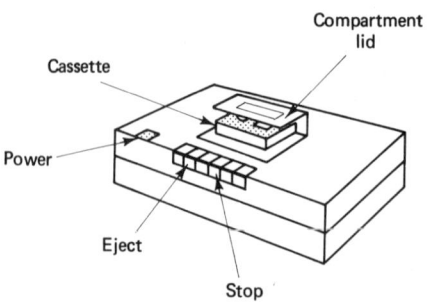

**FIGURE 1-51.** Inserting and removing a video cassette.

ment lid. You then install the cassette in the compartment (usually within a holder) and press down on the compartment lid. The VCR is then ready (in about 3 seconds) to perform any of the normal operations, such as RECORD, PLAYBACK, STOP, REWIND, FAST FORWARD, and so on.

When the cassette compartment lid is pressed down, the supply and takeup reels within the cassette engage the drive mechanisms of the *supply* and *takeup drives* within the VCR. These drives are capable of operating in both directions to accommodate both playback/record and rewind functions. The supply and takeup motors are usually capable of operating at various speeds. Typically, the supply and takeup reel motors receive a low voltage during record and playback modes. This low voltage keeps tension on the tape, but the actual tape drive is provided by the capstan and pinch roller. During fast forward and rewind, the supply and takeup motors receive a higher voltage and provide the tape drive. In the case of VHS, the tape is unloaded and returned back inside the cassette when fast forward and rewind are in operation.

For Beta machines, when the lid is closed, the tape is automatically loaded as described in Sec. 1-8.5. For VHS, tape loading occurs after the lid is closed and the PLAY button is pressed as described in Sec. 1-9.8. Operation of the tape drive motors and loading/unloading circuits is discussed further in Chapters 3, 4, and 5.

To remove a Beta cassette, make sure that the power is turned on and that the VCR is in the STOP mode. On some VCRs, the EJECT button cannot be pressed except in the STOP mode. In other VCRs, the EJECT button can be pressed but will not actuate the circuit unless the VCR is in STOP. Press EJECT, remove the cassette, and close the compartment lid. When a Beta compartment lid is raised by pressing the EJECT button, the tape is automatically unloaded as described in Sec. 1-8.5, and the cassette supply and takeup reels disengage from the tape drive motors.

For VHS, when the STOP button is depressed, the tape is unloaded. The cassette can then be removed by pressing the EJECT button to release the cassette holder.

### 1-10.2 Record Lockout or Malerase Functions

Both Beta and VHS cassettes can be reused to record new program material many times. The erase head automatically erases all recorded material on the tape before the tape reaches the rotating video heads (Fig. 1-19). Both Beta and VHS cassettes have provisions for preventing accidental erasure of recorded material when you want to retain a particular program. These provisions are called "record lockout" or "malerase" or similar term. In a Beta VCR, the record lockout takes the form of a tab located on the bottom of the cassette, as shown in Fig. 1-52a. In VHS, a similar tab or nail is located on the edge of the cassette, as shown in Fig. 1-52b.

With either system, the tabs engage a plunger rod or switch when the

72  Introduction to Videocassette Recorders

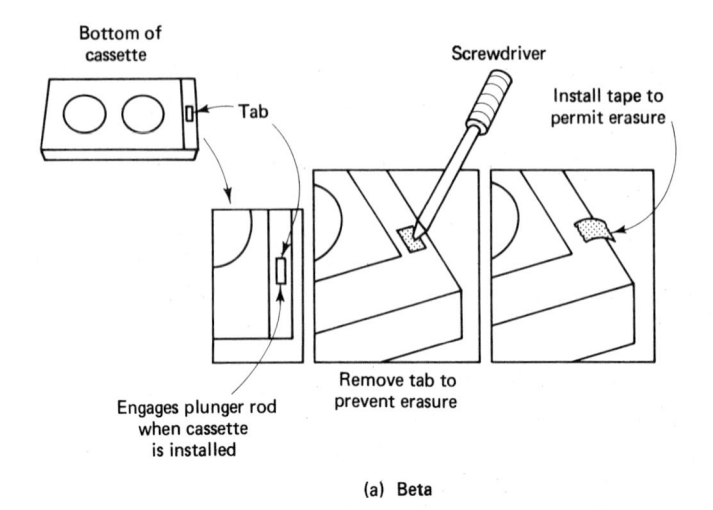

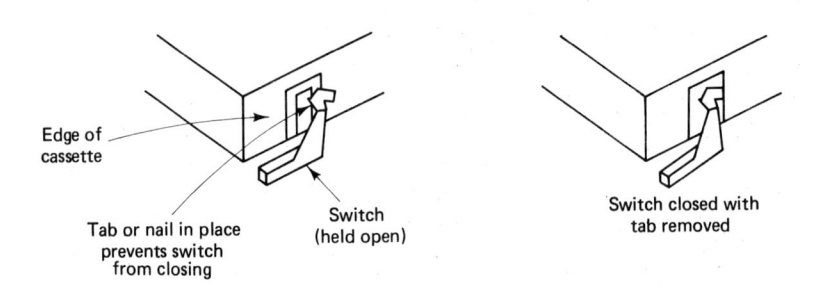

**FIGURE 1-52.** Typical record lockout or malerase function.

cassette is inserted and the compartment lid is closed. In the case of most Beta systems, the RECORD button cannot be pressed unless the rod is pushed down by the tab. In most VHS systems, the tab prevents a switch from closing (closing of the switch disables record operation). If you want to keep a recorded program from being accidentally erased, you break off the tab or nail with a screwdriver. In this way, the plunger is not pushed down (Beta) or the switch is closed (VHS), and the RECORD function is disabled.

If you wish to record on a cassette with the tab or nail removed, cover the hole with a piece of cellophane or vinyl tape. The tape will actuate the plunger rod or hold the switch open, and the RECORD function will be normal. During service, it is sometimes necessary to operate the VCR without a cassette installed (say to observe rotation of the tape drive motor). In this case you can use vinyl tape to hold the plunger rod or switch in place, just as though it was actuated by the tab or nail.

# 2

# VCR Test Equipment and Tools

The test equipment used in VCR service is basically the same as that used in television service and in other fields of electronics. That is, most service procedures are performed using meters, signal generators, color generators, oscilloscopes, power supplies, and assorted clips, patch cords, and so on. Theoretically, all VCR service procedures can be performed using conventional test equipment, provided that the oscilloscopes have the necessary gain and bandpass characteristics, that the signal generators cover the appropriate frequencies, and so on.

However, this condition does not apply to the tools, jigs, and fixtures used in VCR service work. Generally, each VCR requires a special set of tools, available from the manufacturers in the form of kits. Although there are some tools found in all kits, such as tension gauges, there are many special-purpose tools for most VCRs. It is impossible to perform a full set of recommended test and adjustment procedures without all the special tools.

It is not the purpose of this chapter to promote one type of test equipment over another (or one manufacturer over another). Instead, the chapter is devoted to the basic operating principles of test equipment types in common use. Each type of test equipment is discussed in turn. You can then select the type of equipment best suited to your own needs and pocketbook. Although advanced theory has been avoided, the following discussions cover what each type of test equipment does and what signals or characteristics are to be expected from each type of equipment. The discussions describe how the features and outputs found in present-day test equipment relate to specific problems in VCR service.

## 74  VCR Test Equipment and Tools

A thorough study of this chapter will make you familiar with the basic principles and operating procedures for typical equipment used in VCR service. It is assumed that you will take the time to become equally familiar with the principles and operating controls for any particular test equipment you use. Such information is contained in the service literature for the particular equipment. It is absolutely essential that you become thoroughly familiar with your own particular test instruments. No amount of textbook instruction makes you an expert in operating test equipment; it takes actual practice.

It is strongly recommended that you establish a routine operating procedure, or sequence of operation, for each item of service equipment. That will save time and familiarize you with the capabilities and limitations of your own equipment, thus minimizing false conclusions based on unknown operating conditions.

## 2-1  SAFETY PRECAUTIONS IN VCR SERVICE

In addition to a routine operating procedure, certain precautions must be observed during operation of any electronic test equipment during service. Many of these precautions are the same for all types of test equipment; others are unique to special test instruments, such as meters, oscilloscopes, and signal generators. Some of the precautions are designed to prevent damage to the test equipment or to the circuit where the service operation is being performed. Other precautions are to prevent injury to you. Where applicable, special safety precautions are included throughout the various chapters of this book.

The following general safety precautions should be studied thoroughly and then compared to any specific precautions called for in the test equipment service literature and in the related chapters of this book.

There are two standard *international operator warning symbols* found on some test equipment. One symbol, a *triangle with an exclamation point at the center,* advises the operator to refer to the operating manual before using a particular terminal or control. The other symbol, a *zigzag line simulating a lightning bolt,* warns the operator that there may be dangerously high voltage at a particular location, or that there is a voltage limitation to be considered when using a terminal or control. Always observe these warning symbols. Unfortunately, use of the symbols is not universal, particularly on older test equipment.

1. Many service instruments are housed in metal cases. These cases are connected to the ground of the internal circuit. For proper operation, the ground terminal of the instrument should always be connected to the ground of the VCR being serviced. *Make certain* that the chassis of the VCR being serviced is not connected to either side of the a-c line or to any potential above ground, using the leakage current check of Sec. 2-1.1.

2. Remember that there is always danger in servicing VCRs that operate at hazardous voltages, especially as you pull off covers with the power cord connected. Fortunately, most VCR circuits operate at potentials well below the line voltage, since the circuits are essentially solid state. However, a line voltage of 120 V is sufficient to cause serious shock, and possibly death! Always make some effort (such as reading the service literature) to familiarize yourself with the VCR *before* servicing it, bearing in mind that high voltages may appear at unexpected points in a defective VCR.

3. It is good practice to remove power before connecting test leads to high-voltage points. It is preferable to make all service connections with the power removed. If this is impractical, be especially careful to avoid accidental contact with VCR circuits and objects that are grounded. Keep in mind that even low-voltage circuits may present a problem. For example, a screwdriver dropped across a 12-V line in a solid-state circuit can cause enough current to burn out a major portion of the VCR, possibly beyond repair. Of course, this problem is nothing compared to the possibility of injury to yourself! Working with one hand away from the VCR and standing on a properly insulated floor lessens the danger of electrical shock.

4. Capacitors may store a charge large enough to be hazardous. Discharge filter capacitors before attaching test leads.

5. Remember that leads with broken insulation offer the additional hazard of high voltages appearing at exposed points along the leads. Check test leads for frayed or broken insulation before working with them.

6. To lessen the danger of accidental shock, disconnect the test leads immediately after the test is completed.

7. Remember that the risk of severe shock is only one of the possible hazards. Even a minor shock can place you in danger of more serious risks, such as a bad fall or contact with a source of higher voltage.

8. The experienced service technician guards continuously against injury and does not work on hazardous circuits unless another person is available to assist in case of accident.

9. Even if you have considerable experience with test equipment used in service, always study the service literature of any instrument with which you are not thoroughly familiar.

10. Use only shielded leads and probes. Never allow your fingers to slip down to the meter probe tip when the probe is in contact with a "hot" circuit.

11. Avoid vibration and mechanical shock. Most electronic equipment is very delicate. The mechanical portions of a VCR are vulnerable to any kind of shock or vibration. Not only can the mechanical parts be damaged, but they can also be thrown out of adjustment by rough handling. This is especially true of the scanner and video heads. Although the video heads are designed to be rotated continuously at 1800 rpm and to be in constant contact with magnetic tape, they are not designed to be in contact with the tips of screwdrivers, Allen wrenches, and the like.

76    VCR Test Equipment and Tools

12. Study the circuit being serviced before making any test connections. Try to match the capabilities of the instrument to the circuit being serviced. For example, if the circuit under test has a range of measurements to be made (ac, dc, RF, modulated signals, pulses, or complex waves), it is usually necessary to use more than one instrument. Most meters measure dc and low-frequency signals. If an unmodulated RF carrier is to be measured, use an RF probe. If the carrier to be measured is modulated with low-frequency signals, a demodulator probe must be used. If pulses, square waves, or complex waves are to be measured, a peak-to-peak reading meter can possibly provide meaningful indications, but an oscilloscope is the logical instrument.

## 2-1.1  Leakage Current Tests

Before placing a VCR in use (for service or normal home use) it is recommended that you measure possible leakage current. Such leakage indicates that metal parts of the VCR are in electrical contact with one side of the power line. If the leakage problem is severe, it can result in damage to the VCR, or possible shock to anyone touching the exposed metal parts. There are two recommended leakage current tests: *cold check* and *hot check*.

**Cold Check.**   With the a-c plug removed from the 120-V source, place a jumper across the two a-c plug prongs. Turn the a-c or POWER switch ON. Using an ohmmeter, connect one lead to the jumpered a-c plug, and touch the other lead to each exposed metal parts (metal cabinet, screwheads, metal overlays, control shafts, etc.), particularly any exposed metal part having a return path to the chassis. Exposed metal parts having a return path to the chassis should have a minimum resistance reading of 1 MΩ. Any resistance below this value indicates an abnormality that requires corrective action. Exposed metal parts not having a return path to the chassis indicates an open circuit.

**Hot Check.**   Using the diagram of Fig. 2-1 as a reference, measure a-c leakage current with a milliammeter. Leave switch S1 open, and connect the VCR power plug to the test connector. Immediately after connecting the VCR, measure any leakage current with switch S2 in both positions. Set the VCR switches (at least the power switch) to ON when making the leakage current measurements. Now close switch S1, and immediately repeat the leakage current measurements in both positions of switch S2 (and with the VCR switches on). Allow the VCR to reach normal operating temperature, and repeat the leakage current tests. In any of these tests, the leakage current should not exceed about 0.5 mA (for a typical VCR).

If possible, check both the television set and antenna to be used with the VCR for possible leakage currents. Use the same procedure as described for the VCR (Fig. 2-1). To avoid shock hazards, do not connect a VCR to any TV, antenna, cable or accessory that shows excessive leakage current.

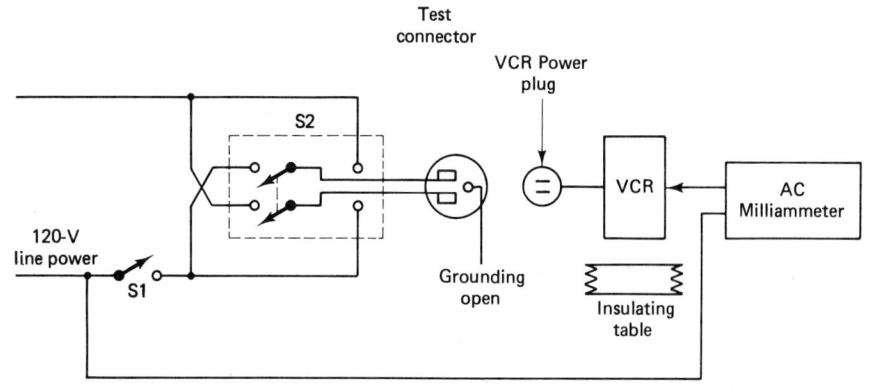

**FIGURE 2-1.** "Hot check" circuit for leakage current tests

### 2-1.2 Basic Handling and Service Precautions

The following precautions apply to all types of VCRs and are to be observed in addition to any precautions described by the service literature.

**Handling and Storage.** Avoid using the VCR in the following places: extremely hot, cold, or humid places; dusty places; near appliances generating strong magnetic fields; places subject to vibration; and poorly ventilated places. Do not block the ventilation openings. Do not place anything heavy on the VCR. Do not place anything that might spill on the top cover of the VCR. Use an accessory cover (if available) to prevent dust and dirt from accumulating on the VCR. Use the VCR in the horizontal (flat) position only. Do not lubricate VCR motors, or any point not recommended for lubrication. Remove any excess lubricant. When reassembling any VCR, always be certain that all the protective devices (nonmetallic control knobs, shield plates, etc.) are put back in place. When service or test is required on any VCR, observe the original lead dress (wire routing, etc.). Pay particular attention to parts that show any evidence of overheating or other electrical/mechanical damage. If you must transport a VCR, avoid violent shocks to the VCR during both packing and transportation. Before packing, be sure to remove the cassette from the VCR.

**Moisture Condensation.** Be very careful to avoid the effects of moisture on a VCR. Moisture condensation on the scanner and video heads (probably the most critical parts of any VCR) will cause damage to the tape. Avoid using a VCR immediately after moving from a cold place to a warm place, or soon after heating a room where it was cold. Otherwise, the water vapor in warm air condenses on the still-cold video head scanner and tape guides, and may cause damage to the tape or VCR, or both.

Because of the moisture condensation problem, most VCRs are provided with a moisture-condensation protection circuit. A moisture sensor (usually called a dew sensor) is mounted near the video heads within the VCR. If moisture exceeds a safe level, the sensor triggers a circuit in the system control and removes power to the tape drive and scanner motors. Operation of various dew sensor circuits is discussed in Chapters 3 and 4.

**Copyright Problems.** Many of the programs broadcast by television stations are protected by copyright, and federal law imposes strict penalties for copyright infringement. Some motion picture companies have taken the position that home recording for noncommercial purposes is an infringement of their copyrights. Until the courts have ruled on the proper interpretation of the law as applied to home video recording, a VCR used to record copyrighted material should be operated at the user's own risk.

## 2-2 SIGNAL GENERATORS

The signal generator is an indispensable tool for practical VCR service. Without a signal generator, you are entirely dependent on signals broadcast by the television station, and you are limited to signal tracing only. This means that you have no control over frequency, amplitude, or modulation of such signals and have no means for signal injection. With a signal generator of the appropriate type, you can duplicate transmitted signals or produce special signals required for alignment and test of all circuits within a VCR. Also, the frequency, amplitude, and modulation characteristics of the signals can be controlled so that you can check operation of a VCR under various signal conditions (weak, strong, normal, abnormal signals).

In addition to conventional *RF generators* and *audio generators,* the signal generators designed specifically for use in television service include the *sweep generator, marker generator, analyst generator, pattern generator,* and *color generator.* These same generators can be used for VCR service. Often, the functions of these generator types are combined. For example, several manufacturers produce a sweep and marker generator. Similarly, the analyst and color generator functions are often combined in a single instrument. Color generators are described in Sec. 2-3. The purpose, operating principles, and typical characteristics of the remaining generators are as follows.

### 2-2.1 Sweep/Marker Generator

At one time, sweep generator and marker generators were manufactured as separate instruments. Today, the sweep and marker functions are usually combined into a single generator. The main purpose of the sweep/marker generator in VCR service is the sweep-frequency alignment of the RF tuner section. A sweep/marker generator capable of producing signals of the appropriate frequency is used in conjunction with an oscilloscope to display the bandpass characteristics of the tuner.

The *sweep portion* of the sweep/marker generator is essentially an FM generator. When the sweep generator is set to a given frequency, this is the *center frequency*. The output varies back and forth through this center frequency. Thus, in simplest form, a sweep generator is a frequency-modulated RF oscillator. The rate at which the frequency modulation takes place is typically 60 Hz. The sweep width, or the amount of variation from the center frequency, is determined by a control, as is the center frequency.

The *marker portion* of the sweep/marker generator is essentially an RF signal generator with highly accurate dial markings that can be calibrated precisely against internal or external signals. Usually, the internal signals are crystal controlled. Marker signals are necessary to pinpoint frequencies when making sweep-frequency alignments. Although sweep generators are accurate in both center frequency and sweep width, it is almost impossible to pick out a particular frequency along the spectrum of frequencies being swept. Thus, fixed-frequency "marker" signals are injected into the circuit together with the sweep-frequency generator output. Usually, this is accomplished by means of a built-in *marker-adder*.

In sweep-frequency alignment, the sweep generator is tuned to sweep the band of frequencies passed by the wideband circuits (RF, IF, video, chroma) in the VCR tuner, and a trace representing the response characteristics of the circuits is displayed on the oscilloscope. The marker generator is used to provide calibrated markers along the response curve for checking the frequency settings of traps, for adjustment of capacitors and coils, and for measuring overall bandwidth of the tuner circuits.

When the marker signal is coupled into the circuit under test, a vertical marker (often called a "pip") appears on the response curve. When the marker is tuned to a frequency within the bandpass limits of the tuner circuits, the marker indicates the position of that frequency on the sweep trace. The tuner circuits can then be adjusted to obtain the desired waveshape, using the different frequency markers are checkpoints.

In addition to the basic sweep and marker outputs, a typical sweep/marker generator will have a number of other special features. For example, a *variable bias* source is provided on some sweep/marker generators. During alignment of most television receivers, it is necessary to disable the AGC circuits. This can be done by applying a bias voltage (of appropriate amplitude and polarity) to the AGC line of the receiver. Some VCR circuits also require a similar voltage.

A *blanking circuit* is another feature found on some sweep/marker generators. When the sweep generator output is swept across its spectrum, the frequencies usually go from low to high and then return from high to low. This makes it possible to view a zero reference line on the oscilloscope during the retrace period.

## 2-2.2 Basic Sweep-Frequency Alignment Procedure

The relationship between the sweep/marker generator and the oscilloscope during sweep-frequency alignment is shown in Figs. 2-2 and 2-3.

If the equipment is connected as shown in Fig. 2-2, the oscilloscope horizontal sweep is triggered by a sawtooth output from the sweep generator. The oscilloscope's internal recurrent sweep is switched off, and the oscilloscope sweep selector and sync selector are set to external.

Under these conditions, the oscilloscope horizontal sweep should represent the total sweep spectrum. For example, as shown in Fig. 2-4, if the sweep is from 10 to 20 kHz, the left-hand end of the horizontal trace represents 10 kHz and the right-hand end represents 20 kHz. Any point along the horizontal trace represents a corresponding frequency. For example, the midpoint on the trace represents 15 kHz. If you want a rough approximation of frequency, adjust the horizontal gain control until the trace occupies an exact number of scale divisions on the oscilloscope screen (such as 10 cm for the 10- to 20-kHz sweep signal). Each centimeter division then represents 1 kHz.

If the equipment is connected as shown in Fig. 2-3, the oscilloscope horizontal sweep is triggered by the oscilloscope internal circuits (both the sweep selector and sync selector are set to internal). Certain conditions must be met to use the test connections of Fig. 2-3. If the oscilloscope is of the *triggered sweep* type (where the horizontal sweep is triggered by the signal applied to the vertical inputs, as discussed in Sec. 2-4), there must be sufficient delay in the vertical input, or part of the response curve may be lost. If the oscilloscope is not of the triggered sweep type, the sweep generator must be swept at the same frequency as the oscilloscope horizontal sweep (usually the line power frequency of 60 Hz). Also, the oscilloscope or the sweep/marker generator must have a *phasing control* so that the two sweeps can be synchronized.

The method shown in Fig. 2-3 is used where the generator does not have a sweep output separate from the RF signal output, or when it is not desired to use the sweep output. The Fig. 2-3 method was popular with older generators,

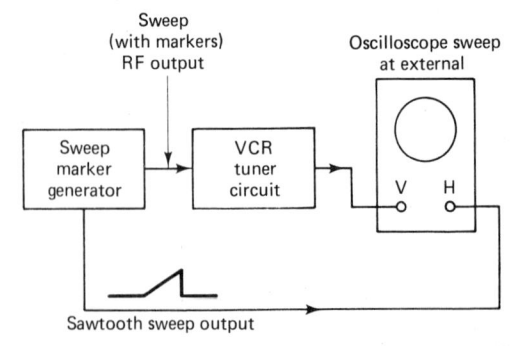

**FIGURE 2-2.** Basic sweep/marker generator and oscilloscope test circuit (using horizontal sweep from generator).

*Signal Generators* 81

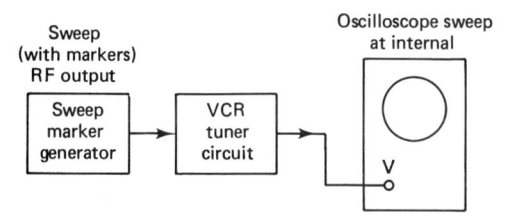

**FIGURE 2-3.** Basic sweep/marker generator and oscilloscope test circuit (using oscilloscope internal horizontal sweep).

although the method is still in use today. Blanking of the trace (if any blanking is used) is controlled by the oscilloscope circuits. The phase of the sweep is adjusted with the phasing control of the oscilloscope (although some sweep generators have phasing controls). If the phase adjustment is not properly set, the sweep curve on the oscilloscope may be prematurely cut off, or the curve may appear as a double or mirror image. These effects are shown in Fig. 2-5.

As shown in Fig. 2-4, the markers are used to provide accurate frequency measurement. On some generators, the marker output frequency can be adjusted until the marker is aligned at the desired point on the trace. The frequency is then read from the marker generator dial. This system has generally been replaced by generators that produce a number of markers at precise, crystal-controlled frequencies. Such fixed-frequency markers are illustrated in Fig. 2-6, which shows the bandpass response curve of a typical VCR tuner IF amplifier. The markers can be selected (one at a time or several at a time) as needed.

The response curve (oscilloscope trace) depends on the VCR tuner circuit under test. If the circuit has a wide bandpass characteristic, the sweep generator is set so that its sweep is wider than that of the circuit. (As shown in Fig. 2-6,

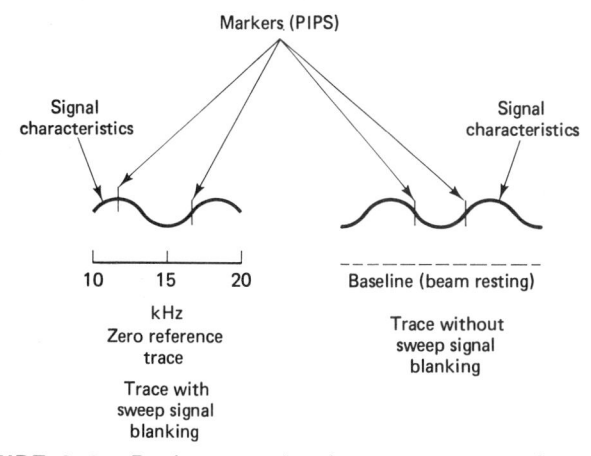

**FIGURE 2-4.** Basic sweep/marker generator and oscilloscope displays.

82   VCR Test Equipment and Tools

**FIGURE 2-5.** Sweep response curves showing effects of improper phasing control adjustment.

the bandpass of tuner circuits is on the order of 6 MHz.) Under these conditions, the trace starts low at the left, rises toward the middle, and then drops off at the right.

The sweep/marker generator-oscilloscope method of alignment tells at a glance the overall bandpass characteristics of the circuit (sharp response, irregular response at certain frequencies, and so on). The exact frequency limits of the bandpass can be measured with the markers.

### 2-2.3  *Direct Injection versus Postinjection*

There are two basic methods for injection or marker signals into a sweep/marker generator-oscilloscope display. With *direct injection,* the sweep generator and marker generator signals are mixed before they are applied to the circuit under test. This method is sometimes called *preinjection,* and has generally been replaced by *postinjection.*

With *postinjection,* as shown in Fig. 2-7, the sweep generator output is applied to the circuit under test. A portion of the sweep generator output is also mixed with the marker generator output in a mixer–detector circuit known as a *postinjection marker adder.* The mixed and detected output from both generators is then mixed with the detected output from the circuit under test. Thus, the oscilloscope vertical input represents the detected values of all three signals (sweep, marker, and circuit output).

Most present-day sweep/marker generators have some form of built-in postinjection marker-adder circuits. The postinjection (sometimes known as *bypass injection*) method for adding markers is usually preferred for VCR serv-

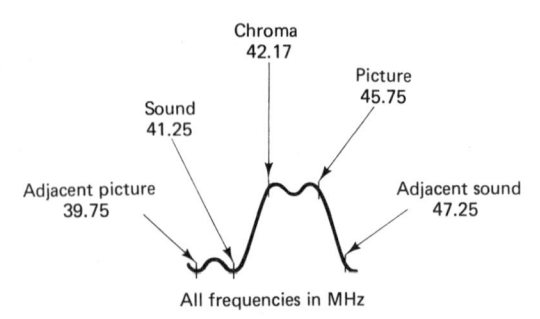

**FIGURE 2-6.** Bandpass response curve of a typical VCR tuner IF amplifier.

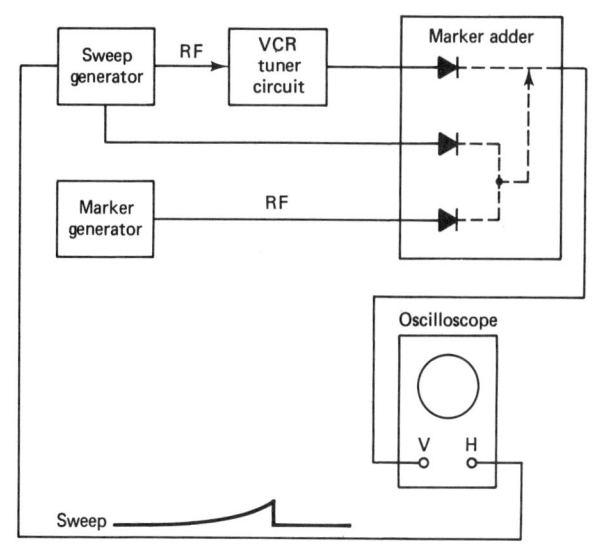

**FIGURE 2-7.** Postinjection (bypass injection) of marker signals.

ice, as it is for television service, because postinjection minimizes the chance of overloading the circuits under test and permits use of a narrow-band oscilloscope. At one time, postinjection marker-adder units were available as separate units and are still in use today.

## 2-2.4 Typical Sweep/Marker Generator

The B&K Precision, Dynascan Corporation Model 415 Sweep/Marker Generator is typical of the sweep/marker generators available for television service. The instrument has a number of features that simplify alignment and troubleshooting of television receivers. These features can also be used effectively in VCR service.

*There are RF outputs,* with equivalents of all IF and chroma markers (Fig. 2-6), available for channels 4 and 10. This makes it possible to connect the RF output to the antenna terminals of the VCR and, without further input reconnections, to evaluate alignment conditions of all tuned signal-processing circuits of the tuner.

*There are 10 crystal-controlled markers* with true postinjection marker adding. All 10 markers can be used simultaneously or individually. The markers shown in Fig. 2-6 are typical.

A *video sweep output* permits direct sweep alignment of the chroma circuits where specified by the manufacturer. With some generators it is necessary to use the IF sweep for signal injection and then monitor the video circuits for response.

The generator includes *pattern polarity reversal* and *sweep reversal* features which permit you to match oscilloscope displays as shown in manufacturers' literature (positive or negative, left- or right-hand sweep, etc.) The *markers can be tilted* to horizontal or vertical positions, thus permitting easy identification. For example, if the sides of a bandpass display are steep (vertical), a horizontal marker is easier to identify. A vertical marker shows up better on the flat top (horizontal) of the same pattern.

The generator also includes a number of features that are primarily of benefit in television service. These features include built-in amplifiers and filters, crystal-controlled marker outputs for spot alignment of traps and bandpass circuits, interconnecting cables, and visual reproduction of idealized alignment curves on the generator front panel to indicate desired marker positions. (When the marker switch is pressed, the corresponding marker light comes on at the position where the marker should appear (on the oscilloscope waveform). This provides a constant reference and minimizes error. Many of these features are quite helpful in VCR service.

### 2-2.5 Analyst and Pattern Generators

An analyst generator is used for *signal substitution* (a form of signal injection). A bench-type analyst generator provides outputs that duplicate all essential signals in a television receiver. Such generators have RF, IF composite video (including sync), sound, audio, separate sync, flyback test, and yoke test signals. All of these signals (except for flyback and yoke test) can be used in VCR service.

The RF, IF, and video signals are usually in the form of a *pattern* (or patterns). On portable-type analyst generators, the patterns are typically lines or bars (horizontal and vertical), dots, crosshatch lines, square pulses, blank rasters, or color bars (color bar generators are discussed in Sec. 2-3). Some typical patterns are shown in Fig. 2-8. Such patterns can be used primarily in troubleshooting. As an example, with a typical portable analyst generator, you can inject a black-and-white crosshatch pattern at some particular VHF channel into the antenna terminal of the VCR, and record the pattern. Then you can repeat the process using the color bar pattern. You then play back the recorded patterns on a known good television monitor. If the display is a clear, sharp crosshatch pattern followed by a good color pattern, you know that the VCR is capable of recording and playing back a good picture. If only black and white is good, you know there are problems in the color circuits of the VCR. If there is no playback, you can play the recorded tape on a known good VCR. If there is still no playback, the problem is in the record circuits of the suspected VCR. If there is a good playback, the problem is in the playback circuits of the suspected VCR. Thus, any form of pattern generator can serve to pinpoint quickly defective circuit areas during VCR troubleshooting.

In addition to the basic analyst generator, there are *test pattern generators* available for bench service. The test pattern generator differs from the basic pattern generator in that the test pattern generator reproduces positive transparen-

*Signal Generators* 85

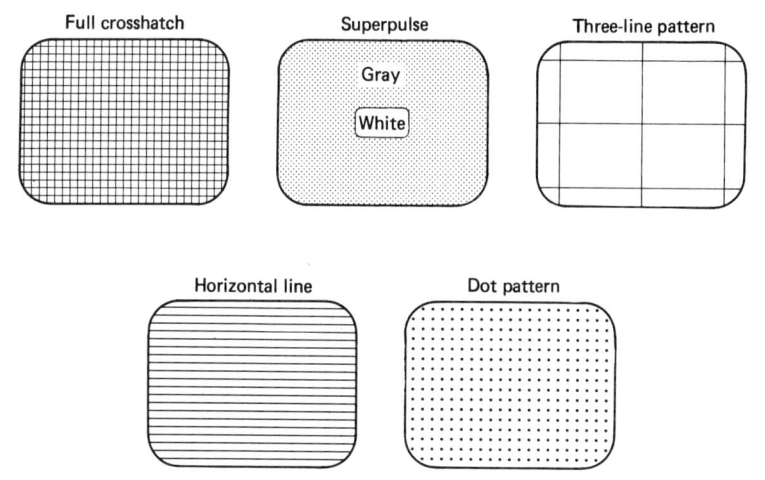

**FIGURE 2-8.** Typical patterns produced by a pattern generator.

cies of various pictures that may be inserted into the generator. In effect, the test pattern generator is a miniature television station, capable of reproducing nonmoving pictures. Figure 2-9 shows two typical black-and-white test patterns.

In operation, the transparency is placed in the generator, the generator output (RF, IF, or video) is connected to the VCR, and the transparency picture is recorded on tape. The recorded pattern can then be played back (as described for the basic analyst-type generator). A test pattern generator is not necessarily a color generator, but most present-day test pattern generators do incorporate some type of color output. Generally, the color output is the *keyed rainbow display* (Sec. 2-3), which is superimposed over the black-and-white test pattern display.

One of the advantages of the test pattern generator is that it provides a conventional "test pattern" presentation that duplicates the test patterns of some television stations. In practically all cases, the test pattern generators also include analyzer functions.

### 2-2.6 Typical Analyst/Pattern Generators

The B&K Precision, Dynascan Corporation Model 1077B Television Analyst is typical of the advanced analyst generators. The instrument produces several patterns, including color at RF, IF, and video frequencies. The instrument generates many signals normally transmitted by a television station, and those produced within a television receiver, for point-to-point television troubleshooting. Many of these signals can be used in VCR service. For example, the instrument generates:

> VHF signals on channels 2, 3, 4, 6, 7, 8, 12, and 13 for testing the RF tuner (and overall performance).

86  VCR Test Equipment and Tools

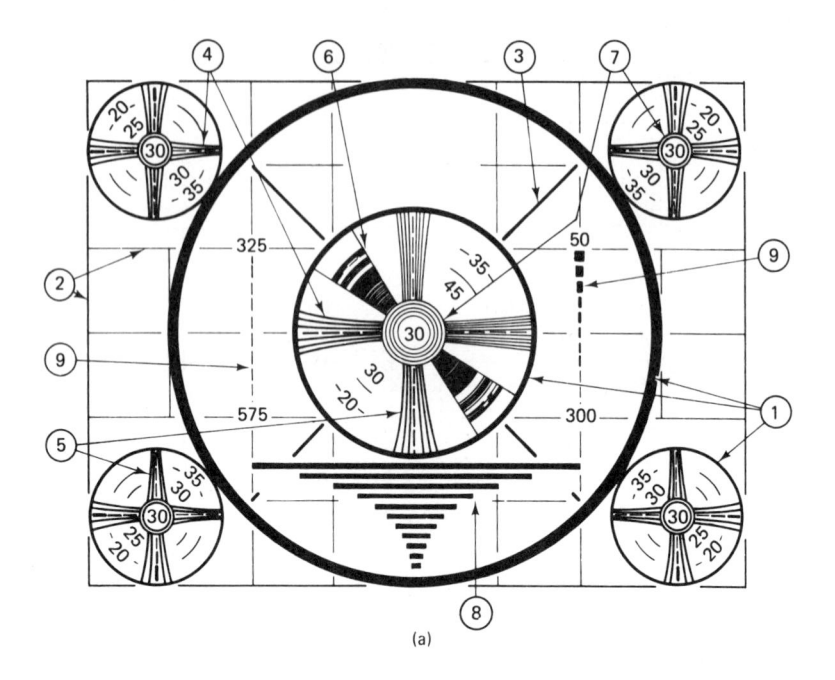

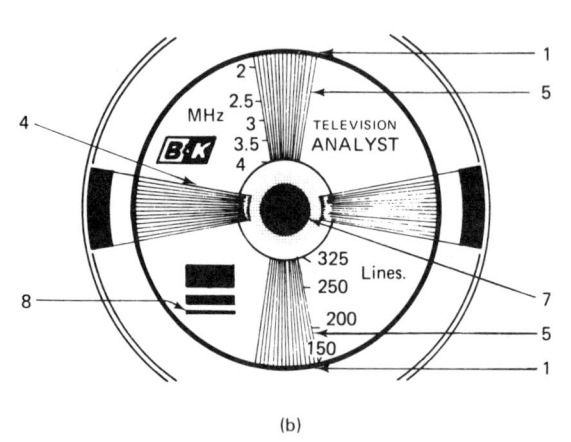

**FIGURE 2-9.** Typical television test patterns.

UHF signals on channels 14 through 83 for testing the UHF tuner (a portable generator usually excludes UHF).

IF signals form 20 to 48 MHz for testing the IF portion of the VCR tuner.

Positive and negative composite video signals (including the sync pulses) for injection into video stages.

A keyed color bar pattern which modulates the RF output (for check of overall color response).

A 4.5-MHz sound channel test signal that is frequency modulated by a 1-kHz audio tone.

A separate 1-kHz audio tone.

A test pattern or other positive transparency of your choice.

## 2-3 COLOR GENERATORS

There are two types of color generators in common use for television service: the *keyed rainbow generator* and the NTSC generator. Both of these can also be used for VCR service. However, the NTSC color generator is a far superior instrument. Many of the test and adjustment procedures described in Chapter 6 require the type of signals found only in NTSC generators. For that reason, the author recommends that you consider an NTSC instrument if you plan to go into VCR service full time.

### 2-3.1 Keyed Rainbow Generators

At one time, the keyed rainbow generator was about the only type of color generator available outside television studios (or at a price that most service technicians could afford). The keyed rainbow generator produces a pattern similar to that shown in Fig. 2-10, where each color of the rainbow display is of uniform spacing and width. Also, there are blanks or bars between each color.

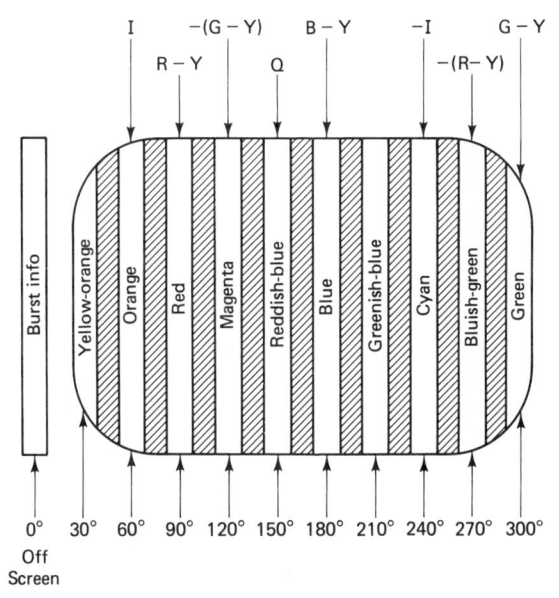

**FIGURE 2-10.** Gated or keyed rainbow display.

The color-bar pattern is produced by gating a 3.56-MHz oscillator at a frequency 12 times higher than the horizontal sweep frequency (15,750 Hz x 12 = 189 kHz). This 189-kHz gating produces color bars that are 30° apart all around the color spectrum. When viewed from the picture tube of a normally operating television receiver, these bars appear as shown in Fig. 2-10.

Note that of the 12 gated bursts, only 10 show on the picture tube as color bars, because one of the bursts occurs at the same time as the horizontal sync pulse and is thus eliminated. The other burst occurs immediately after the horizontal sync pulse and becomes the color sync burst, which is used to control the 3.58-MHz reference oscillator in the television receiver.

**Differences in Keyed Rainbow Generators.** Most present-day rainbow generators are portable (some are battery operated). In the simplest form, a keyed rainbow generator provides a single-channel RF output that can be modulated by color bars, as well as crosshatch, dots, bars, and blank rasters. *Color-gun interrupters* (usually called "gun killers" or similar term) are found on some keyed rainbow generators. These killers or interrupters consist of switches that ground the corresponding (red, blue, green) color gun of the television receiver picture tube through fixed resistors, thus disabling that particular color.

Advanced keyed rainbow generators usually include analyzer functions (Sec. 2-2.5) together with the color signals. Stated another way, most analyzer generator include a keyed rainbow output (either RF, IF, or video, or possibly all three).

### 2-3.2 NTSC Color Generators

At one time NTSC generators were quite heavy and bulky (in addition to being expensive). Today's NTSC generators are almost portable. The unit described in Sec. 2-3.3 weighs about 11 pounds and requires 18 W of power. The NTSC generator differs from the rainbow types in that the NTSC instrument produces single colors, one at a time, in addition to standard NTSC color bar presentations, similar to those transmitted by television stations. A typical NTSC generator provides independent selection of fully saturated colors, plus white, where the phase angles and amplitudes are permanently established in accordance with the NTSC standards.

In a typical NTSC generator, the signals are selected by taps on a linear delay line so that no color adjustments are required. The amplitude of the color signals are also accurately set to NTSC standards, and the color reference burst is placed in its precise NTSC position, closely following the horizontal sync pulse. Thus, the color signal produced by an NTSC-type generator is exactly the same as if the signals were being produced by a television station transmitting color.

## 2-3.3 Typical NTSC Color Generator

The B&K Precision, Dynascan Corporation Model 1250 NTSC Color-Bar Generator is typical of present-day NTSC generators. The instrument generates the *standard NTSC bar pattern* with a -IWQ signal occupying the lower quarter of the pattern, as shown in Fig. 2-11, or will produce the full-screen color pattern, as desired. The generator also produces a *five-step linear staircase pattern,* similar to that shown in Fig. 2-12, with selectable chroma levels. Other patterns available include dot, crosshatch, dot hatch, center cross, and *full color raster.* The following raster colors can be displayed by using three color raster buttons

**FIGURE 2-11.** Standard NTSC bar pattern (75%) with a-IWQ signal occupying the lower quarter of the pattern.

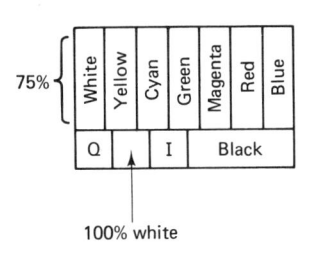

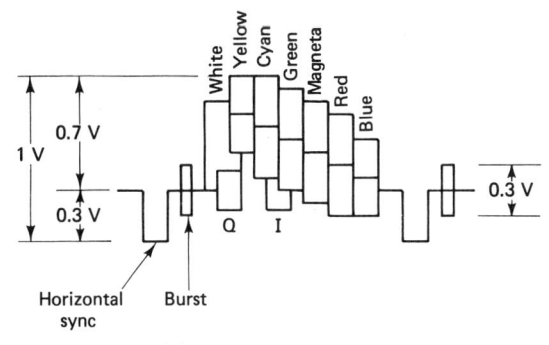

(a) Color bar signal waveform

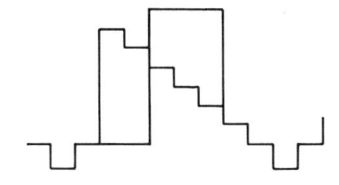

(b) Corresponding five-step linear staircase pattern

**FIGURE 2-12.** Standard NTSC color bar (75%) signal waveform with corresponding five-step linear staircase pattern.

individually or in combination: red, blue, green, yellow, cyan, magenta, white, and black raster. The NTSC bar pattern of Fig. 2-11, and the staircase pattern of Fig. 2-12, are the most often used signals, as discussed in Chapter 6.

The instrument generates an RF output on either channel 3 or 4, or at the standard TV IF frequency of 45.75 MHz. These carriers can be modulated by any of the patterns. In addition to the video patterns, the instrument generates a 4.5-MHz sound carrier with 1- or 3-kHz modulation, or with external modulation, a vertical or horizontal trigger output pulse, and a chroma subcarrier signal of 3.579545 MHz.

## 2-4 OSCILLOSCOPES

A conventional oscilloscope can be used for VCR service. Ideally, the oscilloscope should have a bandwidth of 10 MHz or more, although you can probably get by with a 5- to 6-MHz bandwidth. Minimum sweep time should be 0.1-$\mu$sec per division, although a 0.2-$\mu$sec sweep will probably do the job. The oscilloscope should be capable of a *triggered sweep,* in addition to conventional internal and external sweep synchronization.

With a triggered sweep, the oscilloscope horizontal sweep is synchronized by the signals being applied to the vertical input. The sweep remains at rest until triggered by the signal being observed. This assures that the signals are always synchronized, even when the waveform is of varying frequency. The triggered sweep threshold should be fully adjustable so that the desired portion of the waveform can be used for triggering.

With internal sweep synchronization, the horizontal sweep is set at some fixed rate by selection of controls. With external synchronization, the horizontal sweep is controlled by an external trigger. With some oscilloscopes, it is possible to apply an external sweep signal (not just a trigger) to the horizontal amplifier. Such provisions are helpful when using the sweep-frequency alignment techniques described in Sec. 2-2.2.

A dual-trace oscilloscope will facilitate some measurements (such as timing measurements required in servo adjustment), but is not absolutely required for VCR service. Such refinements as calibrated voltage scales, calibrated sweep rates, Z-axis input (for intensity modulation), sweep magnification, and illuminated scales are, of course, always helpful (as they are in any type of electronic service).

Many oscilloscopes have special provisions for television service. Such features include the *television sync* or *sweep* (usually identified as TV HORIZONTAL and TV VERTICAL, or TVH and TVV, or similar term) and the *vectorscope* provision. Although a vectorscope is of little value in VCR service, the TVH and TVV sweeps can be useful.

## 2-4.1 TVV and TVH Sync and Sweeps

The TVV and TVH functions are usually selected by means of front-panel switches and permit *pairs* of vertical or horizontal sync pulses to be displayed on the oscilloscope screen. As shown in Fig. 2-13, the signal applied to the vertical input is also applied to a sync separator. This separator receives both 60-Hz and 15,750-Hz (vertical and horizontal) sync pulses from the video signals being monitored. The oscilloscope separator operates similarly to a television sync separator and delivers two separate outputs (if both sync inputs are present). If only one sync input is present, only one output is available. Either way, the selected output (60 Hz or 15,750 Hz) is used to synchronize the horizontal sweep trigger.

When TVV is selected, the oscilloscope horizontal sweep is set to a rate of 30 Hz, and the horizontal trigger is taken from the TVV output of the sync separator. With the horizontal sweep at 30 Hz and the trigger at 60 Hz, there are two pulses for each sweep and two vertical sync pulses (or two complete vertical displays) appear on the screen.

When TVH is selected, the oscilloscope horizontal sweep is set to a rate of 7875 Hz, and the trigger (15,750 Hz) is taken from the TVH output of the sync separator. Again, two pulses or displays are presented on the screen since the sweep is one-half the trigger rate.

The TVV and TVH features not only simplify observations of waveforms

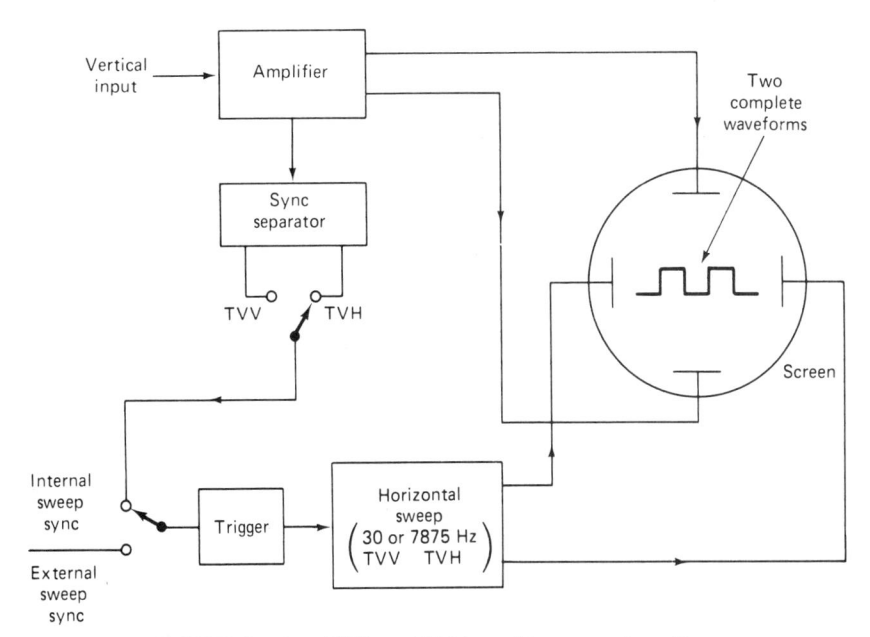

**FIGURE 2-13.** TVV and TVH oscilloscope functions.

(you always get exactly two complete displays), but they can also be helpful in trouble localization. As an example, assume that you are monitoring waveforms at some point where both vertical and horizontal sync pulses are supposed to be present. If you find two steady pictures on the TVV position but not in the TVH position, you know that there is a problem in the horizontal circuits or that horizontal sync pulses are not getting to the point being monitored.

## 2-5 METERS

The meters used for VCR service are essentially the same as for all other electronic service fields. Most tests can be done with the standard VOM (volt-ohmmeter). The VOM may be either digital or moving-needle: the choice is yours. The digital meter is easier to read, but often requires line power and is thus best suited for use in the shop (although there are battery-operated digital meters). The moving-needle VOM is more difficult to read, but operates on internal batteries, and can thus be used in the shop or field. The accuracies of both instruments are about the same.

One problem with digital meters in VCR service is that VCR service literature almost always lists audio signals in terms dB (decibels) rather than in volts. Most digital meters do not have dB scales (whereas most VOMs have at least one dB scale).

Either digital or moving-needle meters can be used to measure both voltages and resistances of all VCR circuits, as required for the test and troubleshooting procedures described in Chapters 6 and 7. When used with the appropriate probe (Sec. 2-7), the meters can be used to trace signals throughout all VCR circuits. Keep in mind that a meter will indicate the presence of a signal in a circuit, and the signal amplitude, but will not show the frequency or waveform.

In addition to accuracy, ranges (both high and low), and resolution or readability, meters are rated in terms of ohms per volt; 20,000 $\Omega$/V (or higher) is recommended. A higher ohms-per-volt rating means that the meter draws less current and thus has the least disturbing effect on the circuit under test. A lower ohms-per-volt rating means more circuit loading, which should be avoided in some critical circuits. For example, the oscillator circuits of some VCRs will not operate properly when loaded with a low ohms-per-volt meter.

One way to avoid the loading problem is to use an electronic voltmeter that has a high input impedance and thus draws little current from the circuit under test. The electronic voltmeter can be a VTVM (vacuum-tube voltmeter), EVM (electronic voltmeter), TVM (transistorized voltmeter), or similar instrument. Most digital meters are electronic meters, and thus draw a minimum of current from the circuit.

One minor problem with some meters is that the frequency range is not sufficient to cover the entire audio circuit range, which is usually considered anything up to about 10 kHz for most VCRs. Some meters will produce indications at the high end of the audio range, but the readings will not be accurate. Some VCR service literature recommends a VTVM, EVM, or TVM that covers the audio range. One practical way to overcome the problem is to use a probe, as discussed in Sec. 2-7.

If you are monitoring the audio circuits of a VCR, and there appears to be a frequency-response problem (such as a drop in amplitude above about 5 to 7 kHz), make certain that the meter is not at fault. This can be done by applying a constant-amplitude signal directly to the meter over the entire audio range. (While you are at it, make certain that the signal source is of constant amplitude over the audio range!)

## 2-6 FREQUENCY COUNTERS

The two most common uses for a frequency counter in VCR service are (1) to check or adjust the various 3.58-MHz oscillators in the chroma record and playback circuits, and (2) measurement of the servo system timing. Most frequency counters have sufficient frequency range to measure the 3.58-MHz signals. There are low-cost portables that will measure up to 30 MHz at an accuracy within 10 Hz. It is the low end of the frequency range that can be a problem.

Many of the servo system signals are in the 30-Hz range (the video head scanner speed, for example). Inexpensive counters do not go down to that frequency. Even if they do cover low frequencies, they are not sufficiently accurate. The accuracy of a frequency counter is set by the *stability of the time base* rather than the readout. The readout is typically accurate to within ± 1 count. The time base of a typical shop counter is 10 MHz, and is stable to within 10 ppm (parts per million) or 100 Hz. The time base of a precision lab counter could be in the order of 4 MHz and is stable to within 1 ppm or 4 Hz. (Accuracy is not to be confused with resolution. The resolution of an electronic counter is set by the number of digits in the readout.)

Obviously, an accuracy of 100 Hz is not sufficient to measure 30-Hz signals. One way to overcome this problem is to use the *period function,* as is often done when measuring turntable speeds. (Unfortunately, many inexpensive counters do not have a period function.) Period is the inverse of frequency (period = 1/frequency). When period is measured on a counter, the unknown input signal controls the counter timing gate, and the time-base frequency is counted and read out. For example, if the time-base frequency is 1 MHz, the indicated count is in microseconds (a count of 333 indicates that the gate has been held open for 333 $\mu$s. In effect, the time base accuracy is divided by the time period. Thus, for 30-Hz signals, where the time period is approximately $\frac{1}{30}$ s, the 100-Hz accuracy is increased to 3.3 Hz (100/30), and the 4-Hz accuracy is increased to 1.3 Hz ($\frac{4}{3}$). Of course, the period count must be divided into 1 to find the frequency.

### 2-6.1 Calibration Check of Frequency Counters

The accuracy of frequency counters used for VCR service should be checked periodically, at least every 6 months. Always follow the procedures recommended in the counter service instructions. Generally, you can send the instrument to a calibration lab, or to the factory, or you can maintain your own frequency standard. (The latter is generally not practical for most VCR service shops!)

No matter what standard is used, keep in mind that the standard must be more accurate, and have better resolution, than the frequency-measuring device, just as the counter must (theoretically) be more accurate than the VCR.

There are two methods for checking frequency counters. One way is to check the counter against the frequency information broadcast by U.S. Government station WWV. The procedure requires a communications-type receiver, preferably with an S-meter, and is described in the author's best-selling *Handbook of Practical CB Service* (Englewood Cliffs, N.J.: Prentice-Hall, Inc., 1978).

A simpler method for checking the counter is to monitor the 3.58-MHz oscillator in a color television receiver. This oscillator is locked in frequency to a color broadcast at a frequency of 3.579545 MHz. The television receiver oscillator remains locked to this frequency, even though the phase and color hue may shift. Therefore, the counter should read 3.579545 MHz when the oscillator frequency is measured. Of course, a seven-digit counter is required to get the full frequency resolution.

## 2-7 MISCELLANEOUS TEST EQUIPMENT

In addition to a good triggered-sweep oscilloscope (with TVV and TVH provisions), a sweep/marker generator, a color generator (preferably an NTSC generator with pattern and analyst features), VOM and digital meters, and a frequency counter (with good low-frequency accuracy and period function), most VCR service can be done with conventional test equipment found in other electronic service fields. These include isolation transformers, testers (transistor and diode), resistance–capacitance substitution boxes, and assorted adapters, clips, and probes.

### 2-7.1 Oscilloscope and Meter Probes

Practically all meters and oscilloscopes used in VCR service operate with some type of probe. In addition to providing for electrical contact with the VCR circuit under test, probes serve to modify the voltage being measured to a condition suitable for display on an oscilloscope or readout on a meter. Typical probes for VCR service include basic, low-capacitance, RF, and demodulator.

**Basic Probe.** In its simplest form, the basic probe is a test prod (possibly with a removable alligator clip). Basic probes work well on VCR circuits carrying a direct current or audio. However, if the circuit contains high-frequency alternating current, or if the gain of the oscilloscope or meter is high, it may be necessary to use a special low-capacitance probe. Hand capacitance in a simple probe can cause hum pickup. This condition is offset by shielding in low-capacitance probes. In a more important problem, the input impedance of the meter or oscilloscope is connected directly to the VCR circuit under test by a simple probe. Such input impedance can disturb circuit conditions.

**Low-Capacitance Probes.** The low-capacitance probe contains a series capacitor and resistors that increase the meter or oscilloscope impedance. In most low-capacitance probes, the resistors form a divider (typically 10:1) between the circuit under test and the meter or oscilloscope input. Thus, low-capacitance probes serve the dual purpose of capacitance reduction and voltage reduction. You should remember that the voltage indications are one-tenth (or whatever value of attenuation is used) of the actual value when low-capacitance divider probes are connected at the inputs of meters or oscilloscopes.

**Radio-Frequency Probes.** When the signals to be measured are at radio frequencies (such as in the tuner, IF, and RF units) and are beyond the capabilities of the meter or oscilloscope, an RF probe is required. RF probes convert (rectify) the RF signals into a d-c output voltage that is equal to the peak RF voltage (or possibly equal to the rms of the RF voltage). The d-c output of the probe is applied to the meter or oscilloscope input and is displayed as a voltage readout in the normal manner.

**Demodulator Probes.** The circuit of a demodulator probe is essentially like that of the RF probe. However, the demodulator probe produces both an a-c and a d-c output. The RF carrier frequency is converted to a d-c output voltage equal to the RF carrier. If the carrier is modulated, the modulating voltage appears as ac (or pulsating dc) at the probe output.

In use, the meter or oscilloscope is set to measure direct current, and the RF carrier is measured. The meter or oscilloscope is set to measure alternating current, and the modulating voltage (if any) is measured. In general, demodulator probes are used primarily for signal tracing, and their output is not calibrated to any particular value. However, this is not always true. Similarly, some RF probes are capable of demodulation (producing outputs equal to both carrier and modulation) and may be described in catalogs as RF/demodulator probes.

## 2-7.2 Industrial Video Receiver/Monitor

If you are planning to go into VCR service on a full-scale basis, you should consider a receiver/monitor such as used in studio or industrial video work. These receiver/monitors are essentially television receivers, but with video and audio inputs and outputs brought out to some accessible point (usually the front panel).

The output connections make it possible to monitor broadcast video and audio signals as they appear at the output of a television receiver IF section (the so-called *baseband* signals, in the range 0 to 4.5 MHz, at 1 V peak to peak for video and 0 db, or 0.775 V, for audio). These output signals from the receiver/monitor can be injected into the VCR at some point in the signal flow past the tuner IF.

The input connections make it possible to inject video and audio signals from the VCR (before they are applied to the RF output unit), and monitor the display. Thus, the baseband output of the VCR can be checked independently from the RF unit.

If you do not want to go to the expense of an industrial receiver/monitor, you can use a standard TV receiver to monitor the VCR. Of course, with a TV receiver, the VCR video signals are used to modulate the VCR RF unit. The output of the RF unit is then fed to the receiver antenna input. Under these conditions, it is difficult to tell if faults are present in the VCR video or in the VCR RF unit. Similarly, if you use an NTSC generator for a video source, the generator output is at an RF or IF frequency, not at the baseband video frequencies.

If you use a TV receiver as a monitor, adjust the vertical height control to *underscan* the picture. In this way you can easily see the video switching point in relation to the start of vertical blanking.

## 2-8 TOOLS AND FIXTURES REQUIRED FOR SERVICE

Figure 2-14 shows some typical tools and fixtures recommended for field service of VCRs. These tools are available from the VCR manufacturer. In some cases, complete tool kits are made available. There are other tools and fixtures used by the manufacturer for both assembly and service of VCRs. These factory tools are not available for field service (not even to factory service centers, in some cases). This is the manufacturer's subtle way of telling service technicians that they should not attempt any adjustments (electrical or mechanical) not recommended in the service literature.

The author strongly recommends that you take this subtle hint! He has heard many horror stories from factory service people concerning the "disaster area" VCRs brought in from the field. Most of these problems are the result of tinkering with mechanical adjustments (although there are some technicians who can destroy a VCR with a simple electrical adjustment). One effective way to avoid this problem is to use only recommended factory tools and perform only recommended adjustment procedures. Always remember the old electronics rule "when all else fails, follow instructions."

| | | | |
|---|---|---|---|
| Driver<br>VJ0002 | Hex wrench<br>M3 | Bending jig<br>VJ0009 | Eccentric screwdriver<br>VJ0018 |
| Alignment tape<br>VJ0037 | Tension gauge<br>for reel<br>VJ0038 | Cassette standard<br>plate gauge<br>VJ0008 | 500 Gram spring<br>tension gauge<br>VJ0014 |
| Eccentric<br>screwdriver<br>VJ0018 | 100 Gram spring<br>tension gauge<br>VJ0012 | Inspection<br>mirror<br>VJ0015 | Gear spacer<br>VJ0084 — Hex wrench<br>VJ0087 |

(a) Beta

Back tension meter

Alignment tape

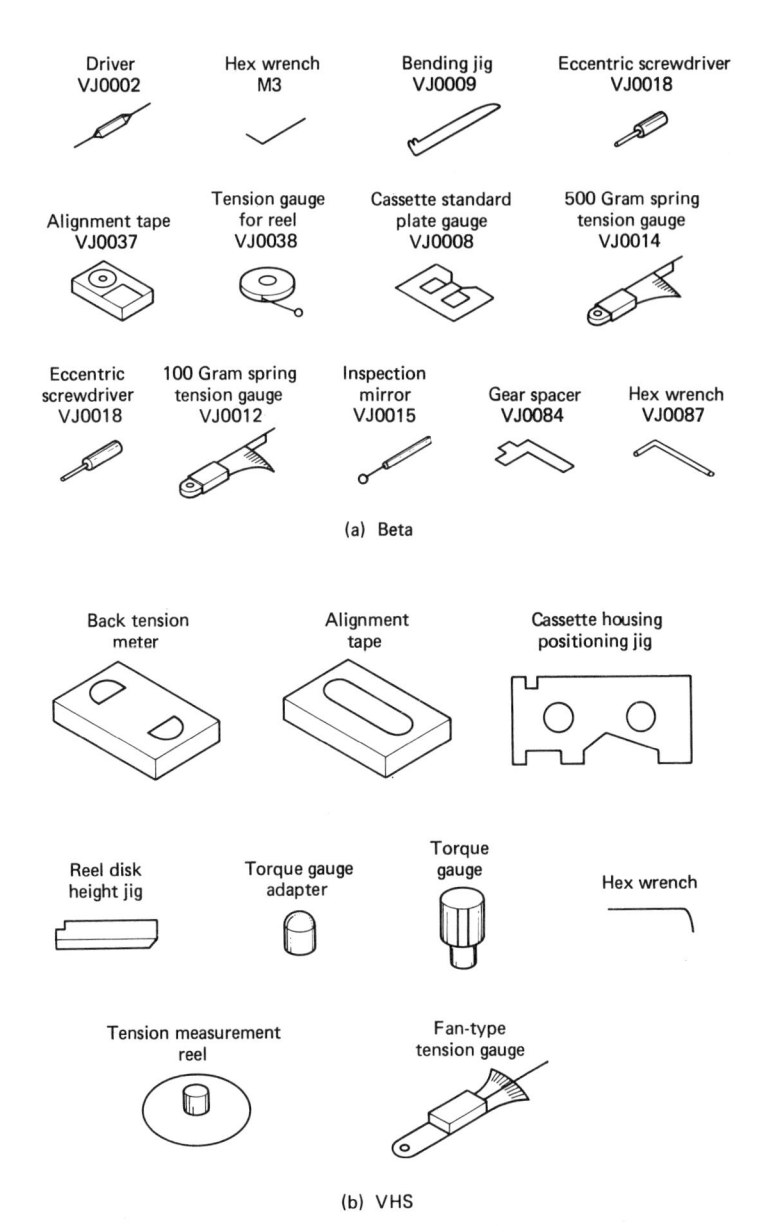

Cassette housing positioning jig

Reel disk height jig

Torque gauge adapter

Torque gauge

Hex wrench

Tension measurement reel

Fan-type tension gauge

(b) VHS

**FIGURE 2-14.** Typical tools, jigs, and fixtures for Beta and VHS mechanical sections.

## 2-8.1 Alignment Tapes

Use of the tools shown in Fig. 2-14 is described in Chapters 6 and 7. One tool merits some discussion here. Most VCR manufacturers provide an alignment tape as part of their recommended tools. An alignment tape is housed within a standard cassette, and has several very useful signals recorded at the factory using very precise test equipment and signal sources. Although there is no standardization, a typical alignment tape contains audio signals (at low and high frequencies, such as 333 Hz and 7 kHz) an RF sweep signal, a black-and-white (monoscope) signal or pattern, and NTSC color bar signals. If you intend servicing one type of VCR extensively, you would do well to invest in the recommended alignment tape.

A typical use for the audio signals recorded on the alignment tape is to check overall operation of the servo speed and phase control systems. For example, if the frequency of an audio playback is exactly the same as recorded (or within a given tolerance), and remains so for the entire audio portion of the tape (as checked on a frequency counter), the servo control systems (both speed and phase) must be functioning normally. If there are any mechanical variations, or variations in servo control, that produce wow, flutter, jitter, and so on, the audio playback will vary from the recorded frequency.

If you do not want to invest in a factory alignment tape, or if you do not want to wear out an expensive factory tape for routine adjustments (alignment tapes do deteriorate with continued use), you can make up your own alignment tape or "work" tape using a blank cassette. The TV stations in most areas broadcast color bars before or after regular programming. (Use the VCR timer for convenience.) These color bars can be recorded using a VCR known to be in *good operating condition*. Any stationary color pattern with vertical lines (such as the white color bar) that extends down to the bottom of the screen) is especially useful. If you have access to a factory tape, you can duplicate it on your own work tape. Of course, make certain to use a *known good* VCR when making the duplication.

## 2-8.2 Miscellaneous Tools

In addition to the special tools described thus far, the mechanical sections of most VCRs can be disassembled, adjusted, and reassembled with common handtools such as wrenches and screwdrivers. Keep in mind that most VCRs are manufactured to Japanese *metric standards,* and your tools must match. For example, you will need metric-sized Allen wrenches and Phillips screwdrivers with the Japanese metric points.

Since VCRs require periodic cleaning and lubrication, you will also need tools and applicators to apply the solvents and lubricants (cleaner sticks for the video heads, etc.). Always use the recommended cleaners, lubricants, and applicators. We describe cleaning and lubrication in Chapter 7.

While on the subject of cleaning, you should be aware of a special cleaning cassette, also known as a *lapping cassette*. Such cassettes contain a nonmagnetic tape coated with an abrasive. The idea is to load the lapping cassette, and run the abrasive tape through the normal tape path (across the video heads, around tape guides, etc.) for *a few seconds*. This cleans the entire tape path (especially the video heads) quite thoroughly. However, prolonged use of a lapping tape can result in damage (especially to the video heads). The author has no recommendation regarding cleaning tape. If lapping cassettes are used, always follow the manufacturer's recommendations, and never use any cleaning tape for more than a few seconds.

# 3

# Typical Beta VCR Circuits

This chapter describes the theory of operation for a number of Beta VCR circuits. (Similar coverage for a cross section of VHS circuits is provided in Chapter 4.) The Beta circuits described here include video, servo, system control, audio, timer, and tuner. Note that the RF unit circuits are not covered. This is because the service literature rarely, if ever, provides any detailed information on the RF unit. At best, only the input, output, and power supply connections are shown. If the RF unit fails for any reason, it must be replaced as a unit, preferably at a factory service facility. This is because the RF unit is essentially a miniature TV broadcast station or transmitter. Any improper adjustment can produce interference with normal television broadcasting.

By studying the circuits found in this chapter, you should have no difficulty in understanding the schematic and block diagrams of similar Beta VCRs. This understanding is essential for logical troubleshooting and service, no matter what type of electronic equipment is involved. No attempt has been made to duplicate the full schematics for all circuits. Such schematics are found in the service literature for the particular VCR. Instead of a full schematic, the circuit descriptions are supplemented with partial schematics and block diagrams that show such important areas as signal flow paths, input/output, adjustment controls, test points, and power source connections. These are the areas most important in service and troubleshooting. By reducing the schematics to these areas, you will find the circuit easier to understand, and you will be able to relate circuit operation to the corresponding circuit of the VCR you are servicing.

Note that, as shown in the illustrations of this chapter, many circuit parts are contained within *integrated circuits* (ICs). These ICs carry the reference designation Q (Q101, Q102, etc.), as do individual transistors outside the ICs. Both the ICs and the transistors are mounted on printed circuit boards (PC boards), as are other individual or discrete circuit parts. The PC boards carry the reference designation of W. This arrangement, as presented in the illustrations of this chapter, is typical for many present-day VCRs.

## 3-1 INTRODUCTION TO BETA CIRCUITS

The majority of the circuits described in this chapter are part of the Sanyo VCR5000. This VCR, described as "Betacord" or "Betavision" by the manufacturer, uses the conventional Beta color-under recording system described in Sec. 1-7 and the Beta high-density PI (phase inversion) described in Sec. 1-8. However, the VCR5000 has two tape formats, called BII (Beta 2) and BIII (Beta 3). The VCR uses a BII/BIII switching method which permits recording of the conventional BII system, in addition to the extremely high density BIII recording system. Figure 3-1 shows the major differences between the BII and BIII systems.

As shown in Fig. 3-1a, the tape formats are essentially the same for BII and BIII, except in video track pitch and tape speed. (The slower tape speed permits longer playing time for a given amount of tape.) However, strictly speaking, the BIII system records on a true zero guard band, while the BII system records with a guard band of about $2\mu$m.

Another difference between BII and BIII is alignment in the color PI system. As described in Sec. 1-8.3, the Beta system uses 1H phase reversal of the chroma signal (recorded at 688 kHz). As shown in Fig. 3-1c, the BIII system uses a 0.5H alignment of phase reversal, rather than the 0.75H alignment for BII (shown in Figs. 3-1b and 1-28).

As shown in Fig. 3-1c, when BIII is selected, the burst of the recording 688-kHz chroma signal is amplified by 6 dB. This makes the recording current of the burst about the same level as a signal with high color saturation, and improves the signal-to-noise (S/N) ratio. By improving the S/N ratio of the color burst signal, the jitter cancel effect of the APC circuit at the time of playback (Sec. 1-8.4) is improved considerably. This improvement is required since the recording density of BIII is much greater than for BII, and the effects of jitter are more noticed. After the BIII playback color burst has been used by the APC circuit, the amplitude of the burst is restored to normal by a burst deemphasis circuit.

## 3-2 BETA VIDEO CIRCUITS

Figure 3-2 is an overall block diagram of the video circuits. As shown, the video signal system uses four ICs (Q1 a CX187, Q2 a CX134A, Q3 a CX188, and Q4 a CX196). The video circuits process both the luminance (Y) and chroma (C)

*102  Typical Beta VCR Circuits*

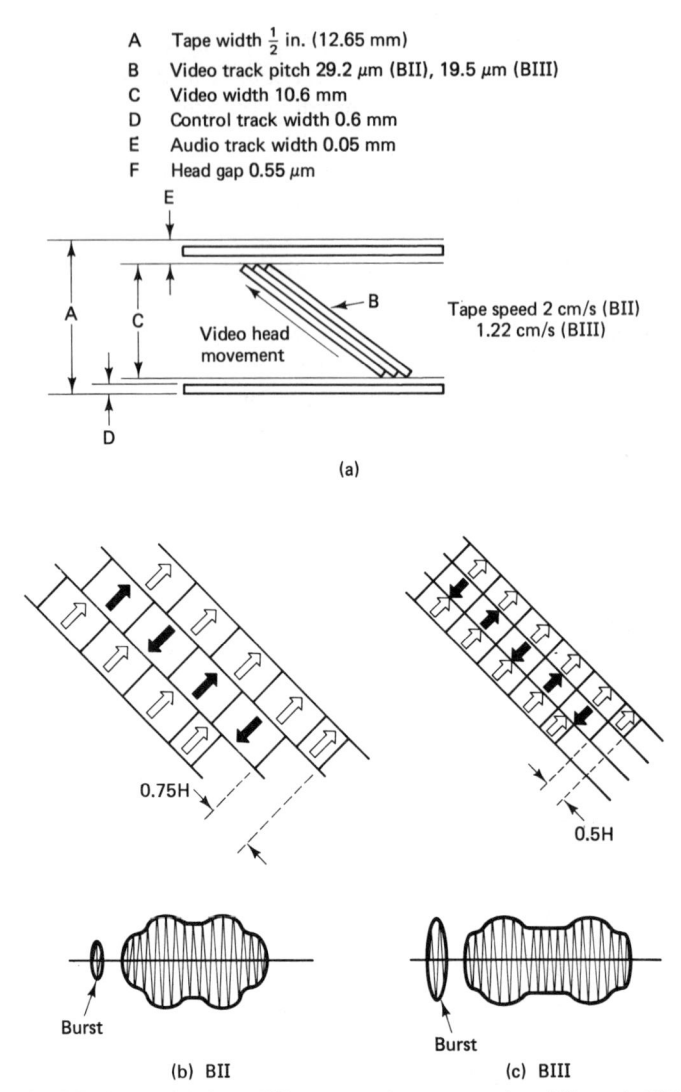

```
A    Tape width 1/2 in. (12.65 mm)
B    Video track pitch 29.2 µm (BII), 19.5 µm (BIII)
C    Video width 10.6 mm
D    Control track width 0.6 mm
E    Audio track width 0.05 mm
F    Head gap 0.55 µm
```

Tape speed 2 cm/s (BII)
1.22 cm/s (BIII)

(a)

(b) BII

(c) BIII

Burst

**FIGURE 3-1.** Major differences between the BII and BIII systems.

signal during record and playback. Most of the processing takes place within the ICs, as is described in the following paragraphs.

### 3-2.1  Luminance Signal System Circuits

The luminance signals are processed mainly by ICs Q1 and Q2. IC Q1 has both recording and playback functions, as shown in the simplified blocks of Fig. 3-3 and 3-4. IC Q1 is used primarily in playback, as shown in Fig. 3-5.

During record, the video signal from the tuner is applied through the record/playback changeover switch in Q1 to the AGC (automatic gain control) system. Note that the AGC system consists of the SYNC AGC in Q1 and the PEAK AGC (Q10 to Q14) outside the IC.

**AGC Circuit and Y-FM Modulator.** Two AGC systems are used to accommodate video signals with different sync/signal ratios. In the case of an input signal with a sync/signal ratio of about 30% or more, the SYNC AGC provides a fixed sync signal level. When the sync/signal ratio is less than 30%, the AGC output can become excessive with only the SYNC AGC. This causes overmodulation of the FM modulator and RF unit, resulting in buzz, and so on. When the sync/signal ratio is low, the PEAK AGC circuit takes over to prevent an excessive AGC output. With either circuit, the AGC signal output is applied to the comb filter circuit through pin 24 of Q1.

When the input signal is color, the signal from pin 24 passes the comb filter and low-pass filter. The luminance signal (with color removed) is returned to Q1 at pin 22. When the input is black and white, the signal is applied directly to pin 23 by a switch in Q1. Either way, the output from pin 23 is applied through C11 and pin 2 to a sync tip clamp within Q1. The output of the sync tip clamp is applied to an external emphasis network through pin 1. The emphasis network contains preemphasis, white clip, and black clip circuits, as well as several adjustments. VR15 is used for sync tip modulation frequency adjustment. VR17 is used for frequency shift adjustment, whereas VR16 controls the half-H adjustment. The white clip and black clip circuits are adjusted by VR18 and VR19, respectively. The output of the emphasis network is applied to the FM modulator.

The output of the FM modulator is applied through a 688-kHz trap and a phase compensation circuit to the Y record amplifier Q21. The trap is controlled by a color killer signal from pin 17 of Q1. When the video signal is black and white, the trap removes the 688-kHz color signal.

**Record Amplifier.** The recording amplifier for the Y-FM signal consists of Q21–Q24. The Y-FM signal (at the specified recording current) is applied to the video heads through T1 and T2. VR6 is used for adjustment of the recording current value. VR5 is used to adjust the frequency characteristic of the recording current. Q25 is the recording amplifier for the color signal (after being converted to 688 kHz). This color signal is superimposed on the Y-FM signal (at the specified level), and both signal currents are supplied to the video heads.

**Playback Amplifier and DOC.** The playback amplifier and DOC circuits are shown in Fig. 3-5. Operation of these circuits is similar to that described in Sec. 1-8.7.

The low-amplitude playback signal from the video heads passes step-up transformers T1 and T2 and is fed to the first-stage amplifiers Q16 and Q19

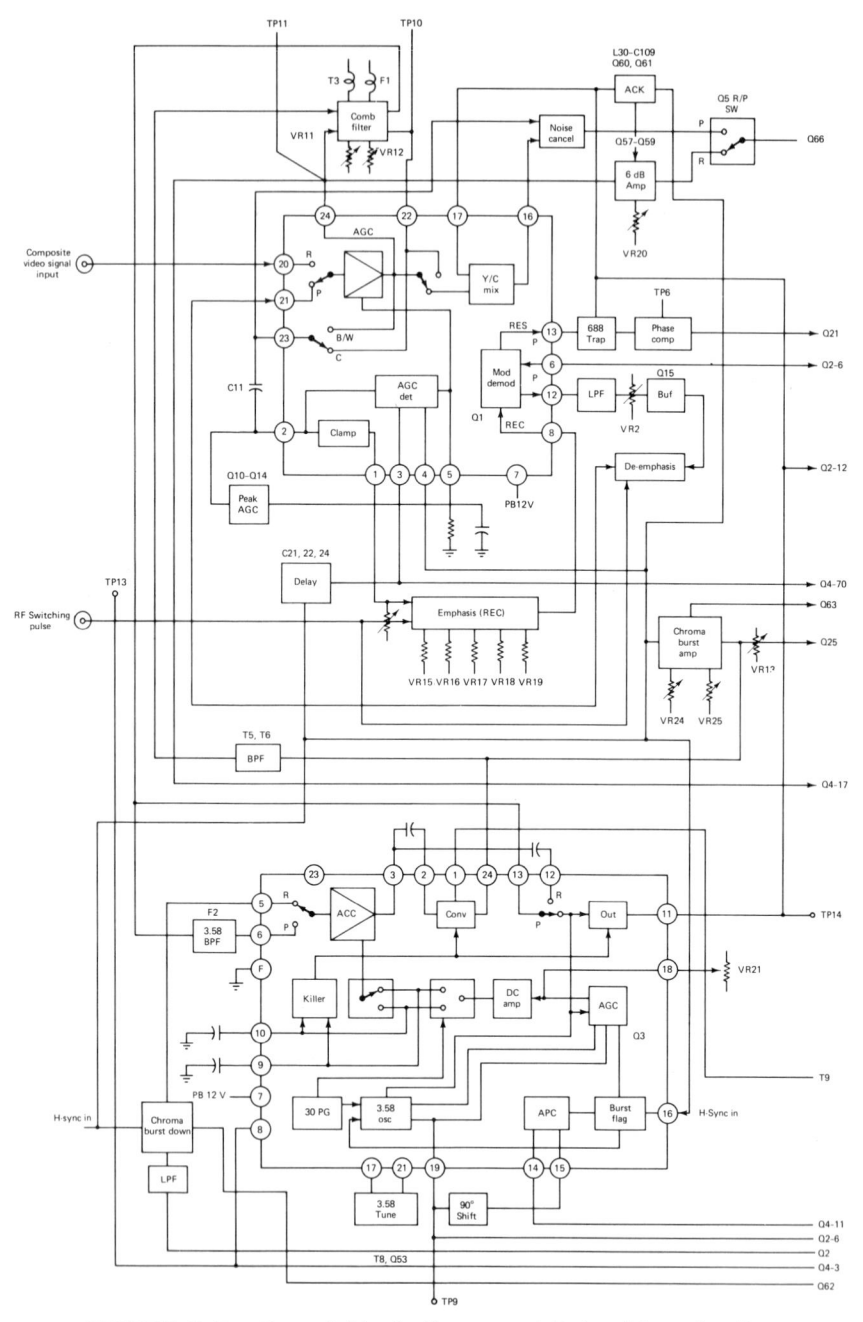

**FIGURE 3-2.** Overall block diagram of Beta video circuits

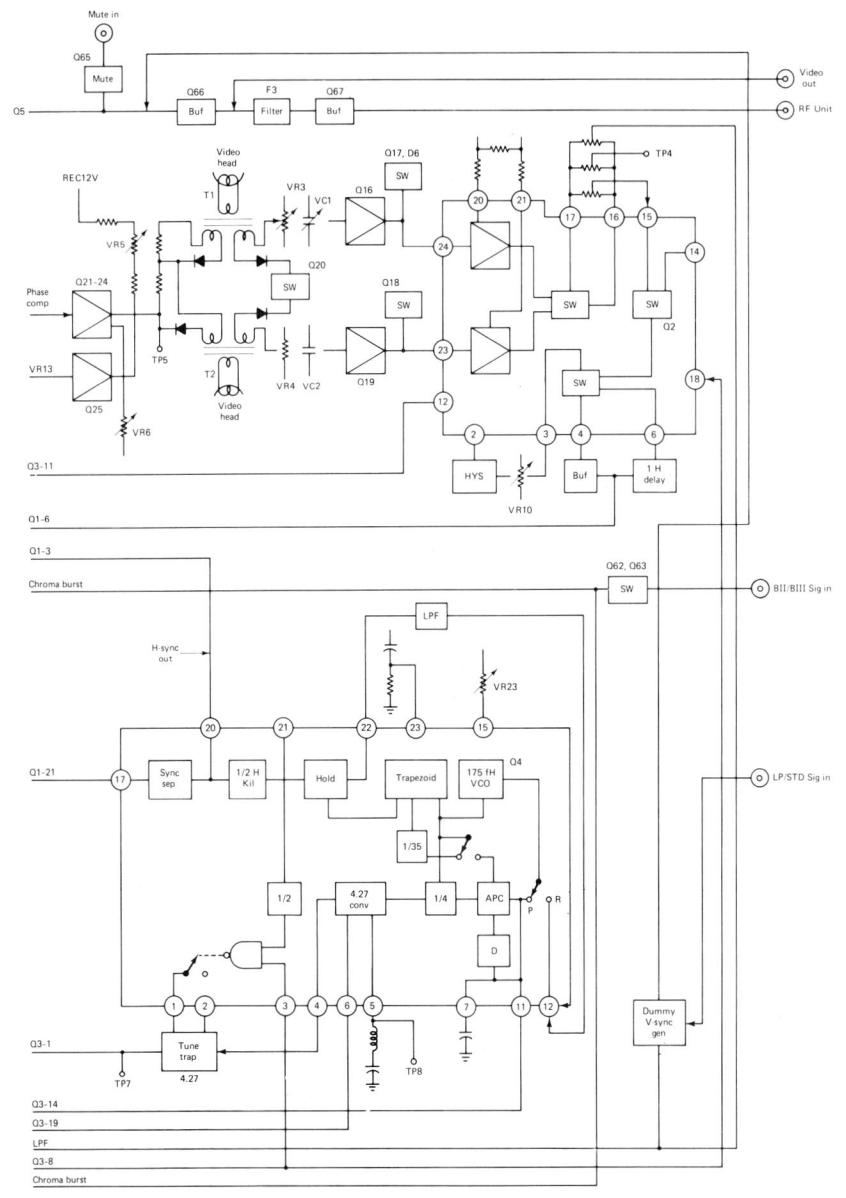

**FIGURE 3-2.** (continued)

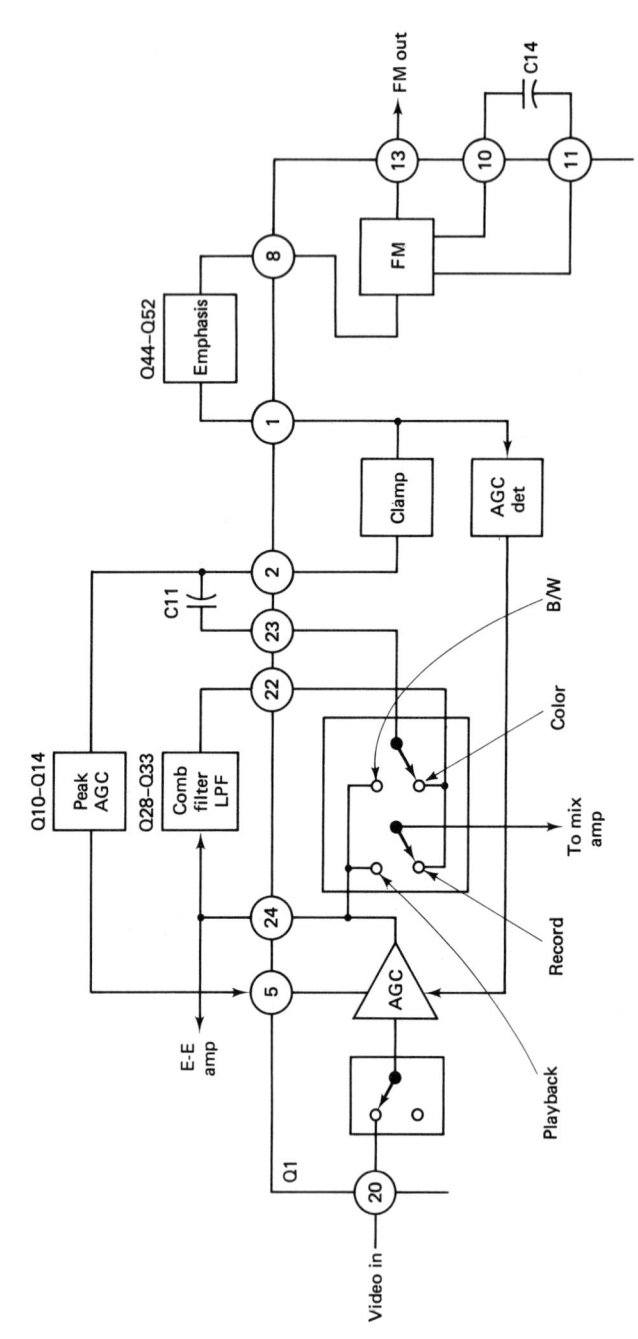

**FIGURE 3-3.** Typical Beta luminance signal circuit.

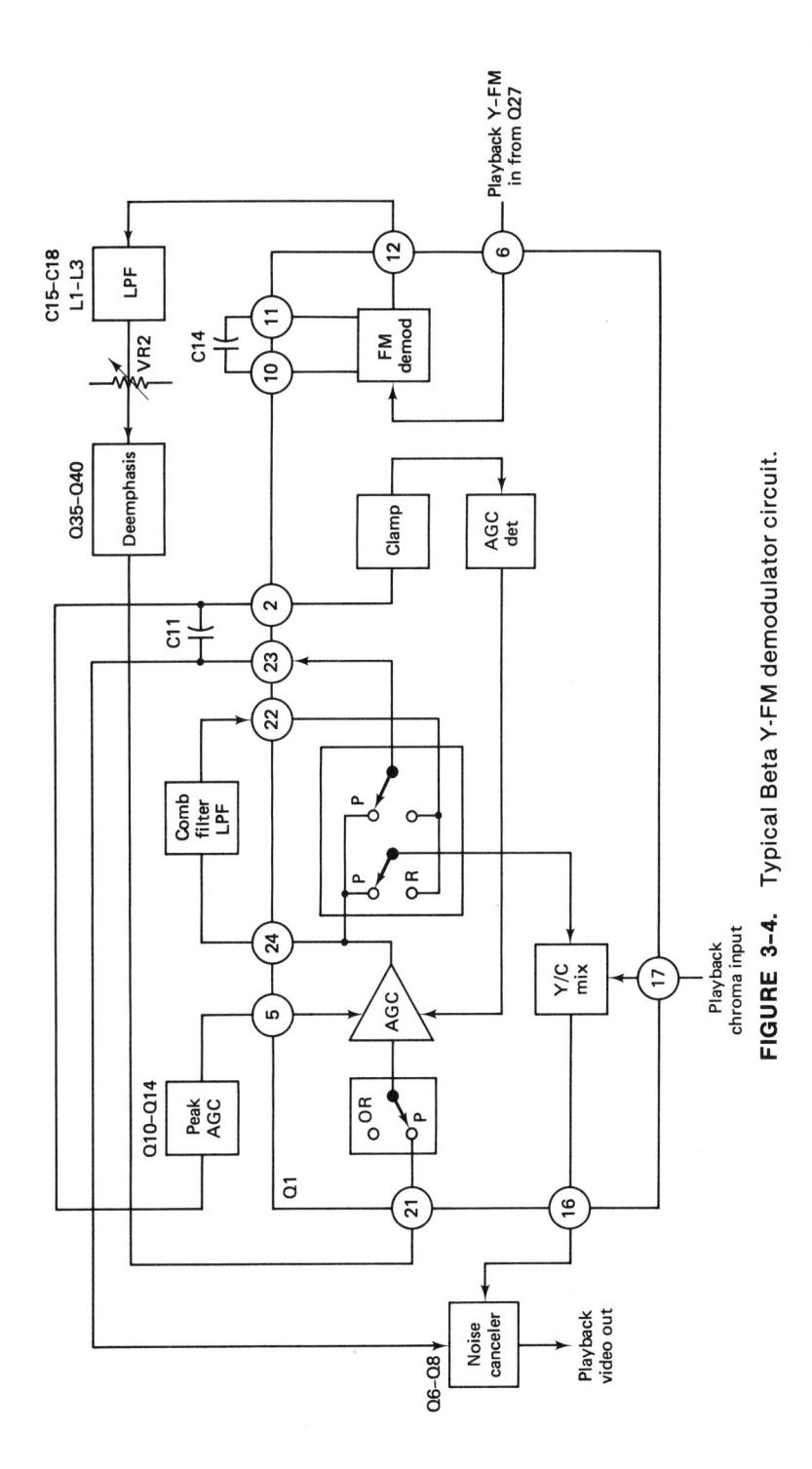

**FIGURE 3-4.** Typical Beta Y-FM demodulator circuit.

108  *Typical Beta VCR Circuits*

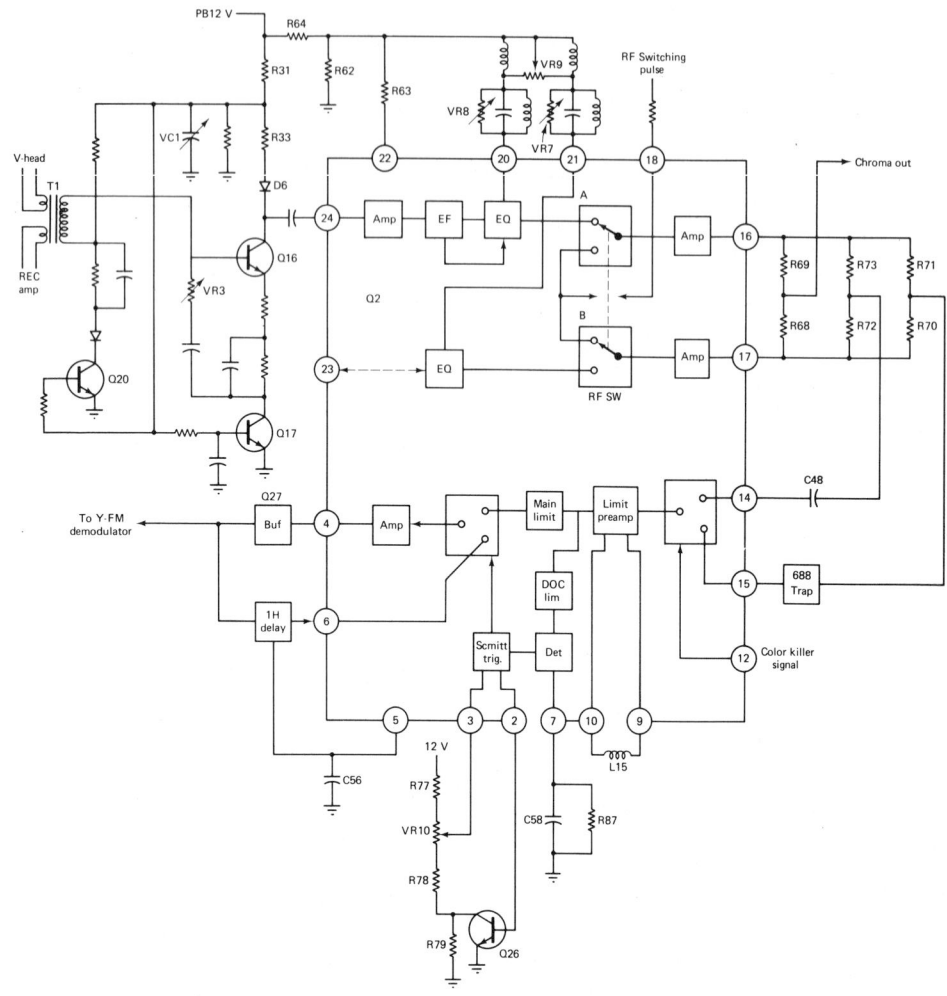

**FIGURE 3-5.** Typical Beta playback amplifier and DOC circuit.

(which operate as low-noise amplifiers). The outputs of Q16 and Q19 are applied to cascade amplifiers within Q2. (Note that T1 and T2 operate as input/output transformers, supplying recording current to the video heads during record, and stepping up the low-amplitude playback signals during playback.)

The frequency characteristics of the playback signals are adjusted by VC1, VR3, and VR8 for channel A, and by VC2, VR4, and VR7 for channel B. The balance between channels A and B is adjusted by VR9. The frequency-compensated or adjusted playback signal is switched by the RF switching pulse at pin 18, and appears as an output from Q2 at pins 16 and 17. The chroma signal is taken from this output at the junction of R68 and R69 and is applied to the chroma processing circuits (Sec. 3-2.2). One Y signal is taken from the output at the junction of R72 and R73, and is applied through C48 to pin 14 of Q2. Another Y signal is taken from the junction of R70 and R71 and is applied through a 688-kHz trap (C50-L11) to pin 15 of Q2. Both Y signals are applied to a color/black and white switch within Q2. This switch is operated by a color killer signal applied at pin 12. When the playback contains color, the signal at pin 14 is used in the normal manner. In the absence of color, the switch selects the signal from pin 15.

The output from the color/black and white switch is applied to the DOC circuit, which operates as described in Sec. 1-8.7. The limiter suppresses any waveform distortion and passes the video signal to the detector and adjustable Schmitt trigger. The adjustment control VR10 sets the firing point of the Schmitt trigger and thus sets the point at which dropout compensation occurs. If dropout is present, the Schmitt trigger switches the signal so that the preceding 1H delayed signal (passed through the 1H delay line, external to Q2) appears at the output. The compensated signal at pin 4 is applied through emitter-follower buffer Q27 to the Y-FM demodulation circuit.

**Y-FM Demodulation Circuit.** The circuits of Q1, used at the time of recording, operate as a demodulator at the time of playback. This is shown in Fig. 3-4.

The signal from the playback amplifier and DOC circuit is applied to pin 6 of Q1 during playback. This signal is demodulated by circuits within Q1, and is applied through a low-pass filter (C15-C18, L1-L3) to a deemphasis circuit (Q35-Q40). This deemphasis circuit has the reverse frequency characteristics from the preemphasis at the time of recording. The deemphasis circuit is set for optimum characteristics by VR2.

After passing through the deemphasis circuit to obtain the correct frequency characteristics, the playback Y signal is returned to Q1 at pin 21 and is applied to the same AGC circuit used during record (Fig. 3-3). Reprocessing by the AGC (at playback) helps restore the playback signal to the same level that existed at the time of record. After passing the AGC, the Y signal is mixed with the chroma signal (at pin 17) and appears as the playback composite video signal at pin 16. The Y signal is also applied through pin 23 to a noise canceler circuit (Q6-Q8). The noise canceler output is combined with the composite video signal at pin 16. In the noise canceler, the high-frequency noise component is removed, the phase is inverted, and with reverse phase and the same amplitude, the processed video signal is added back to the composite video signal, thus canceling the noise. (This sequence is shown in Fig. 1-37.)

**Record/Playback and Video Output Circuit.** The playback composite video signal from pin 16 of Q1 is applied to the RF converter (and to a VIDEO OUT terminal) through a record/playback switching system and various video output circuits, as shown in Fig. 3-2. Operation of these circuits is controlled by an IC analog switch Q5, shown in Fig. 3-6. IC Q5 provides the necessary changeover for the E-E and playback signals. (As discussed in Sec. 1-8.6, when

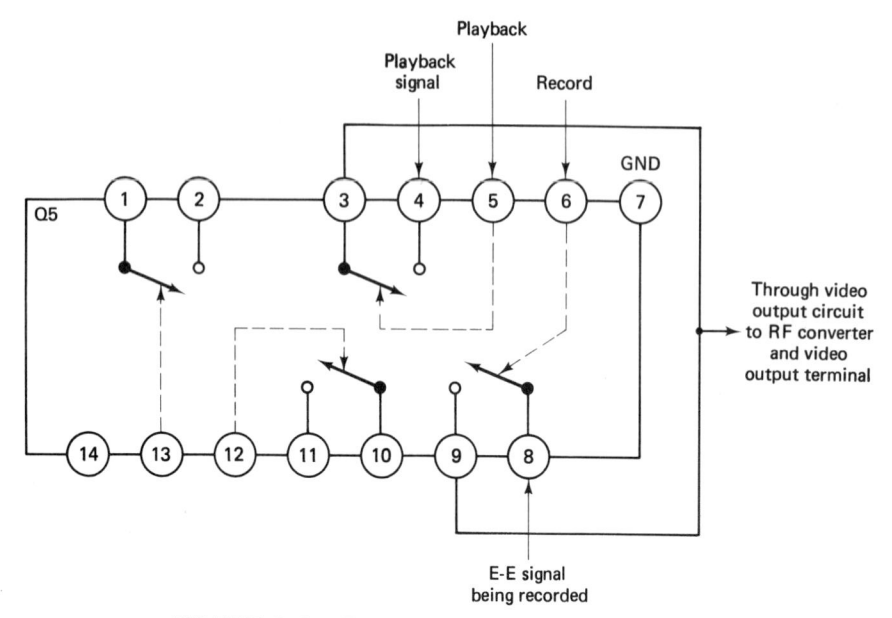

**FIGURE 3-6.** Record/playback switching system.

in the record mode, the record output circuit is connected to the playback input so that the video signal can be monitored on a TV set.)

As shown in Fig. 3-6, operation of switches within Q5 is controlled by signals at pins 5 and 6. During record, a signal is applied to pin 6, closing the switch across pins 8 and 9. This connects the E-E signal (signal being recorded) to the RF converter and VIDEO OUT terminals. During playback a signal is applied to pin 5, closing the switch across pins 3 and 4. This connects the playback signal to the RF converter and VIDEO OUT.

As shown in Fig. 3-2, the video signal from pin 24 of Q1 is applied to a 6-dB amplifier (Q57–Q59) and is then applied to pin 8 of Q5 (Fig. 3-6). VR20 is used to adjust the E-E level from the 6-dB amplifier. The ACK (automatic color killer) circuit (Q60–Q61) functions to remove the burst signal from the E-E video if the signal is black and white or if the color signal being recorded is extremely attenuated. The ACK accomplishes this burst removal by inserting a 3.58-MHz trap when a color killer signal is applied. (This is the same color killer signal applied to pin 12 of Q2, Fig. 3-5.)

After passing the noise canceler circuit, the playback video signal is applied to pin 4 of Q5. Either the playback signal (pin 4) or the E-E signal (pin 8) is selected by operation of the switches in Q5. The output from Q5 (pins 3 and 9, tied together) is applied through the video output circuit to the RF converter and VIDEO OUT terminal.

The video output circuit is composed of transistors Q65, Q66, Q67, and filter F3. Q65 operates as a muting circuit for the video signal. During playback,

Beta Video Circuits   111

if there is no control signal output from the servo system because of a malfunction (or if the CTL signal was not properly recorded), the video signal is muted (cut off at Q65). This produces a black raster on the monitor TV screen. Transistor Q66 functions as an emitter follower and provides a 75-Ω output impedance for the VIDEO OUT terminal. Transistor Q67 operates as a buffer between filter F3 and the RF converter.

### 3-2.2 Chroma Signal System Circuits

The chroma or color signals are processed mainly by ICs Q3 and Q4. The relationship of Q3 and Q4 to the remaining video circuits is shown in Fig. 3-2. Figures 3-7 and 3-8 show the Q3 and Q4 circuits in greater detail. Before discussing the operation of these circuits, let us review the overall functions of the chroma circuits.

As discussed in Sec. 1-8 and shown in Fig. 3-9, the main function of the chroma circuits during record is to convert the 3.58-MHz color burst signal to a 688-kHz signal and to record this signal on tape. The 688-kHz signal is locked in phase and frequency to the broadcast 3.58-MHz and H-sync signals by AFC and

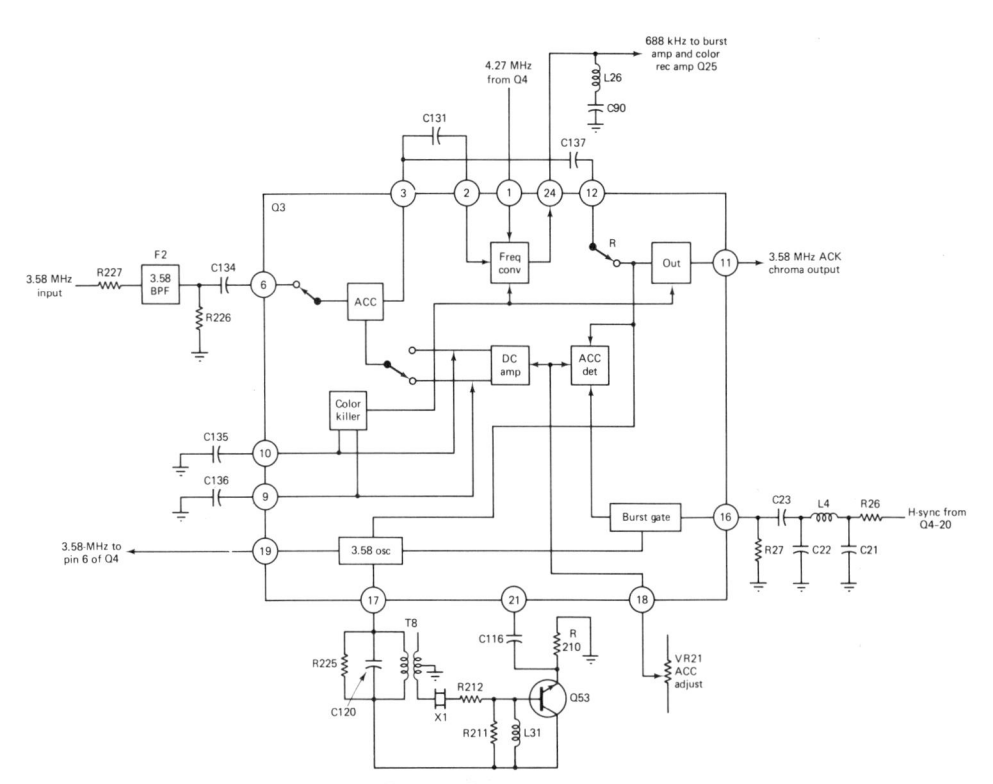

**FIGURE 3-7.** Typical Beta record mode operation.

**FIGURE 3-8.** Generating the 4.27-MHz signal.

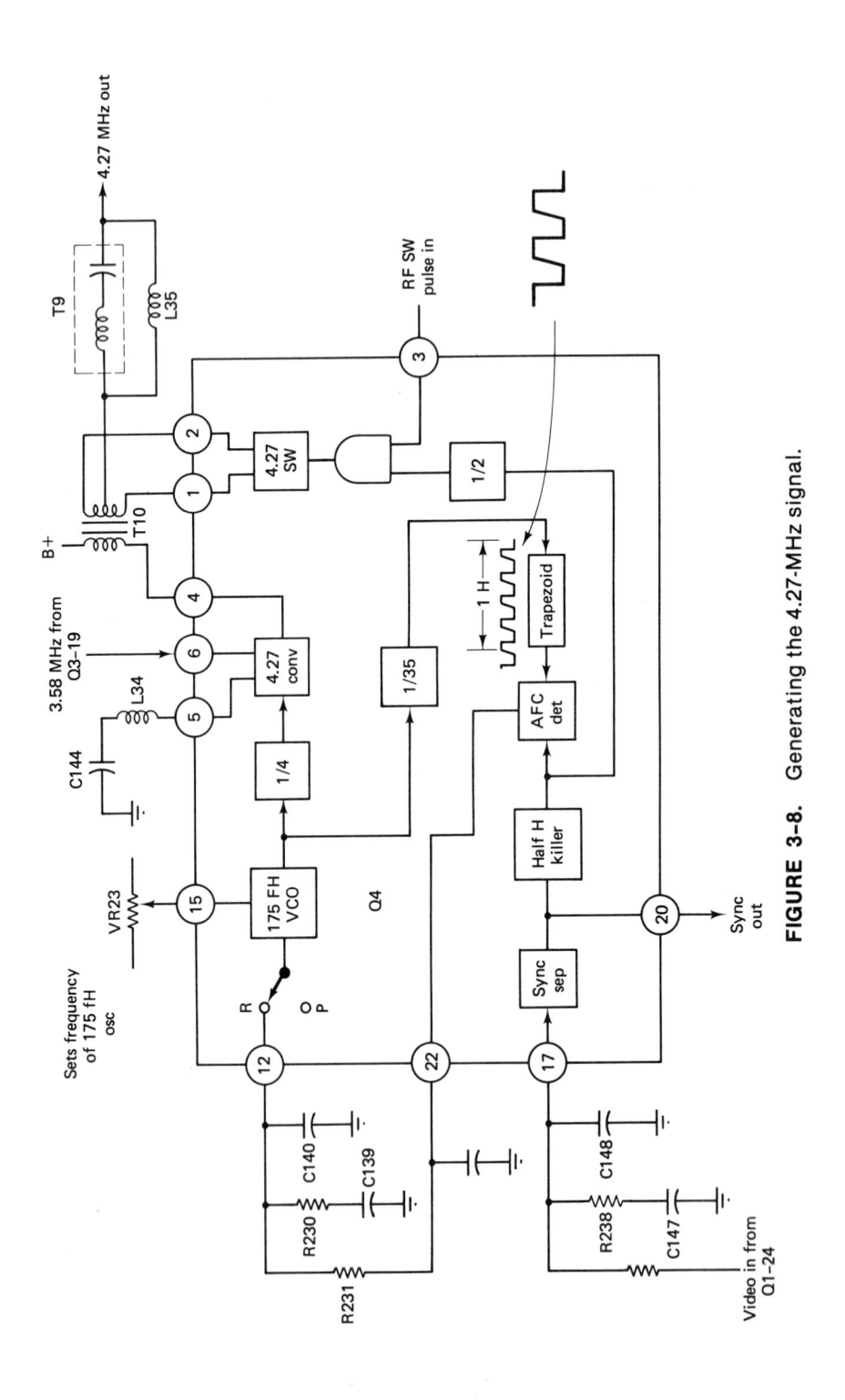

Beta Video Circuits    113

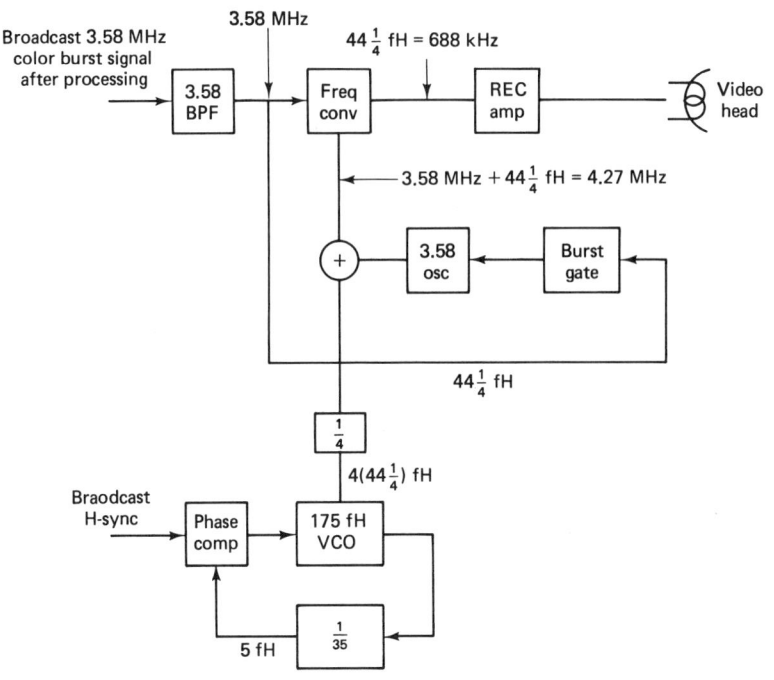

```
Broadcast 3.58 MHz        3.58 MHz
color burst signal                     44¼ fH = 688 kHz
after processing   ┌──────┐   ┌──────┐      ┌──────┐
   ──────────────▶│ 3.58 │──▶│ Freq │─────▶│ REC  │──────▶ Video
                  │ BPF  │   │ conv │      │ amp  │        head
                  └──────┘   └──────┘      └──────┘

                        ◀──── 3.58 MHz + 44¼ fH = 4.27 MHz ────

                         ┌───┐   ┌──────┐   ┌──────┐
                         │ + │◀──│ 3.58 │◀──│ Burst│◀──────┐
                         └───┘   │ osc  │   │ gate │       │
                                 └──────┘   └──────┘       │
                                                           │
                                    44¼ fH                 │
                         ┌───┐                             │
                         │ ¼ │                             │
                         └───┘                             │
                         4(44¼) fH                         │
Broadcast        ┌──────┐   ┌──────┐                       │
H-sync           │Phase │   │175 fH│                       │
──────────────▶  │ comp │──▶│ VCO  │───────────────────────┘
                 └──────┘   └──────┘
                     ▲         │
                     │  ┌────┐ │
                     └──│1/35│◀┘
                   5 fH └────┘
```

**FIGURE 3-9.** Basic functions of chroma circuits during record.

APC circuits. Because of zero guard band recording, the 688-kHz signal is recorded with a phase reversal of 1H on every other track to remove crosstalk. As shown in Fig. 3-10, the function of the chroma circuits during playback is to convert the 688-kHz signal back to a 3.58-MHz color burst. This 3.58-MHz signal is locked in phase and frequency to the playback 688-kHz and H-sync signals. The 1H phase reversal is removed, and the 3.58-MHz signal is combined with the luminance signal to produce a composite video signal identical (hopefully) to the broadcast video signal.

**Record Mode.** The composite video signal at the output of the AGC circuit within Q1 (Fig. 3-2) is taken from pin 24 and applied to pin 6 of Q3 through a comb filter and a 3.58-MHz bandpass filter F2. T3, VR11, and VR12 set the level of the video signal passing through the comb filter. Filter F2 passes only the 3.58-MHz signal applied to pin 6 of Q3 (Fig. 3-7). The H-sync is applied to the burst gate circuit within Q3 through pin 16. The output of the 3.58-MHz oscillator, locked to the burst gate signal, appears at pin 19 of Q3. This signal is applied to a 4.27-MHz converter within Q4. Simultaneously, the 3.58-MHz signal at pin 6 of Q3 is applied to the frequency converter within Q3 through the ACC circuit and C131. The frequency converter also receives a 4.27-MHz signal from Q4 through pin 1 of Q3. The combination of the two signals at the fre-

## 114  Typical Beta VCR Circuits

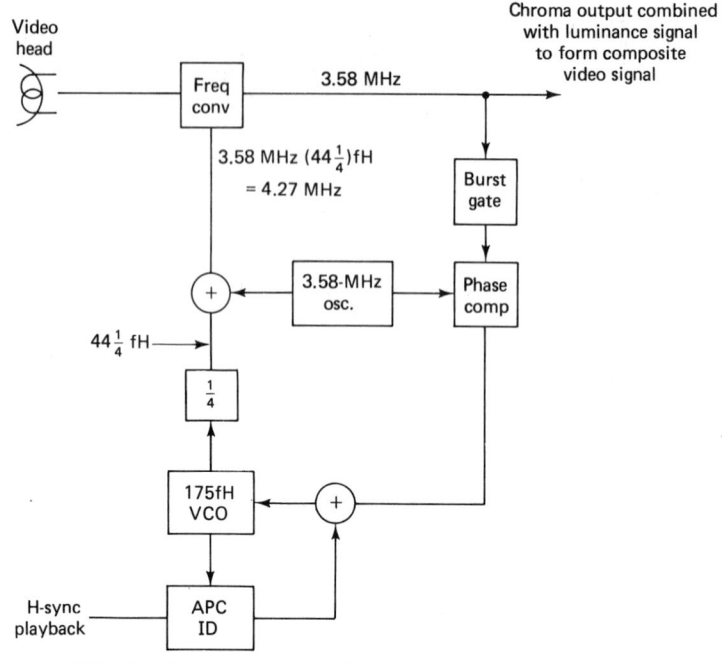

**FIGURE 3-10.** Basic functions of chroma circuits during playback.

quency converter produces a 688-kHz signal which is applied (through pin 24 and a burst amplification circuit) to the color record amplifier Q25. Since the 4.27-MHz signal is phase reversed at 1H on every other track, the 688-kHz color signal is recorded with the phase inversion.

During record, the 3.58-MHz signal at pin 3 of Q3 is applied through C137 and circuits within Q3 to appear as a 3.58-MHz ACK (automatic color killer) output at pin 11 of Q3. In the absence of a 3.58-MHz signal (black and white, or badly attenuated color) the ACK circuits function to cut off the color circuits (such as removal of the E-E video as described in Sec. 3-2.1).

Q3 also contains ACC (automatic color control) circuits which maintain the level of the color signal. VR21 sets the level of the color signal at 1V.

**BIII Burst Amplification Circuit.** As discussed in Sec. 3-1, the 688-kHz signal to be recorded is amplified by 6 dB when the BIII mode is selected. This is accomplished by the burst amplification circuit shown in Fig. 3-11. As shown, a burst gate pulse (H-sync) is applied to the base of Q42, and a BII/BIII switching control signal is applied to Q72 and Q73. The switching control signal is high for BIII and low for BII. VR13 sets the record signal level during both BII and BIII.

Although the circuit is called a burst amplifier, the circuit function is more

Beta Video Circuits 115

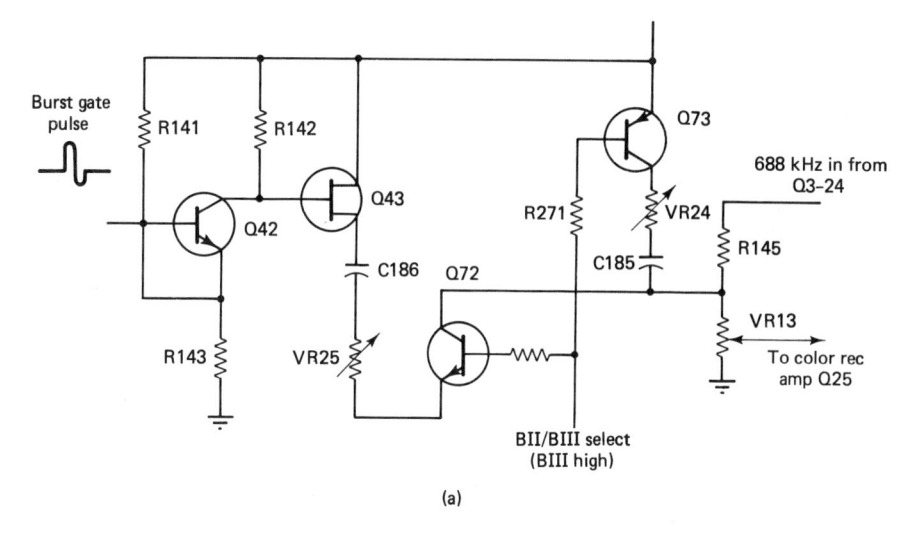

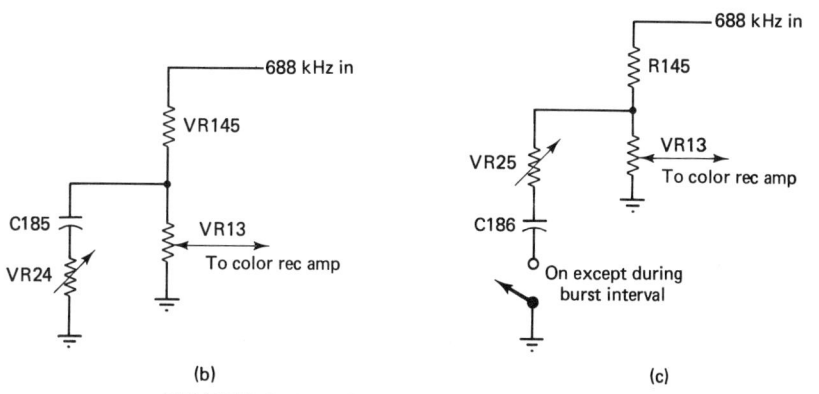

**FIGURE 3-11.** BIII burst amplification circuit.

like that of an attenuator. When BII is selected, Q73 is ON and Q72 is OFF. During BII, the circuit appears as shown in Fig. 3-11b, where part of the 688-kHz signal to be recorded is passed through C185 and VR24 to ground. This attenuates part of the signal. The level of the recorded signal during BII is set by adjustment of VR24.

When BIII is selected, Q72 is ON and Q42 is OFF (during the burst interval) and ON at other times. During BIII, the circuit appears as shown in Fig. 3-11c, where part of the 688-kHz to be recorded is passed through VR25 and C186 to ground, at all times except during the burst period (when the burst gate is present). This attenuates the signal, except during the burst period. The net result is that the burst level is increased *in relation* to the remainder of the signal. VR25 sets the level of record during BIII.

**Generation of 4.27-MHz Signal during Record.** As shown in Fig. 3-8, the 4.27-MHz signal used in the color circuits is generated in Q4. IC Q4 receives a composite video signal (containing H-sync signals) at pin 17, RF switching pulses at pin 3, and a 3.58-MHz signal from Q3 at pin 6. The 4.27-MHz output is at pins 1 and 2. The H-sync pulses at pin 17 are separated by the sync separator circuits within Q4 and appear as an output at pin 20.

During record, the output of the 175fH VCO (voltage-controlled oscillator) is counted down $\frac{1}{35}$ to 5 fH by circuits within Q4, and produces a trapezoidal wave. Simultaneously, the H-sync separated from video (at pin 17) is compared in phase with the trapezoidal wave, and the resultant error voltage is applied to the 175fH VCO through pins 22 and 12 and a filter circuit (C139, C140, R230, R231). This filter circuit smooths out the error voltage which locks the 175fH VCO to H-sync. The exact frequency of the 175fH VCO is set by VR23.

The 175fH signal is divided by 4 and the resulting 44 $\frac{1}{4}$fH signal (locked to the H-sync) is applied to a frequency converter within Q4. This frequency converter combines the 44 $\frac{1}{4}$ fH with the 3.58-MHz signal from Q3 to produce a 4.27-MHz signal. Since the 44 $\frac{1}{4}$ fH signal is locked to H-sync, and the 3.58-MHz signal is locked to the broadcast 3.58-MHz color signal, the 4.27-MHz signal from Q4 is thus locked to both signals. The 4.27-MHz signal is applied through transformer T10 to the converter in Q3.

Transformer T10 acts as a carrier-phase inverter. The phase of the 4.27-MHz signals passing through T10 is reversed every 1H by operation of circuits within Q4. These phase inverter circuits receive both RF switching pulses and H-sync pulses, and function to alternately short pin 1 or pin 2 of Q4 to ground for each H-sync pulse. This shorts either end of the T10 secondary every 1H, but only during track A. The RF switching pulses override the H-sync pulses when track B is being traced by head B. Each time one end of the T10 secondary is grounded, the center-tap output is inverted in phase.

**Playback Mode.** The basic function of the video circuits during playback is shown in Fig. 3-10. Figure 3-12 shows operation of the Q3 circuit during playback. As shown, the 688-kHz output from the preamplifier in Q2 is applied to pin 5 of Q3 through a low-pass filter (C130, C133, L33) at the time of playback. As discussed in Sec. 3-1, when BIII is selected, the burst is amplified 6 dB by a separate circuit (Fig. 3-11). During BIII playback, the burst signal is restored to a normal level by a burst deemphasis circuit (described in the following paragraphs).

In either BII or BIII modes, the 688-kHz signal at pin 5 of Q3 is passed through the ACC circuit. Note that the ACC circuit of Q3 has an independent d-c amplifier circuit for each field to improve transient response. Fields A and B are switched by the RF switching pulses applied at pin 8. The 688-kHz signal from the ACC circuit is applied to a frequency converter through C131. The frequency converter also receives a 4.27-MHz signal generated in Q4 through pin 1. The

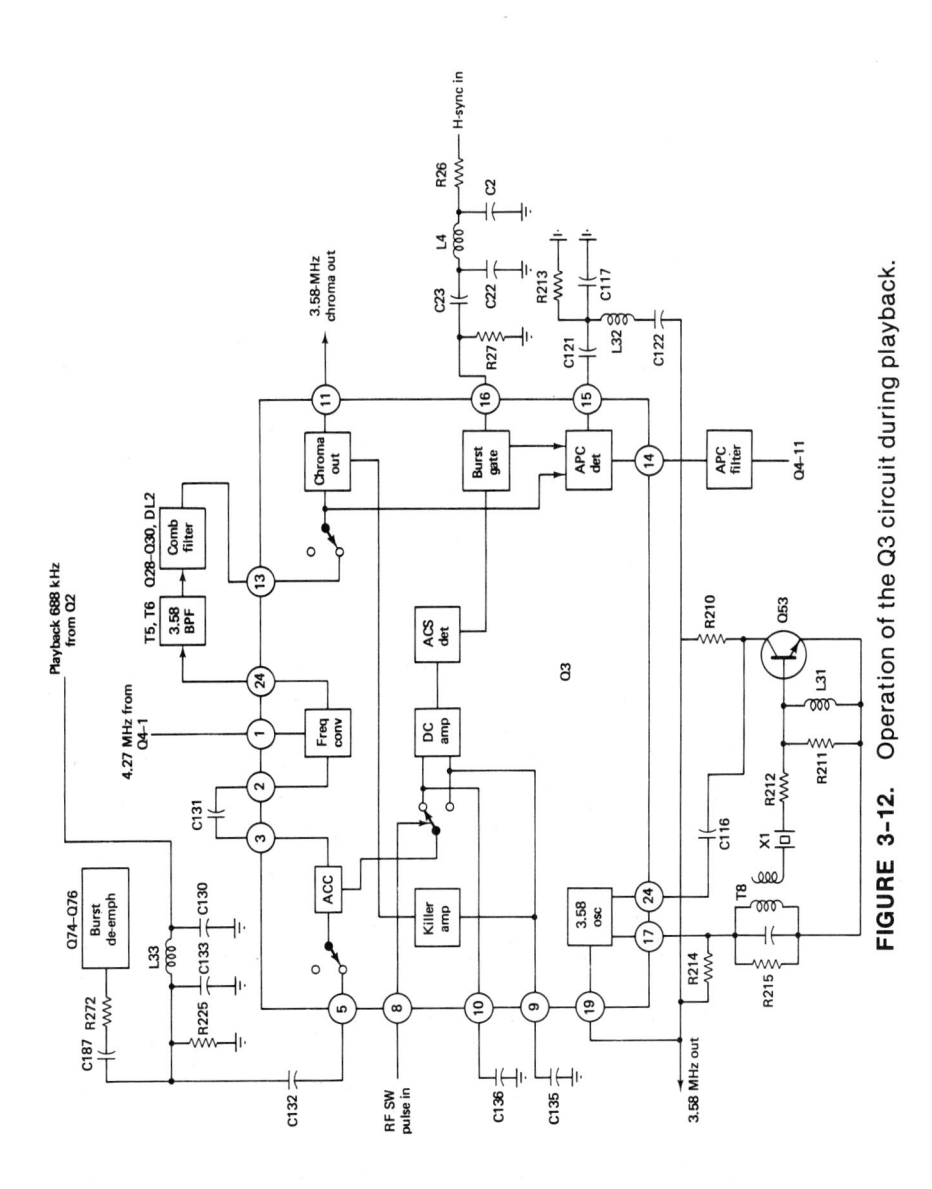

**FIGURE 3-12.** Operation of the Q3 circuit during playback.

*118 Typical Beta VCR Circuits*

resultant 3.58-MHz signal is applied through a bandpass filter (T5, T6) and a comb filter (Q28–Q30, DL2) to a chroma output amplifier in Q3. The comb filter restores the phase reversal introduced by Q4 and T10 (Fig. 3–8). The 3.58-MHz playback signal from the chroma output amplifier is applied through pin 11 and the ACK circuits to pin 17 of Q1, where the chroma signal is mixed with the luminance signal.

As shown in Fig. 3–12, the 3.58-MHz signal at pin 13 of Q3 is also applied to an APC (automatic phase control) circuit. This APC circuit also receives H-sync signals through a burst gate (pin 16) and 3.58-MHz signals from pin 19 (shifted 90° by C117, L32, and R213). The burst gate signals turn on the APC detector during the time of the color burst. The 3.58-MHz crystal reference signal is compared with the phase of the 3.58-MHz playback signal. Any difference in phase produces an error signal at pin 14. This error signal is applied to pin 11 of Q4 to control the phase of the 175fH VCO.

**Generation of 4.27-MHz Signal during Playback.** At the time of recording, Q4 produces $44\frac{1}{4}$ fH by a one-fourth countdown of the 175fH VCO locked to the H-sync of the video signal. Similarly, the 4.27-MHz signal is produced from the 3.58-MHz signal locked to the input burst. At the time of playback, the 175fH VCO is controlled by the APC circuit (Fig. 3–12). With control only by the APC, it is possible for a mislock to occur. Such a mislock condition is prevented by an APC-ID (APC identification) circuit within Q4.

Figure 3–13 shows operation of the APC-ID circuits. As shown, the 175fH VCO output is applied through a gate to a counter. This counter is switched open, and then reset, every 4 H by pulses from a flip-flop (FF) locked to the playback H-sync. If the 175fH frequency is correct, the counter will count 700 pulses before reset (175 x 4 = 700). If the 175fH VCO is off-frequency, the count will not be 700 and an error voltage is applied to the input of the 175fH VCO (to correct the frequency). Thus, the 175fH VCO is locked to both the playback 3.58-MHz (through the APC, Fig. 3–12) and the playback H-sync (through the APC-ID, Fig. 3–13).

**Burst Deemphasis Circuit.** Details of the burst deemphasis circuit of Fig. 3–12 are shown in Fig. 3–14. This circuit operates only during playback, when BIII is selected, and functions to attenuate the 688-kHz signal passing to pin 5 of Q3. As shown, a burst gate pulse (H-sync) is applied to the base of Q75, and a BII/BIII switching control signal is applied to the base of Q76. The switching control signal is low for BIII and high for BII. A high during BII turns Q76 ON, shorting the collector to emitter, and thus bypasses the burst gate pulse to ground. Under these conditions, Q75 and Q74 are OFF. This turns Q74 ON only during BIII and only during the time of the color burst. Thus, the signal is attenuated only during this time period.

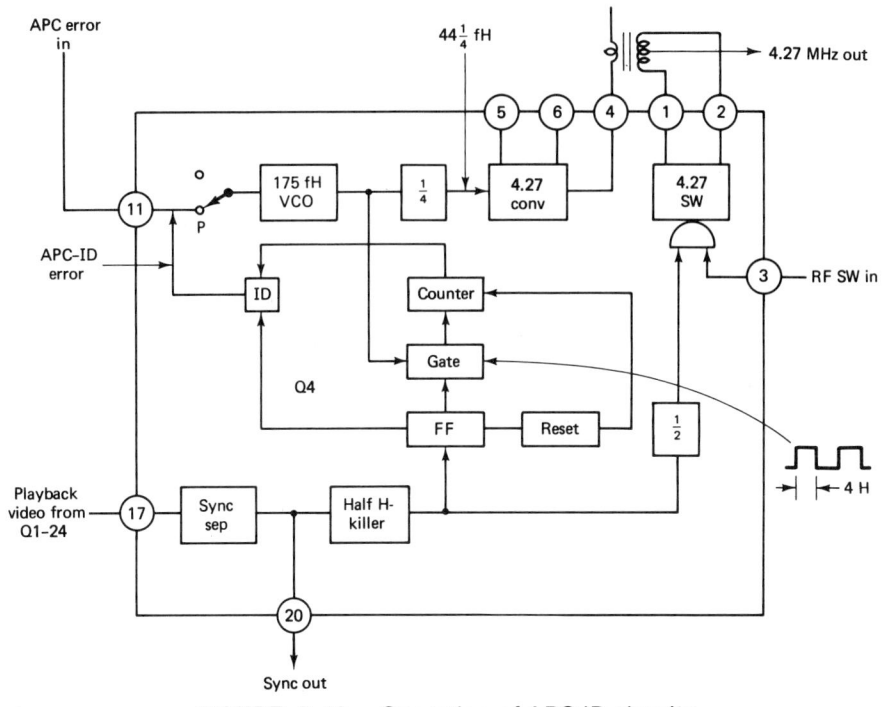

**FIGURE 3-13.** Operation of APC-ID circuits.

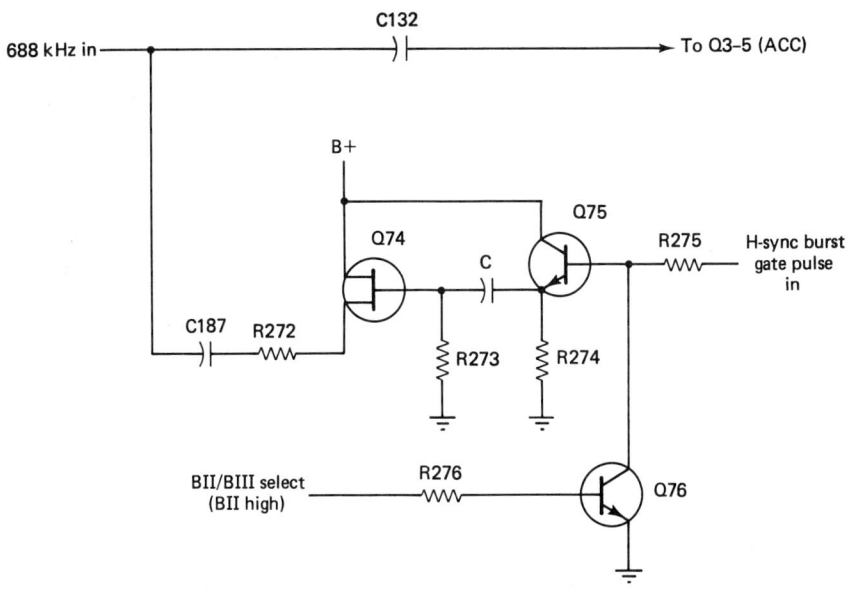

**FIGURE 3-14.** Operation of burst deemphasis circuit.

*119*

## 3-3 BETA SERVO CIRCUIT I

Figure 3-15 is an overall block diagram of servo circuit I. Note that there are two servo circuits in the VCR5000. Servo circuit II is discussed in Sec. 3-4. As shown, servo circuit I uses two ICs (Q416, a CX143, and Q404, a CX186), and a drum motor driver PC board (W8). Most of the servo control functions take place within the ICs and PC board, as described in the following paragraphs. Note that servo circuit I includes a drum servo, a capstan servo, a picture search circuit (part of the drum servo), and a loading/unloading circuit (part of the capstan servo).

The *drum servo signals* are processed mainly by the circuits of Q404. These circuits provide error signals to the drum motor driver circuits on W8. The circuits also provide switching and control signals to the capstan motor control IC Q416. The signals generated by the drum PG coils (A and B) and applied to pin 1 of Q404 are used for both speed and phase control of the drum motor. The signals at pin 18 (the drum error signal) contain both speed and phase information. The speed information is derived from the PG signals, whereas the phase information is based on a comparison of the PG signal phase with composite video V-sync signals (during record) or a servo reference signal (during playback). The servo reference signal is obtained by countdown of a 31.468-kHz signal on the W11 PC board (part of the servo II circuit, Sec. 3-4). Either the V-sync or servo reference signals are applied at pin 12 of Q404. The *picture search circuit* (pin 24 of Q404) varies drum speed so that the H-sync signal of the playback video is set to become 15.734 kHz during picture search.

The *capstan servo signals* are processed by the circuits of Q416 as well as by several other discrete component circuits. The capstan servo circuit has three major functions: (1) keeping the tape speed stable, (2) selecting the tape speed for either BII (2 cm/s) or BIII (1.33 cm/s), and (3) operation as a tracking servo during playback. The capstan is belt driven by a coreless d-c motor. The tape is held against the capstan by a pinch roller, and the tape is pulled from one cassette reel to another. The cassette reel motor is provided for rewind, fast-forward, and take-up of the tape onto the cassette. The reel motor is controlled by the system control circuit (Sec. 3-5). The tape is also driven by the reel motor during picture search. The system control circuit provides tape speed control during picture search. The *loading/unloading* functions are controlled, in part, by the capstan servo circuits and by the system control circuit.

### 3-3.1 Drum Servo Circuit

Figure 3-16 shows the arrangement of the PG coils and video heads within the drum. As shown, the magnets rotate with the video heads and produce switching pulses when the magnets pass over the PG coils. The video heads, magnets, and PG coils are arranged so that the switching pulses are generated when the video heads are about 8° ahead of the switching position. The PG-A and PG-B pulses (which are of opposite polarity) are applied to pin 1 of Q404. The PG pulses are then amplified by a PG-AB detector and are converted into 30-Hz square waves by FF 1. These square-wave pulses are supplied to a SW position delay B circuit, and to the speed servo system. Figure 3-17 shows the relationships of the pulses at pins 1, 2, 3, and 4 of Q404.

The pulses at the output of FF 1 pass through delay B and delay A circuits to become the RF switching pulses (also known as the video head switching pulses) at pin 4 of Q404. Each of these delay circuits is provided with an electrical adjustment which determines the start and stop time (or switching time) of the RF switching pulses. Thus, the relationship of the RF switching pulses to the physical location of the video heads can be adjusted by VR401 (pin 2) and VR402 (pin 3). As discussed in Chapter 6, VR401 and VR402 are adjusted so that the phase difference between the trailing edge of the RF switching pulse and the front edge of the V-sync signal is 7H.

**Controlling Drum Motor Speed.** Drum motor speed is controlled by an error signal from pin 18 of Q404 applied through switching 3 transistor Q414 to Q1608 on the drum motor driver circuit W8 (Fig.3-15). The error signal is developed from the PG signals taken from FF1. Figure 3-18 shows the relationship of the pulses involved.

The PG-A pulse is delayed by about one-fourth of a cycle with the speed MM1 circuit, and trapezoidal waveforms are produced by the slope 2 circuit. A gate pulse is formed from the PG-B pulse by the gate pulse circuit, and the trapezoidal waveforms are gated by gate 2 and slope 2 circuits. The gated voltage is held by the hold 2 circuit. After the current is amplified, the resultant error signal is sent to the drum motor drive circuit from pin 18.

If the drum motor speed is low, there is a wide gap between the PG-A pulse and the PG-B pulse and the gated voltage increases. Conversely, when motor speed is high, the gap between the two pulse narrows and the gated voltage decreases. If the voltage fed out to pin 18 is high, the motor speed increases. Thus, if the drum motor speeds up for any reason, the pulse gap narrows and the gated voltage decreases, decreasing the pin 18 voltage to slow the motor. The opposite occurs if the drum motor tends to slow down.

Drum motor speed is also affected by adjustments VR404, VR408, and VR409, which change the time constant of the Q407 switch 2 circuit connected to the speed MM1 circuit through pin 24. By changing this time constant, the gated voltage at pin 18 (and thus motor speed) is changed. VR409 is used during normal forward record and play modes. VR404 is used during the F-search (forward picture search) mode. VR408 is used during the R-search (reverse picture search) mode.

When there is no need to rotate the video heads, the voltage that controls the drum motor drive circuit is grounded and the switch 2 circuit Q407 causes the

## 122 Typical Beta VCR Circuits

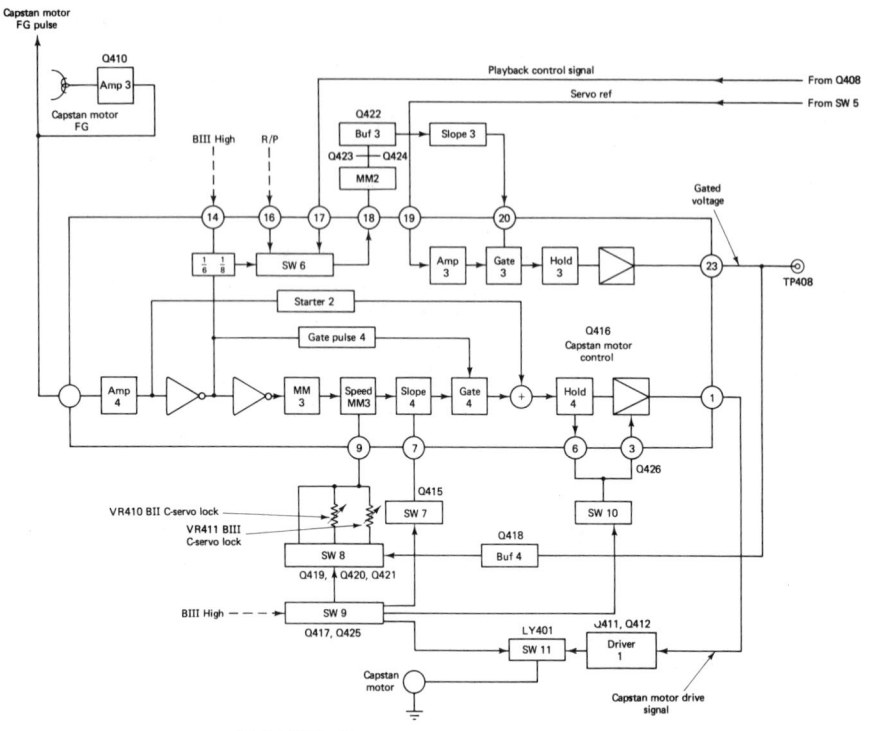

**FIGURE 3-15.** Beta servo circuit I.

drum motor to stop rotating. This control signal is sent from the system control circuit (Sec. 3-5).

**Separating the Vertical Sync Signal.** The rotary phase of the video heads during record must be synchronized with the video signals being recorded. The V-sync separate circuit (Fig. 3-15) serves to extract the V-sync signal from the composite video input. The V-sync signal is then used to control the drum motor phase during record. Figure 3-19 shows the relationship of the pulses involved.

The composite sync signal is current amplified by emitter follower Q401, and the H-sync signal is filtered out by a low-pass filter. The V-sync signal is taken from the filter and is converted into square waves by Q402. The shaped V-sync signal passes through switching 1 diode D402 and is applied to pin 12 of Q401. After the V-sync signal noise immunity has been enhanced by MM1, the 60-Hz signal is converted into 30-Hz square waves by the $\frac{1}{2}$-countdown circuit. The 30-Hz signals at pin 13 are applied through the slope 1 circuit to the drum phase control circuit, through the amp/switching circuit to the control head, and through the switching 5 circuit to the capstan servo.

During playback, the Q402 power supply is cut off so that the V-sync signal

Beta Servo Circuit I / 123

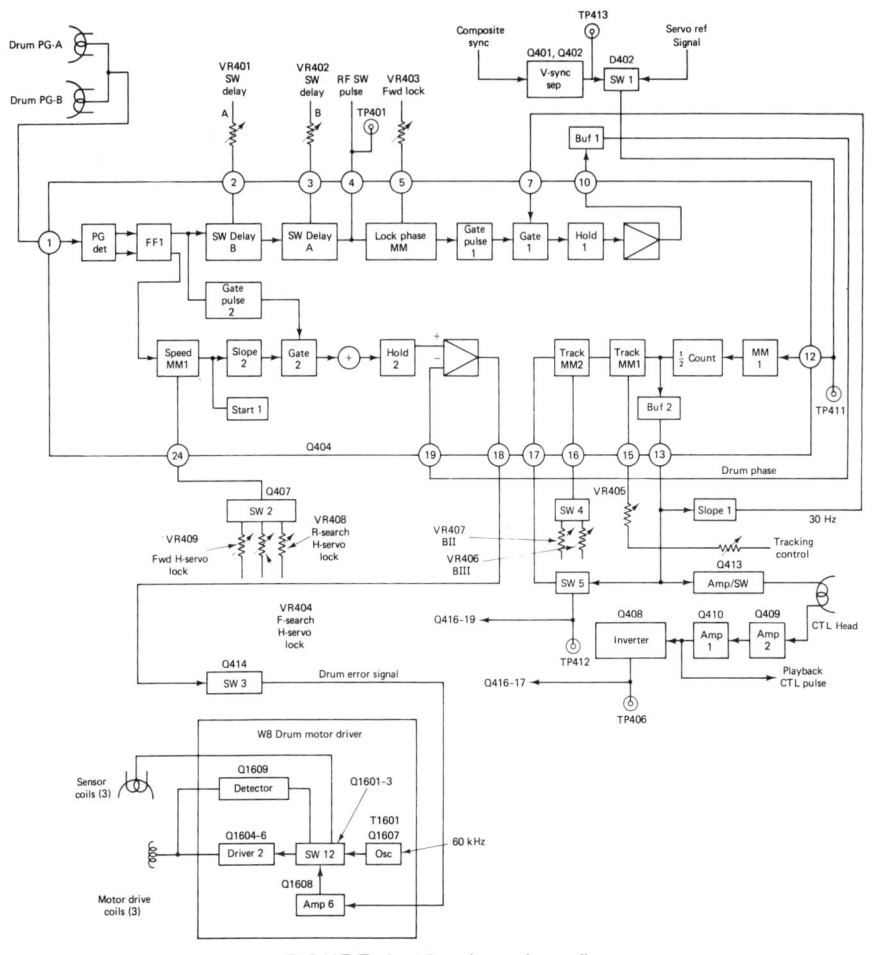

**FIGURE 3-15.** (*continued*)

cannot pass. Instead, the signal generated by the reference signal generator circuit on the W11 PC board is applied to pin 12 of Q404 through D401.

**Controlling the Drum Motor Phase.** During record, the time at which the video head starts to scan is recorded at the bottom of the tape by the CTL head so that the rotary phase of the video heads can be synchronized with the video signal. The position at which the video head starts scanning on the tape is 7H before the leading edge of the video V-sync signal. During playback, phase control (together with speed control) is applied to the drum motor in order to start the scanning of video head A, so that the video head is made to trace accurately on the tape pattern made during record.

Since the tape is driven by the capstan, the phase control signal applied to

124  *Typical Beta VCR Circuits*

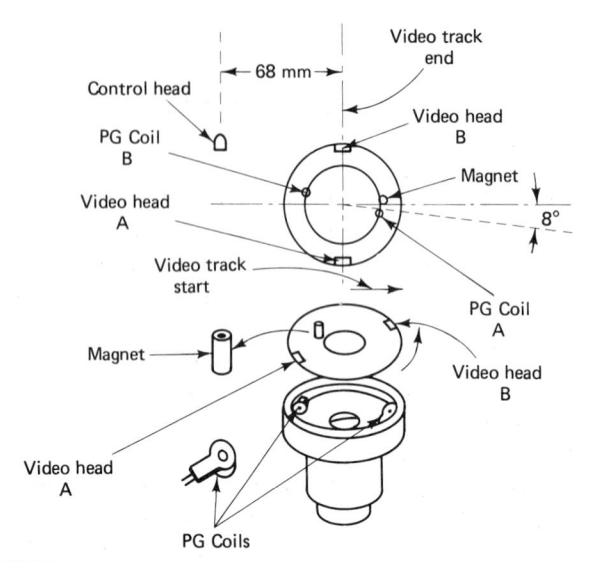

**FIGURE 3-16.** Arrangement of the PG coils and video heads within the drum.

the capstan servo is also used by the drum servo for phase control (although the signal paths are different). During record, the phase control signal is developed by a comparison of the RF switching pulses with the V-sync pulses (which are also being recorded on tape by the CTL head, after a conversion from 60 Hz to 30 Hz). Since the RF switching pulses are developed by the PG pulse, the RF switching pulses represent the speed and phase of the drum. During playback, the phase

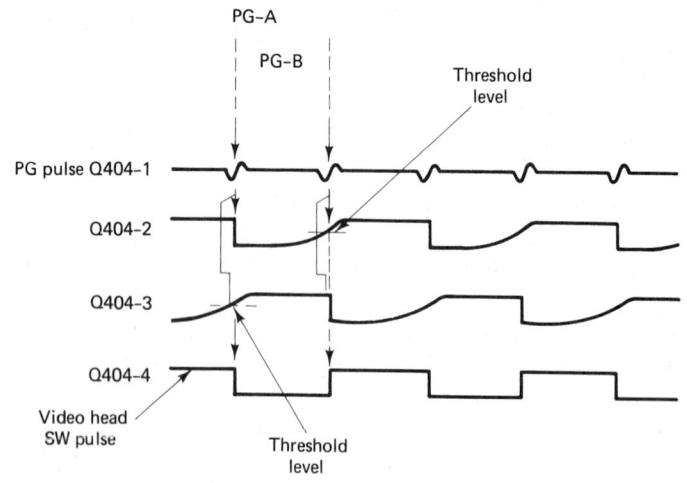

**FIGURE 3-17.** Relationships of pulses in drum servo circuits.

Beta Servo Circuit I 125

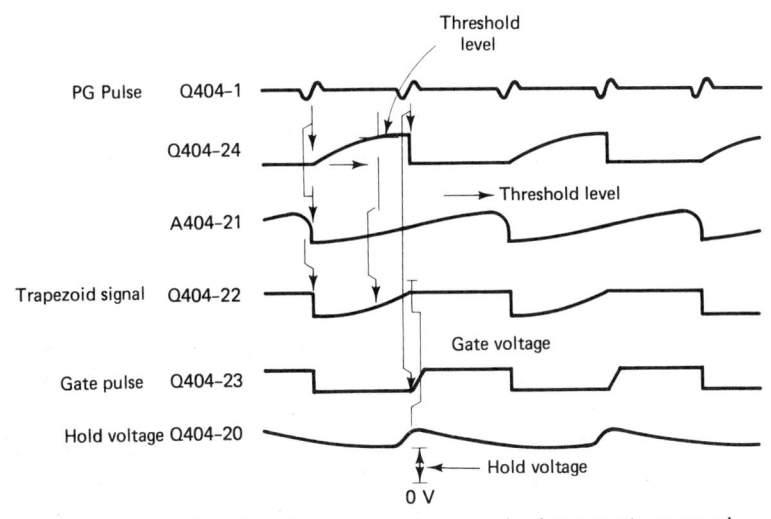

**FIGURE 3-18.** Relationships of pulses in drum motor speed controlling circuits.

control signal is developed by a comparison of the recorded CTL pulses with the servo reference signal (obtained by a countdown of a 31.468-kHz signal in the servo II circuit).

The signal that indicates the rotary phase of the video heads is the fall part

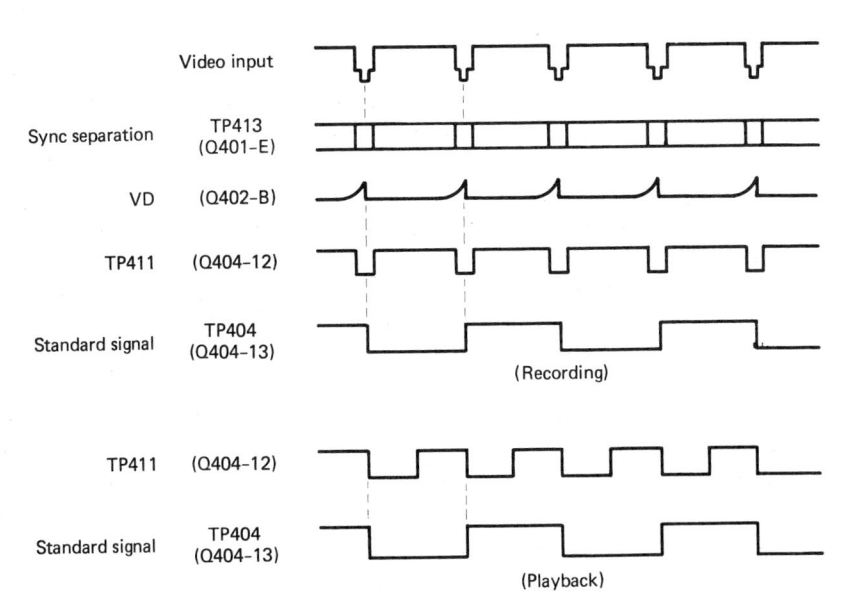

**FIGURE 3-19.** Relationships of pulses in vertical sync signal separation circuits.

## 126  Typical Beta VCR Circuits

of the RF switching pulse, which corresponds to the PG-A pulse. Figure 3-20 shows the relationship of the pulses involved. The RF switching pulse is delayed by the lock phase MM, turned into a gate pulse by gate pulse 1 and then sent to gate 1. The amount of delay introduced by lock phase MM is set by VR403 connected through pin 5 of Q404 (Fig. 3-15).

Gate 1 also receives a 30-Hz trapezoidal wave through pin 7. This wave originates at pin 12 (either from V-sync during record or from the servo reference signal during playback), is taken from pin 13, and is converted to a trapezoidal wave by slope 1. The trapezoidal wave is gated in gate 1 by the gate pulse from gate pulse 1. The gated wave is held, current amplified, and fed through pin 10, buffer Q406, and pin 19 to be mixed with the drum speed control signal in a comparator circuit. The combination of the two signals (speed and phase) controls drum rotation.

If the drum rotation speed increases, the phase control signal applied to the comparator increases and causes the comparator output to decrease, decreasing the pin 18 voltage to slow the drum motor. The opposite occurs if the drum motor tends to slow down. The amount of phase control signal increase or decrease is set by the amount of delay introduced by lock phase MM, which, in turn, is controlled by VR403. As discussed in Chapter 6, VR403 is adjusted so that the video head starts to scan 7H before the leading edge of the V-sync signal.

**Driving the Drum Motor.**  As shown in Fig. 3-15, the three-phase, four-pole drum motor is controlled by two circuits on the W8 PC board. One circuit switches the current flowing in the phase coils, while the other circuit controls the

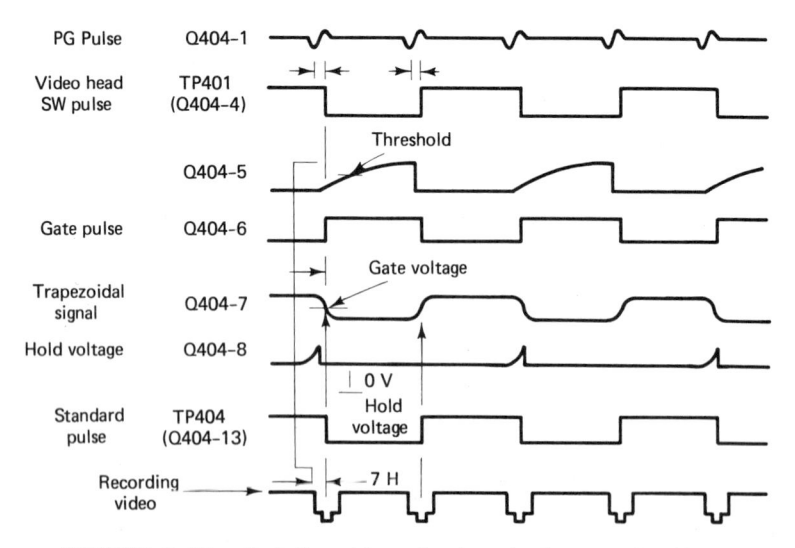

**FIGURE 3-20.** Relationships of pulses in drum motor phase controlling circuits.

current. These circuits control drum motor rotation based on the error voltage supplied by the drum servo circuit (from pin 18 of Q404, through switching transistor Q414). The drum motor circuits include an oscillator and switching circuits. Unlike the other servo circuits, the W8 PC board power supply is 15 V. The remaining servo circuits use 12 V.

A Hartley oscillator composed of Q1607 and T1601 produces signals at a frequency of approximately 60 kHz. These signals are sent to sensor coils mounted on the drum motor stator as shown in Fig. 3-21. The sensor coils are arranged so that the phase of the drum motor stator coil and rotor magnet can be detected. The oscillator signals applied to the sensor coils produce magnetic oversaturation at the magnet's south pole. When the rotor's south pole comes near the sensor coil, there is no change since the coil is oversaturated. However, when the north pole approaches, the sensor coil inductance is reduced. As a result, the oscillation waveforms are reduced (and the approaching north pole is detected).

The output of the three sensor coils (representing the phase relationship of the three-phase motor windings) is applied through switching circuits (Q1601, Q1602, Q1603) to control drum motor speed. The motor speed is changed as necessary (by switching action) if the three phases are not 120° apart. (When the motor is properly synchronized, the motor goes through one rotation while the stator's polarity changes 12 times.)

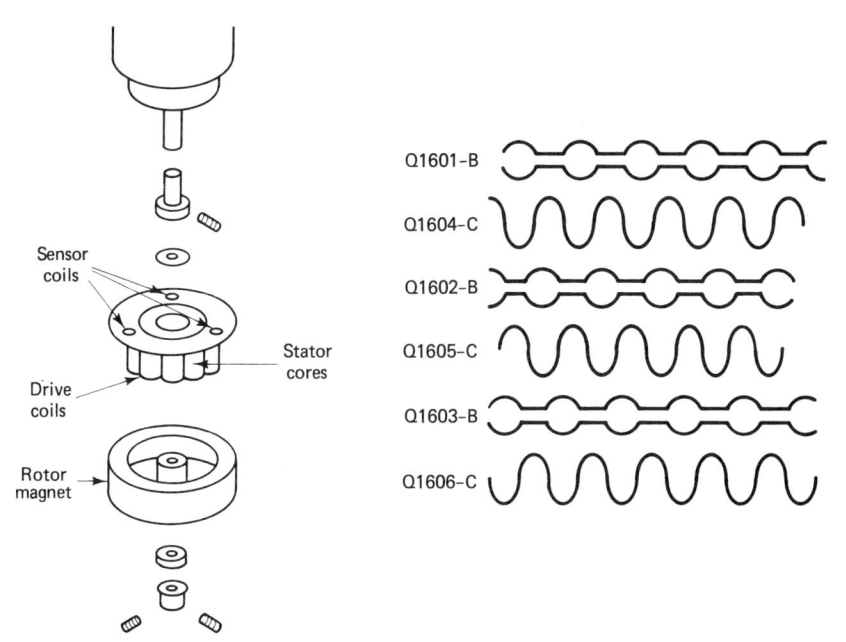

**FIGURE 3-21.** Arrangement of drum motor stator and rotor drive circuits (with related waveforms).

128  *Typical Beta VCR Circuits*

The switching circuits also receive signals from a detector (D1601, Q1609), which senses any irregularity in the current flowing through the three motor windings. Again, the motor speed is changed as necessary if the current is not the same through the three motor windings.

Keep in mind that the purpose of the W8 PC board circuits is to maintain the drum motor at a speed and phase determined by the error signal from the drum servo. If the error signal remains constant, drum motor speed must remain constant.

**Changing the Drum Motor Speed during Picture Search.** The VCR is capable of a picture search function which permits the tape to be run through at high speeds (to search for a particular picture recorded on tape). The tape can be run through in either forward or reverse (F-search or R-search) at speeds 10 to 20 times normal tape speed. In either direction, the video heads straddle a number of tape patterns, as shown in Fig. 3-22. Also, the CTL pulses recorded on tape move by at increased speeds. (Keep in mind that the CTL pulses are recorded at 30 Hz, or 262.5H.)

When the video heads straddle several tape patterns for tracing, the signal played back in a single trace is equal to $[262.5 - 0.5(m - 1)]$H, where $m$ is the tape speed, which is compared with normal tape speed ($m$ is negative with review). For example, if $m$ is 21, only signals equivalent to 252.5H will be played back in a single trace. Since a single trace is equivalent to the V-sync period, this means that the H-sync pulses within the reduced V-sync period are also reduced by 10 (or about 4%).

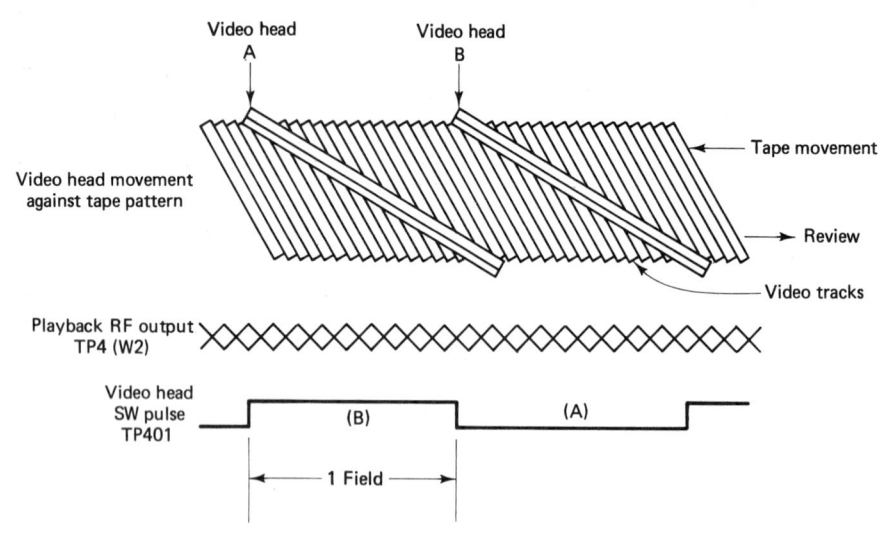

**FIGURE 3-22.** Relationships of video head movement, playback RF, and video head switching pulses.

The H-sync range of a typical TV set is about 3%. Thus, if the H-sync is reduced by 4%, the TV set will probably not be locked in horizontal sync (the pictures will move continuously). For this reason, the video head speed during picture search must be changed so that the H-sync deviation is less than 2% (well within the capabilities of most TV sets). The video head speed (drum motor speed) is changed during picture search by changing the servo reference signal developed in servo circuit II (Sec. 3-4). By changing the servo reference signal countdown, video head speed is increased by about 2.5% during F-search, and when the tape speed is about 15 times the normal rated speed, the H-sync frequency deviation is reduced to almost zero. In the R-search mode, tape speed is reduced by about 2.9%, and when tape speed is about 14 times the rated speed, the H-sync frequency deviation is brought to nearly zero.

### 3-3.2 Capstan Servo Circuit

As shown in Fig. 3-15, the capstan servo circuit receives two inputs from the drum servo circuits. One input (applied at pin 19 of Q416) is the servo reference signal recorded on tape by the CTL head. The other input (applied at pin 17 of Q416) is the playback control signal from the CTL head. The capstan servo circuits also receive FG pulses (applied at pin 13 of Q416) that indicate capstan motor speed and phase. Outputs from the capstan servo circuits are applied to the capstan motor to control both speed and phase.

**Recording the Control Pulse.** During record, the servo reference signal is processed by inverting amplifier Q413. During the time at which video head A starts scanning, the reference signal is recorded at the bottom of the tape by the CTL head. Figure 3-23 shows the relationships of the pulses involved in recording and playback of the CTL signals.

**Amplifying the Playback Control Signal.** During playback, the phase of the CTL head is inverted by Q413 and D408. (A voltage is applied to the anode side of D408, Q413 is set to ON, and the recording side of the CTL head is grounded.) The playback output level is about 1 mV (peak to peak). Amplifier Q409 serves to amplify the very low amplitude playback signal by about 60 dB. Further amplification is provided by amplifier Q410, and the output signal is converted to a square wave. The square wave is further amplified by Q410. The output from Q410 is applied to the muting circuit (part of the system control circuit, Sec. 3-5) as the signal that detects whether or not the tape speed detection circuit (BII/BIII) and playback signals have been recorded. The output from Q410 is also inverted by Q408 and applied to pin 17 of Q416.

130  Typical Beta VCR Circuits

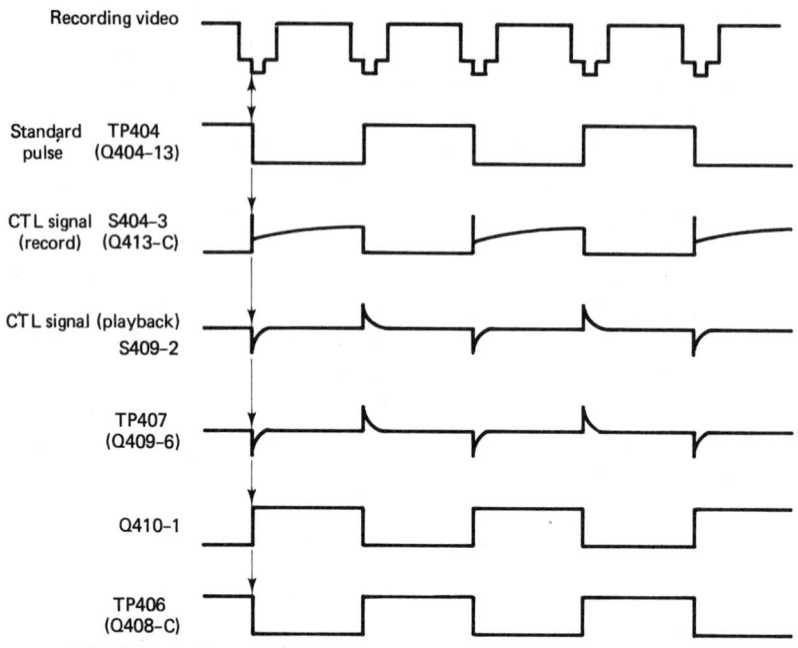

**FIGURE 3-23.** Relationships of pulses involved in recording and playback of CTL signals.

**Delaying the Reference Signal by 360°.** If there is a difference in the mounting position (distance from the video track writing end position and the control head) of the control heads in a recording VCR and a playback VCR (when the recorded tape is played back on a different VCR), the playback time of the control signal deviates by an amount equivalent to the mounting position deviation. When the deviation is large, the playback will not track. The tracking MM 1 and tracking MM 2 circuits within Q404 function to compensate electrically for these mechanical errors. The amount of compensation is determined by the time constants of tracking MM 1 and tracking MM 2. The time constants of the tracking MM 1 circuit are set by VR405 and the TRACKING control. (VR405 is an internal center-set adjustment for the front-panel TRACKING control.) The time constants of the tracking MM 2 circuit are set by VR407 (for BII) and VR406 (for BIII). Figure 3-24 shows the relationships of the pulses involved.

If the control head position of the recording VCR deviates to a negative degree from the position in the playback VCR, the playback time of the control signal is delayed. This playback time must be advanced to align the phases (to restore proper tracking). Although the tracking MM 1 and MM 2 circuits do not actually advance the time of the control signal, they are adjusted (by adjusting the time constants) so as to produce less delay and thus restore tracking. The out-

Beta Servo Circuit / 131

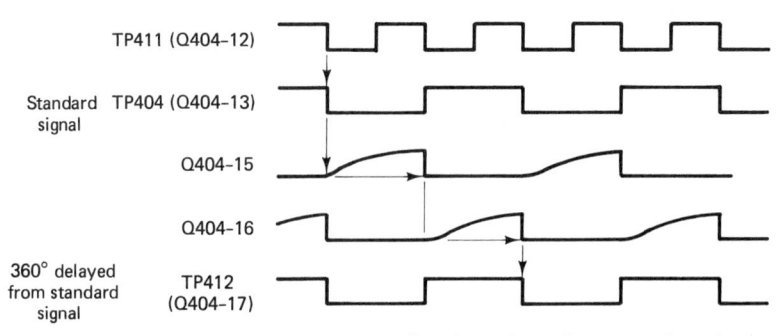

```
TP411 (Q404-12)

Standard   TP404 (Q404-13)
signal

           Q404-15

           Q404-16

360° delayed
from standard   TP412
signal         (Q404-17)
```

**FIGURE 3-24.** Relationships of pulses in reference signal delay circuits.

put of the tracking MM 1 and MM 2 circuits is applied to the capstan servo circuit at pin 19 of Q416.

**Generating the Capstan Motor FG Pulse.** The capstan FG coils convert the speed of the capstan motor into electrical signals. Figure 3-25 shows the mechanical details of the capstan motor FG coils. As shown, the internal gear with 30 teeth is attached to the rotary shaft of the motor. An external gear with 30 teeth is attached to the motor body, opposite the internal gear. The FG coil windings, magnet, and yoke are also mounted as shown in Fig. 3-25.

When the capstan motor rotates to drive the capstan, the magnetic field changes each time the internal and external gear projections face one another. These changes produce voltages in the FG coil, virtually in sine-wave form. Since there are 30 teeth, a total of 30 pulses are produced for each rotation of the capstan. The motor rotates 12 rps during BII and 8 rps for BIII, producing frequencies of 360 and 240 Hz, respectively.

**Amplifying the FG Pulse.** The FG pulses are amplified and converted into square waves by amplifier Q410. The square waves are supplied to pin 13 of Q416 in the capstan servo circuit, and to the tape speed selector or BII/BIII switching of the servo II circuit (Sec. 3-4). During playback, the FG signal serves to detect the tape speed (BII or BIII) of the recorded tape.

**Controlling the Capstan Motor Speed.** The speed control circuit for the capstan motor is essentially the same as for the drum servo (Sec. 3-3.1). The fall of the FG pulse fed to pin 13 of Q416 is delayed by about a half-cycle by MM 3 and speed MM 3, and trapezoidal waves are formed by slope 4 in Q416. Figure 3-26 shows the relationships of the pulses involved. When the trapezoidal waves coincide with the fall of the next FG pulse, the waves are gated with a pulse provided by gate pulse 4, held, and current amplified. The resultant signals are then fed from pin 1 of Q416 to the drive circuit of the capstan motor.

There are two tape speeds (2 cm/s for BII and 1.33 cm/s for BIII). Also, the

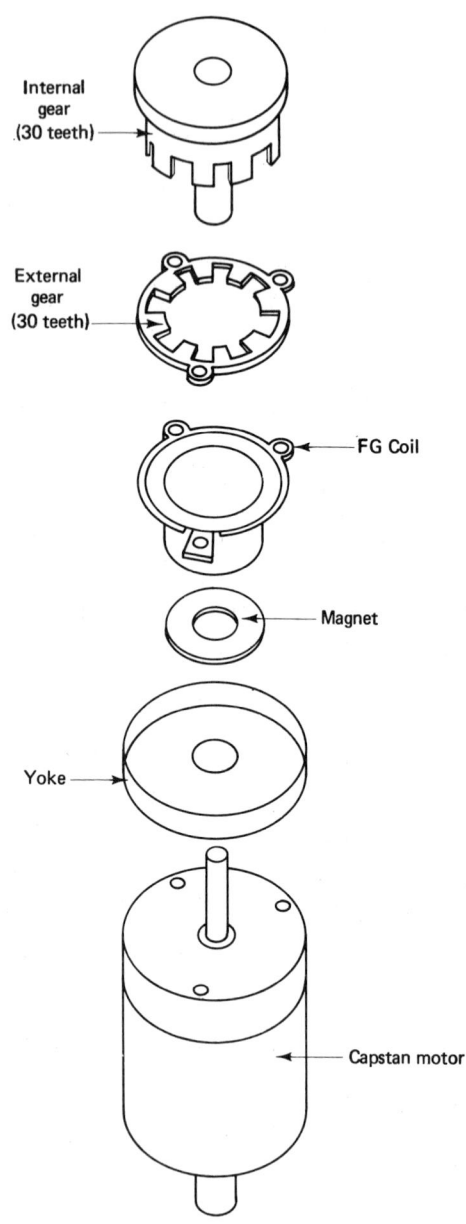

**FIGURE 3-25.** Mechanical details of capstan motor FG coils.

*Beta Servo Circuit I*  133

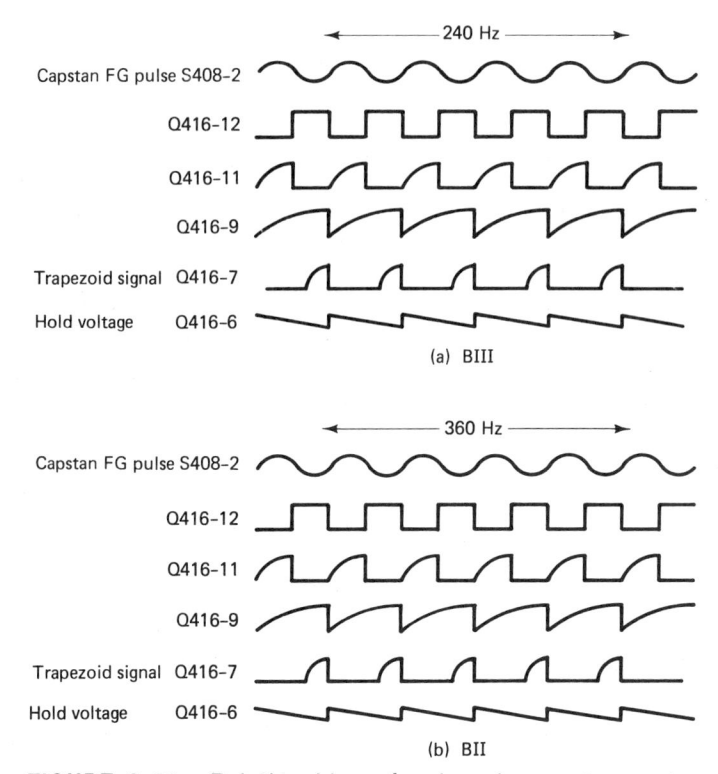

**FIGURE 3-26.** Relationships of pulses in capstan motor speed controlling circuits.

loading and unloading of the loading ring are controlled by the speed control circuit. Thus, a total of four speed control functions are required. These four times are provided by switching the time constants of speed MM 3 (in Q416) using the switching 8 circuits (external to Q416).

With BIII, Q419 goes on, and the time constants of speed MM 3 are delayed as necessary. VR410 sets the amount of time constant delay during BII.

With BIII, Q420 goes on, and the time constants of speed MM 3 are delayed, as determined by the setting of VR411.

With loading and unloading, Q421 goes on, and the motor speed is determined by two sets of time constants (one for loading and one for unloading).

The signal that sets Q419, Q420, and Q421 to ON is provided by the switching 9 circuit. In turn, the switching 9 circuit receives signals from the tape speed selection circuits (Sec. 3-4), as well as the loading and unloading circuits (Sec. 3-5).

The switching 7 circuit (connected to slope 4 through pin 7 of Q416) changes the inclination of the slope 4 trapezoidal waves (as determined by BII or BIII). With BIII, Q415 goes on, and the inclination of the waves becomes

## 134  Typical Beta VCR Circuits

gentler. This change in the trapezoidal wave is necessary since motor response and rotation with BIII is slower than with BII.

**Dividing the FG Pulse Frequency.** During record, there is no need to determine the rotary phase of the capstan motor. However, when the phase of the capstan motor is controlled, it is possible to control the motor speed accurately. Therefore, the servo reference signal (produced by a one-half countdown of the V-sync signal during record, and used by the video head servo, Sec. 3-3.1) is also used by the capstan servo. The capstan motor rotation is synchronized by a comparison of the servo reference signal and the capstan FG pulses.

So that the capstan FG pulses will be the same frequency (30 Hz) as those of the servo reference signal, the capstan FG pulses must be divided down to 30 Hz from 360 Hz (for BII) or from 240 Hz (for BIII). The division by 12 (BII) or 8 (BIII) is accomplished by the MM 2 circuits and the (1/6, 1/8) circuits within Q416. Figure 3-27 shows the relationships of the pulses involved.

When BIII is selected, a "high" voltage is applied to pin 14 of Q416, causing the (1/6, 1/8) circuit to function as a 1/8 divider. This causes the 240-Hz BIII FG signals to be divided down to 30 Hz.

When BII is selected, a "low" voltage is applied to pin 14 of Q416, causing the (1/6, 1/8) circuit to function as a 1/6 divider. This causes the 360-Hz BII FG signals to be divided down to 60 Hz.

The MM 2 time constants are designed so that when BIII is selected, the output from Q423 to Q422 is simply shaped and remains at 30 Hz. When BII is selected, the MM 2 time constants are changed so that a single cycle of the output waveforms is masked, resulting in an output of 30 Hz. Thus, the output from Q423 to Q422 is divided down to 30 Hz with both BII and BIII.

During record, a "low" voltage is applied to pin 16 of Q416, and the output

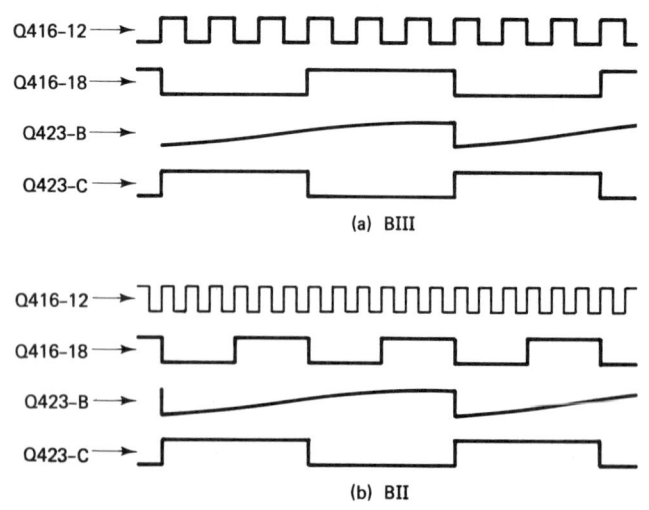

**FIGURE 3-27.** Relationships of pulses in FG pulse frequency-division circuits.

of the (1/6, 1/8) circuit passes through the switching 6 circuit to pin 18. During playback, pin 16 is set to "high" and the signal fed into pin 17 (playback control signal) is fed out at pin 18.

**Controlling the Capstan Motor Phase.** Operation of the circuits for capstan motor phase control is essentially the same as that for drum motor phase control (Sec. 3-3.1). Figure 3-28 shows the relationships of the pulses involved.

During playback, the playback control signal (shaped into square waves) passes through the switching 6 circuit from pin 17 of Q416. The signal then enters MM 2 from pin 18 of Q416, is processed by inverting amplifier Q422, is turned into a trapezoidal wave, and enters the gate 3 circuit from pin 20. The gate 3 circuit also receives the servo reference signal through pin 19 of Q416 and the amplifier 5 circuit.

As discussed, the servo reference signal is delayed about 360° by the tracking MM 1 and tracking MM 2 circuits within Q404. The gated-voltage output from gate 3 (representing both the playback control signal and servo reference signal) is held and current amplified by circuits in Q416. The gated-voltage is applied through pin 23 of Q416, buffer Q418, and the switching 8 circuits to speed

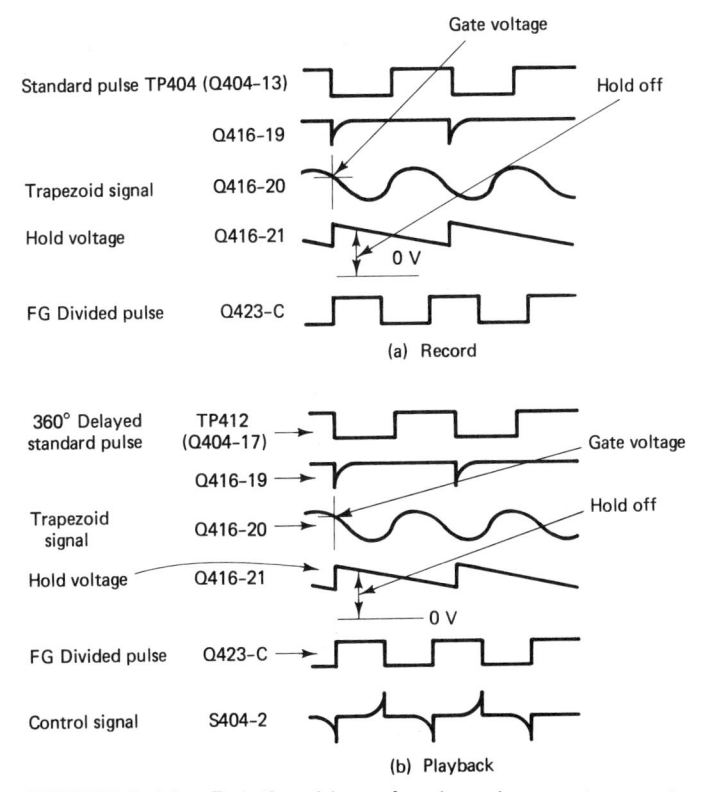

**FIGURE 3-28.** Relationships of pulses in capstan motor phase control circuits.

MM 3 within Q416. As discussed, the capstan motor drive signal at pin 1 of Q416 is determined (in part) by the speed MM 3 circuit. The capstan drive signal thus represents the combined effects of the servo reference signal, playback control signal, and capstan FG pulses.

**Driving the Capstan Motor.** The capstan motor drive signal is current amplified by driver Q412 and Darlington Q411, and is applied to the capstan motor. Both the phase and speed of the capstan motor (and thus the phase and speed of the tape being driven by the motor) are controlled by the three control signals (servo, playback, and capstan FG). When the cassette is not loaded, or when the VCR is set to the stop or still modes, power is removed from the capstan motor and the capstan brake is applied.

## 3-4 BETA SERVO CIRCUIT II

Figure 3-29 is an overall block diagram of servo circuit II. As shown, servo circuit II consists of four separate but interrelated circuits: timer, reference signal generator, BII/BIII switching, and power supply or voltage selector.

The timer circuit sends an automatic shutoff signal to the system control circuit when the VCR is left in the pause or still mode for about 5 to 10 minutes. Under these conditions, the pause or still mode is canceled, and the VCR goes into the stop mode.

The reference signal generation circuit produces the 60-Hz reference signal for the drum and capstan servos (Sec. 3-3). The circuit consists of a 31.468-kHz crystal oscillator and a frequency divider. At the time of normal playback and still modes, a frequency division of 1/525 is made, converting the 31.468 kHz to 60 Hz (actually to 59.94 Hz). When the forward search and reverse search modes are selected, the division ratio is changed as necessary (Sec. 3-3.1).

The BII/BIII switching circuit generates the signals for operation of the video circuits (Sec. 3-2), servo circuits (Sec. 3-3), and audio circuits (Sec. 3-6). The output level from the circuit is low (0 V) for BII and high (about 11 V) for BIII. This signal is generated (high or low) according to the BII/BIII changeover switch (during record), or according to the recorded control signal (during playback).

The power supply or voltage selector circuit produces 12 V signals for control of various circuits.

### 3-4.1 Timer Circuit

When the VCR is locked in pause or still modes, the timer circuit generates a pulse after about 5 to 10 minutes. This pulse returns the VCR to the stop mode (shutoff condition). The timer circuit is composed of Q1818, Q1803, Q1822, and Q1819. When either pause or still modes occur, an RC oscillator (composed of inverters Q1818, Q1803) is started. The RC oscillator signals are counted by an 11-stage frequency-division circuit. When a count of 1024 is reached, a negative

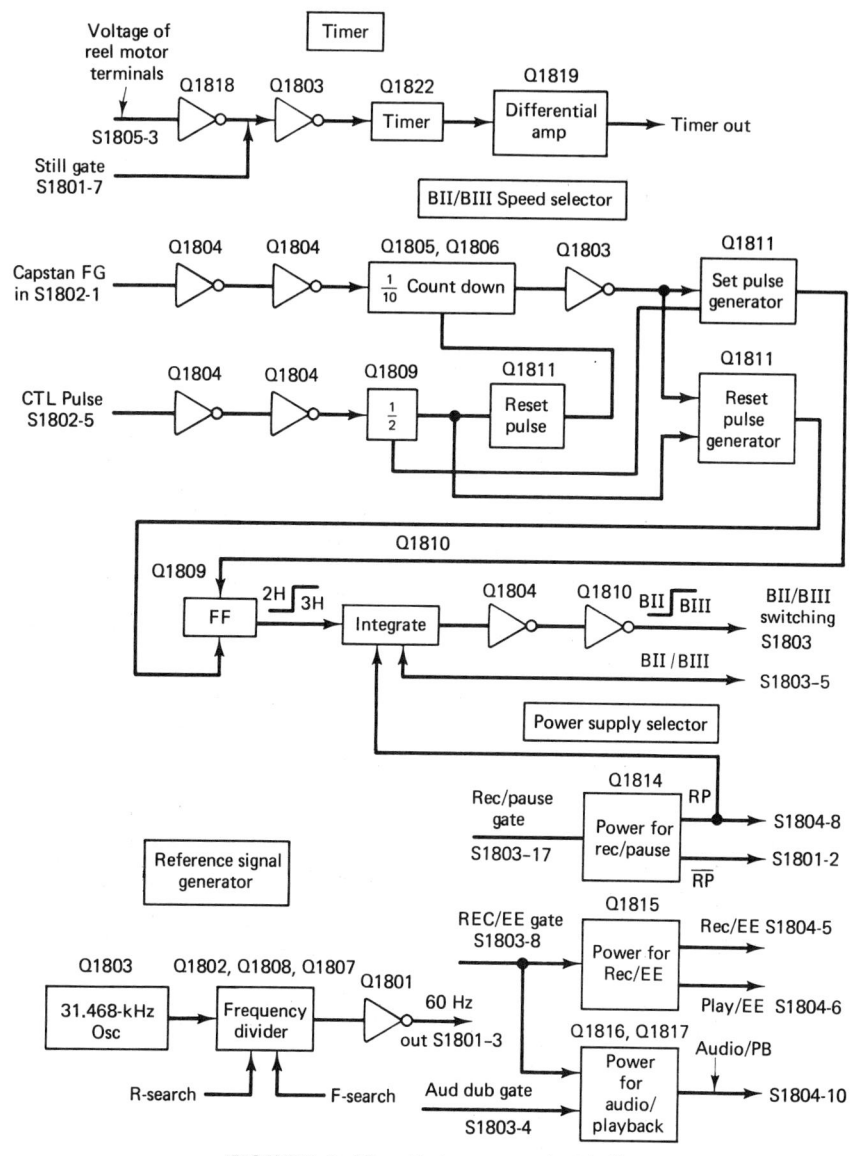

**FIGURE 3-29.** Beta servo circuit II.

polarity pulse is sent from differential amplifier Q1819 to the system control as a shutoff signal.

### 3-4.2 Reference Signal Generation Circuit

The basic signal frequency of 31.468 kHz is produced by a crystal oscillator composed of two inverters within Q1803. This signal frequency is divided down to approximately 60 Hz by a frequency-division circuit consisting of 12-stage binary counter Q1802, AND gates within Q1808, bilateral switch Q1807, and inverter Q1801. Figure 3-30 shows the frequency-division circuit.

During standard playback or still playback, the 31.468-kHz signal is divided down to 59.94 Hz by operation of the binary counter Q1802 and the AND circuit Q1808 connected in a classic "divide-by" circuit. In such divide-by circuits, the selection of AND-gate connections determines the exact countdown frequency. In the circuit of Fig. 3-30, the selection of AND-gate connections (and thus the countdown frequency) is determined by the setting of switches within Q1807. In turn, the Q1807 switches are set by signals from the system control (Sec. 3-5).

As an example, during R-search, the AND output of FF stages 3, 4, 5, and 10 in Q1802 is taken from pin 1 of Q1808 and applied to pin 9 of Q1807. Also during R-search, a high input is applied to pin 6 of Q1807, closing the switch between pins 8 and 9 of Q1807. The output from pin 8 of Q1807 is applied to pin 11 of Q1802, resetting Q1802 as necessary to produce a 540 countdown. This results in a 58.27-Hz reference signal which is sent to the servo systems (Sec. 3-3) via the inverter of Q1801. When F-SEARCH is selected, the frequency-division circuits produce a 61.42-Hz reference signal for the servo systems.

### 3-4.3 BII/BIII Switching Circuit

The BII/BIII switching circuit discriminates between recordings made with BII and BIII at the time of playback. The circuit, composed of Q1804, Q1805, Q1806, Q1809, and part of Q1804, receives CTL pulse and capstan FG pulse inputs and produces a BII/BIII switching pulse used by the servo system (Sec. 3-3). The BII/BIII switching pulse is also used to mute the audio at the time of switching between BII and BIII. The overall BII/BIII switching circuit shown in Fig. 3-29 is composed of a 10-count circuit, a 10-count reset circuit, a discrimination signal generation circuit, and an integration circuit.

**10-count Circuits.** The 10-count circuit, shown in Fig. 3-31, receives capstan FG pulses and reset pulses as inputs, and produces a 10-count output. The reset pulses are generated by counting down the playback CTL pulses as shown in the 10-count reset circuit of Fig. 3-32. In the case of BII, the capstan motor speed is 12 rps and the number of FG teeth is 30, producing 360 FG pulses per second, or 12 pulses between each control pulse. Therefore, for BII, a 10-count occurs before the next CTL pulse. During BIII, the capstan motor speed is 8 rps, producing 240 FG pulses per second, or 8 pulses between each CTL pulse, and a 10-count occurs after the next CTL pulse. In this way the circuit can discriminate between BII and BIII.

**Discrimination Signal Generation Circuit.** The discrimination circuit, shown in Fig. 3-33, receives the 10-count pulse (applied at Q1811, pins 9 and 12) and produces a BII/BIII switching pulse at pin 1 of Q1809. The switching pulse is

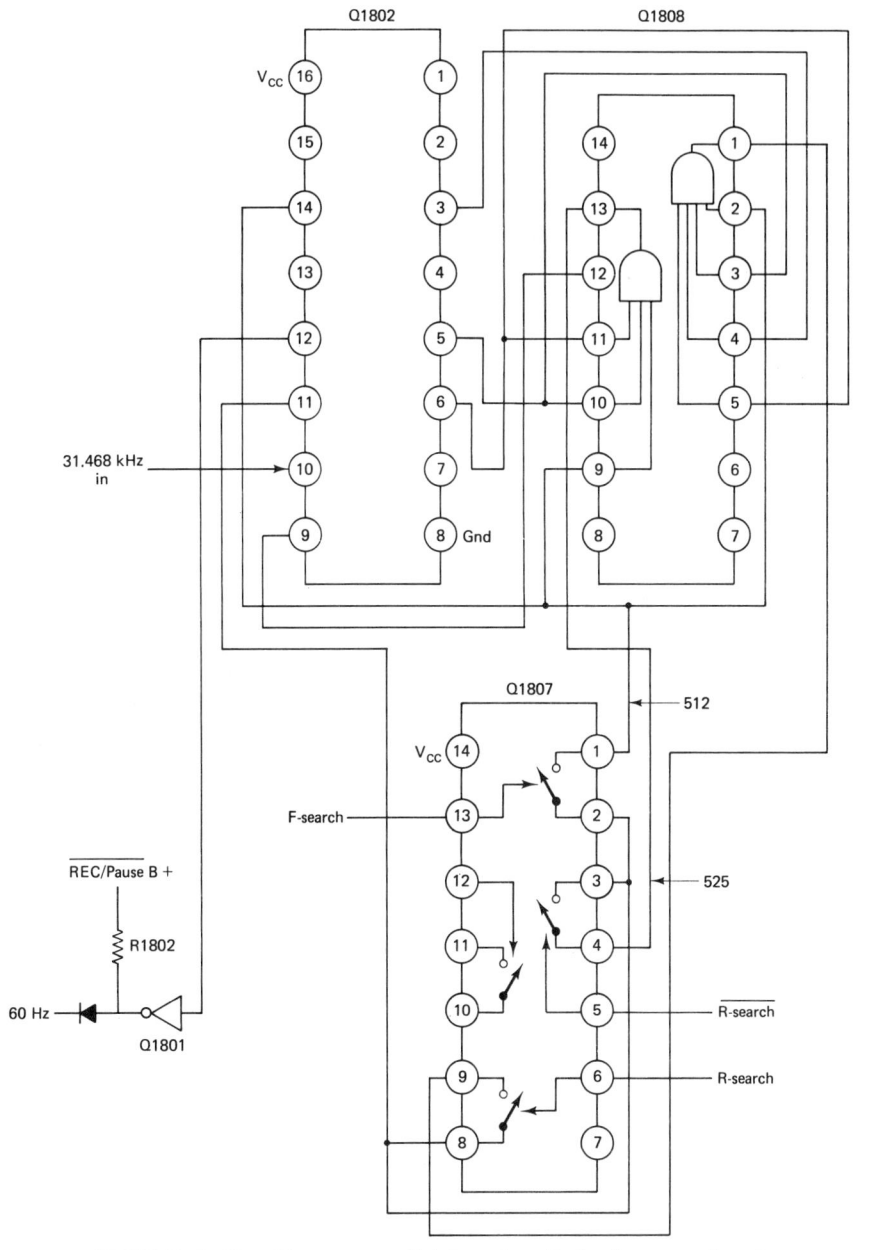

**FIGURE 3-30.** Frequency-division circuit of reference signal generator.

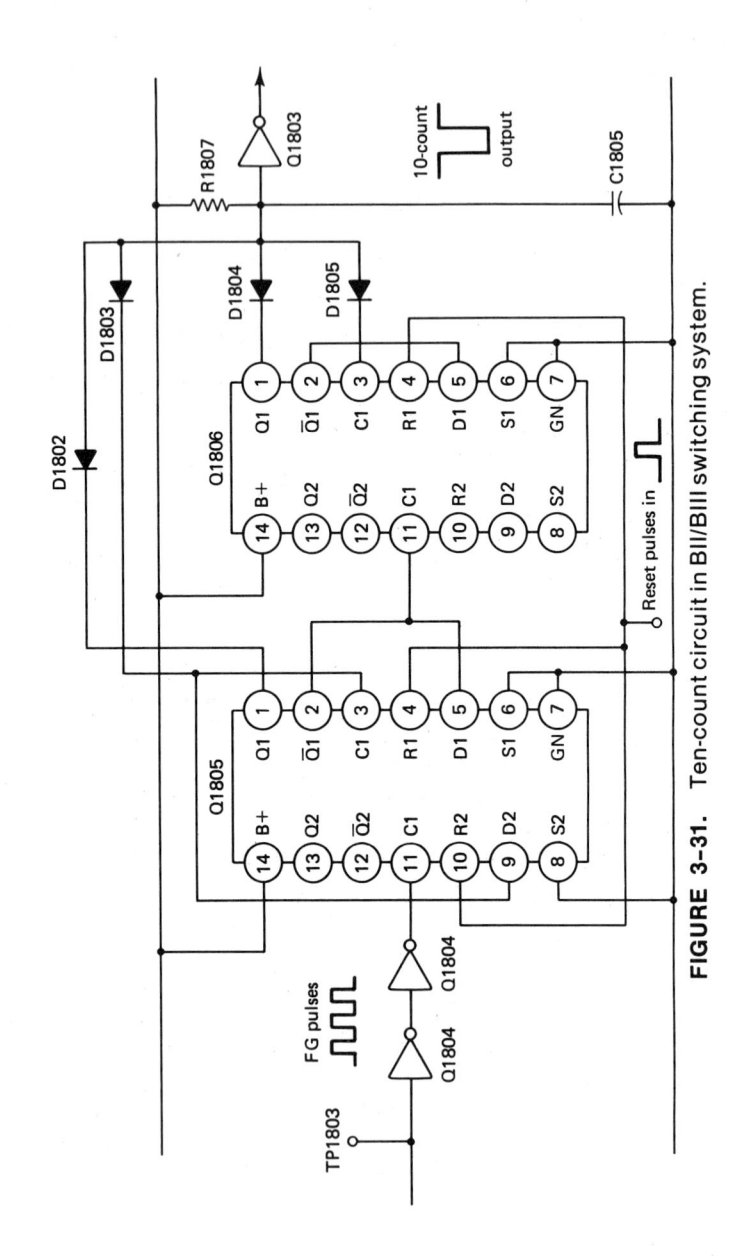

**FIGURE 3-31.** Ten-count circuit in BII/BIII switching system.

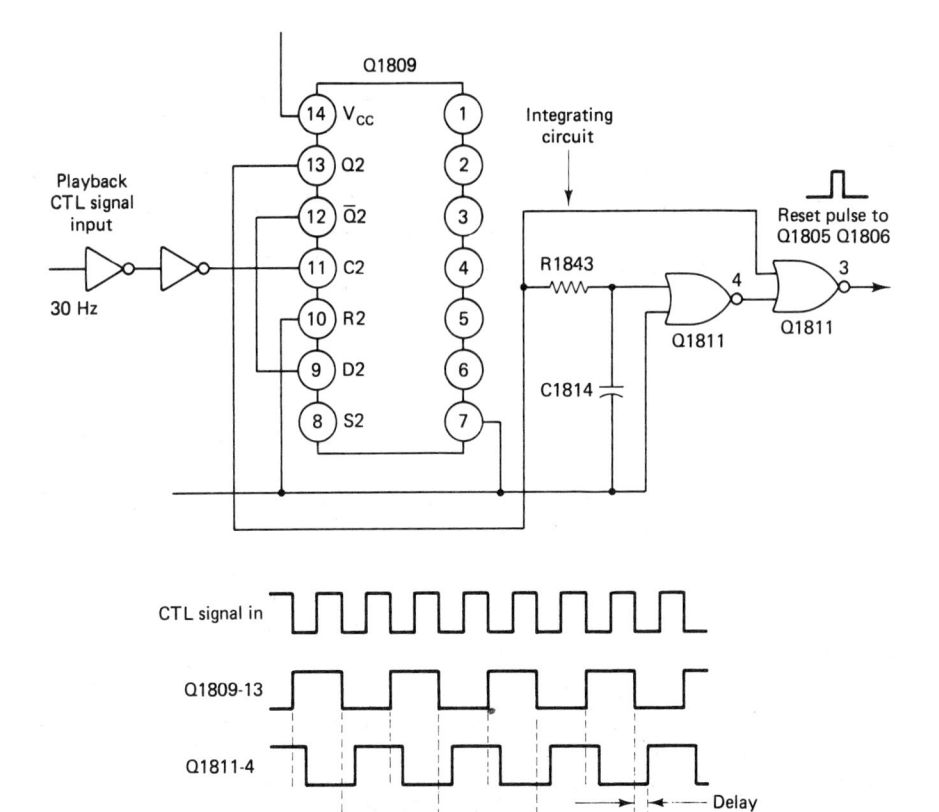

**FIGURE 3-32.** Ten-count reset circuit in BII/BIII switching system.

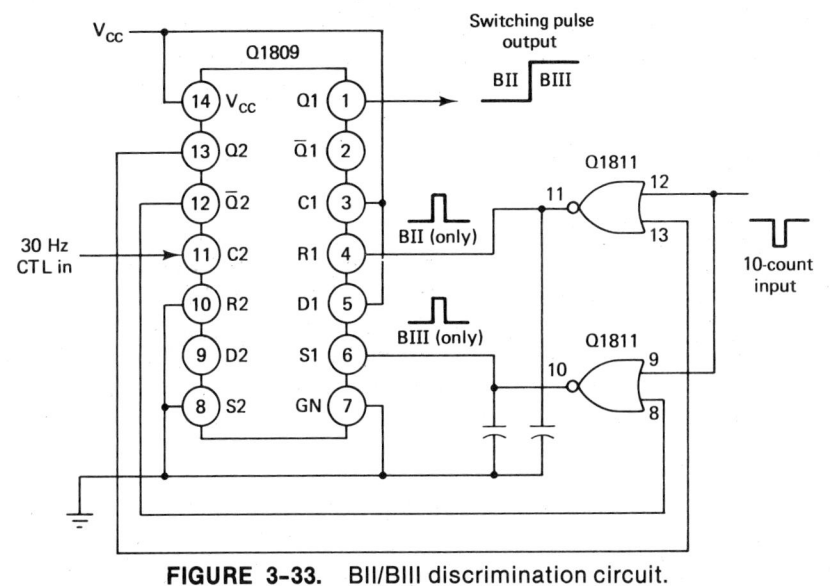

**FIGURE 3-33.** BII/BIII discrimination circuit.

141

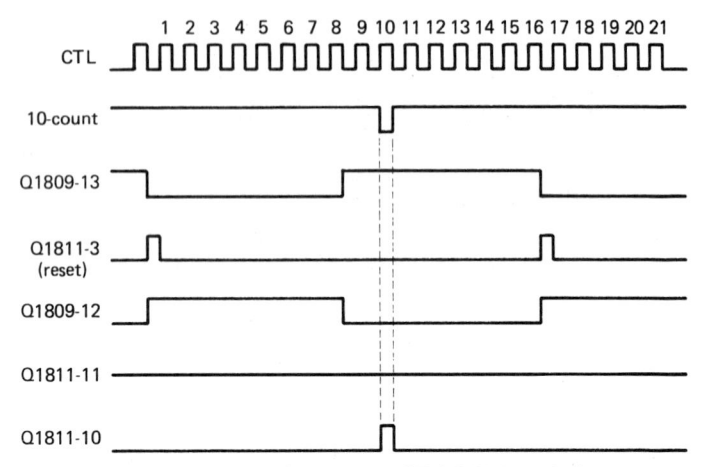

**FIGURE 3-34.** BIII mode pulses of BII/BIII discrimination circuit.

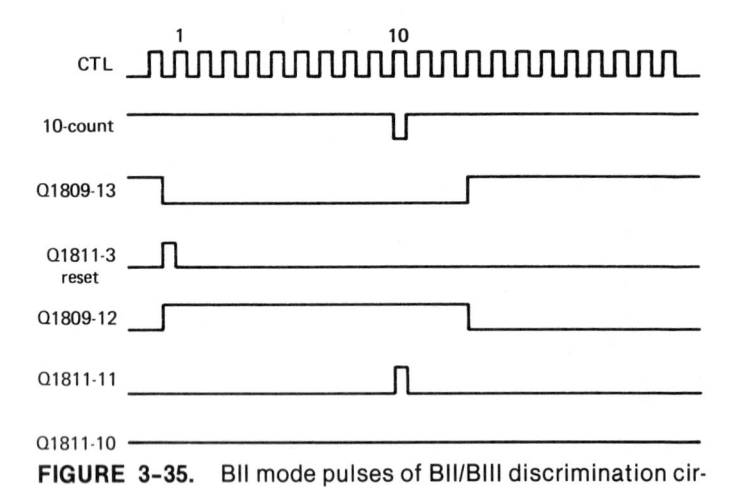

**FIGURE 3-35.** BII mode pulses of BII/BIII discrimination circuit.

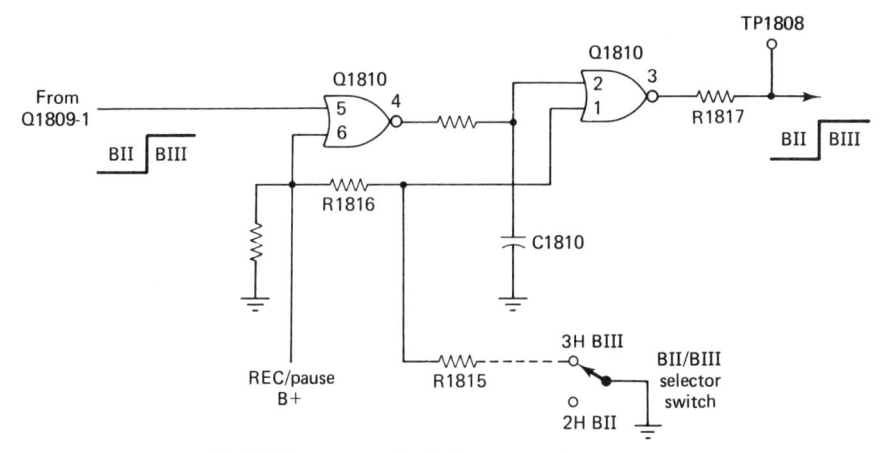

**FIGURE 3-36.** BII/BIII integration circuit.

low for BII and high for BIII. The timing diagrams of Figs. 3-34 and 3-35 show the pulse relationships of the discrimination circuit for BIII and BII, respectively. As shown, the reset pulse is sent from pin 11 of Q1811 for BII recordings, and from pin 10 of Q1811 for BIII recordings.

**Integration Circuit.** The integration circuit, shown in Fig. 3-36, receives an input from pin 1 of Q1809 (low for BII and high for BIII) and produces the BII/BIII switching pulse, which is applied to the servo system through inverters Q1804 and Q1813. The integration circuit also receives a B+ signal during record/pause modes, and a control signal (ground) from the BII/BIII selector switch. The integration circuit absorbs noise associated with BII/BIII switchover and prevents momentary erroneous operation of the Q1809 BII/BIII switching signal that might result from a count mistake or similar occurrence.

During record, record/pause B+ is applied to pin 6 of Q1810, and the output of pin 4 goes low, no matter what the condition of pin 5. As voltage is applied to pin 1 of Q1810 via R1816, pin 1 of Q1810 goes high or low according to the BII/BIII selector switch. For example, when BIII is selected, pin 1 of Q1810 goes low, and a high (indicating BIII) is obtained at the output (TP1808). The opposite occurs when BII is selected.

During playback, the record/pause B+ is no longer applied, and the state of Q1810 (pin 4) depends on the state of the input signal (BII or BIII) at Q1810 (pin 5). Similarly, during playback, the output of the integration circuit at TP1808 depends on the state of the input from pin 1 of Q1809.

### 3-4.4 Power Supply Circuit

The power supply or voltage selection circuit (Fig. 3-29) produces three control signal voltages in response to three control signals. A high input from the record/pause gate to Q1814 produces a record/pause B+ voltage. A high input from the record/E-E gate to Q1815 produces a record/E-E B+, whereas a low to Q1815 produces a playback/E-E B+. A high input from the audio dub gate, during the time of playback, to Q1816/Q1817 produces an audio playback B+.

## 3-5 BETA SYSTEM CONTROL CIRCUIT

Operation of the system control circuit is determined primarily by a microprocessor. The use of such microprocessors for system control is typical for many present-day VCRs. The microprocessor accepts logic control signals from the VCR operating controls (feather-touch pushbuttons) and from various tape sensors. In turn, the microprocessor sends control signals to video, audio, servo, and power supply circuits, as well as drive signals to solenoids and motors.

We will not go into operation of microprocessors here since such information is beyond the scope of this book (and each VCR has its own particular microprocessor applications). However, we do discuss the circuit inputs and outputs to and from the microprocessor, since such circuits are typical for many VCRs. If you want a thorough discussion of microprocessors, your attention is invited to the author's best-selling *Handbook of Microprocessors, Microcomputers, and Minicomputers* (Englewood Cliffs, N.J.: Prentice-Hall, Inc., 1979). Note that many system control functions are closely related to mechanical operation of the VCR, as described in Chapter 5. Therefore, it is essential that you study the related sections of Chapter 5 when reviewing system control operation.

### 3-5.1 *Overall System Control Circuit Operation*

Figure 3-37 is an overall block diagram of the system control circuits. As shown, the main PC board W4 is composed primarily of a microcomputer Q1001, and I/O (input/output) expanders. There is also a control pulse detector, a voltage regulator, and an LED driver circuit on the W4 board. The W7 PC board contains the circuits for the 10 VCR operating control pushbuttons. When the buttons are pushed, control signals are applied to the microprocessor. In turn, the microprocessor sends control signals to the servo, video, audio, and power circuits, as well as the control circuits on the W6 board. The W6 board also has sensors and circuits which provide control signals to the microprocessor.

The microprocessor performs its functions in 13 modes. The STOP, REC, PLAY, AUDIO DUB, PAUSE, STILL, F-SEARCH, R-SEARCH, F FWD and REW modes are selected by the corresponding one of the 10 operating control pushbuttons. The loading, unloading, and cassette UP modes are selected (by the corresponding switches) when the cassette is loaded or ejected.

The W6 PC board is composed of the following circuits:
The *camera pause circuit* generates pulses at the ON time and OFF time of

**FIGURE 3-37.** Beta system control circuit.

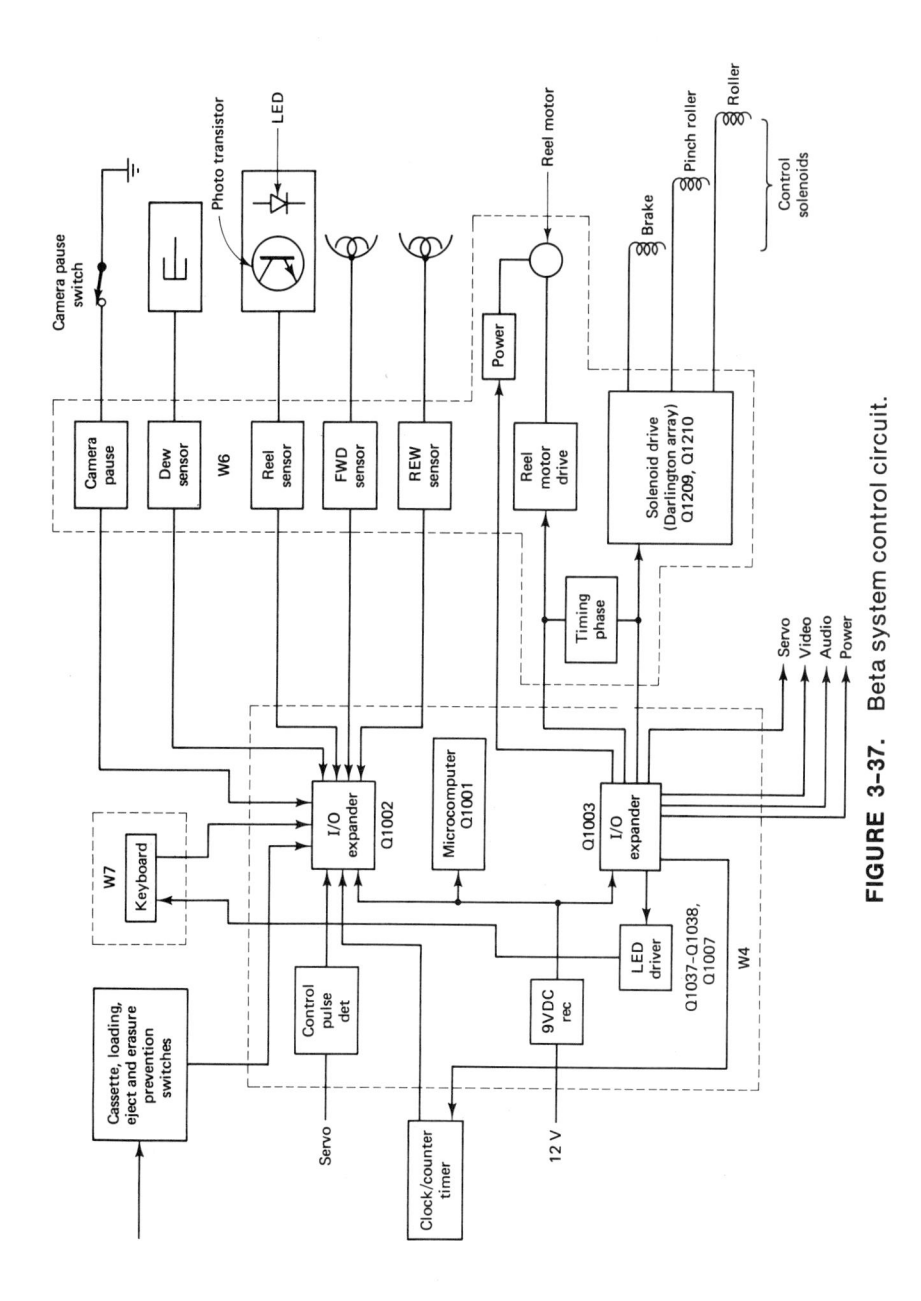

the remote PAUSE switch. These pulses are applied to the microcomputer instead of the control signal during the PAUSE mode.

The *dew sensor circuit* senses when dew or moisture forms on the video head drum surface. The dew sensor resistance (part of the feedback network of an oscillator) decreases when moisture is present, causing the oscillator to produce signals. These oscillator signals cause the microcomputer to place the VCR in the STOP mode.

The *reel base sensor circuit* detects the pulses generated when grooves cut into the take-up reel intermittently shut off a photocoupler. This determines whether the reel base has stopped or not.

The *forward sensor circuit* is part of an oscillator. When aluminum foil attached to the start of the tape approaches this sensor, the Q of the sensor coil decreases, as does the oscillation signal output (indicating that the tape is at the start position).

The *rewind sensor circuit* operates in the same way as the forward sensor circuit, except that the rewind circuit oscillator signal output drops when the aluminum foil at the end of the tape passes the oscillator coil (placing the VCR in the STOP mode, and indicating that the tape must be rewound).

The *reel motor drive circuit* supplies voltage to the reel motor according to the respective mode. In the REC, PLAY, and AUD DUB modes, a voltage of about 3 V is supplied to the reel motor. A voltage of about 8 V is supplied in the F FWD, REW, loading, and unloading modes.

The *solenoid drive circuit* controls ON and OFF of the three solenoids (part of the mechanical functions described in Chapter 5). The three solenoids are: the pinch roller solenoid, which pushes the pinch roller against the capstan; the brake solenoid, which pushes the brake against the supply reel; and the roller solenoid, which switches the tape travel between forward and rewind directions.

The *timing phase circuit* drives the pinch roller solenoid and the reel motor in synchronization with the head motor. This is required so that control pulses are recorded in line on the tape control track (when recording stops temporarily and starts again).

## 3-5.2 Relation between Timer Circuit and System Control

When the timer REC (record) signal is applied to the system control circuits (while the POWER switch is set to TIMER), the power supply turn on and the timer recording condition occurs. The timer recording condition is stopped in the following cases: (1) when the recording time expires, and (2) when the VCR power signal goes low (which occurs when the dew, forward, or rewinding sensors operate, or when the cassette is up in the automatic stop condition). Operation of the timer is discussed further in Sec. 3-7.

### 3-5.3  9-V Power Supply Circuit

The power supply circuit consists of transistor and switching circuits that produce an instantaneous 9 V when the 12-V input is applied. This rapid rise of the 9 V is required to properly initialize the microprocessor.

### 3-5.4  LED Driver Circuit

This circuit provides for the LED (light-emitting diodes) installed on the operation circuit board (above the operation control pushbuttons.) The maximum output from the Q1003 expander is about 7 mA, too low a current for LED operation. The current is increased as necessary for the LEDs by Darlington array Q1007 and transistors Q1037-Q1038.

### 3-5.5  Dew Sensor Circuit

The dew sensor and related circuit are shown in Fig. 3-38. When changes of temperature and humidity cause condensation of dew on the surface of the video head drum, this is detected by the dew sensor, and the STOP mode is produced to prevent damage to the tape and mechanism. When the dew sensor operates, only EJECT operation is possible (no other mode can be selected).

As shown in the circuit of Fig. 3-38, Q1201 is an IC op-amp connected as a feedback oscillator. Resistor R1204 is in the negative feedback loop, while the dew sensor resistance and resistors R1201-R1205 are in the positive feedback

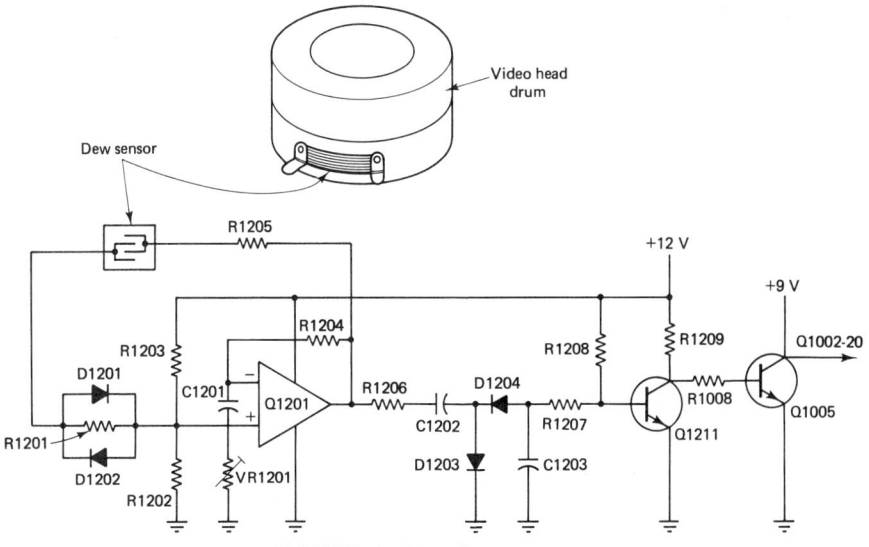

**FIGURE 3-38.**  Dew sensor.

loop. When dew occurs (moisture accumulates on the dew sensor resistance), the dew sensor resistance decreases and positive feedback increases. This sets Q1201 into oscillation at a point determined by the time constants of C1201 and VR1201. Thus, the oscillation point of Q1201 can be adjusted by VR1201. Diodes D1201 and D1202 provide hysteresis for oscillation start and stop, so the circuit is not affected by very small changes in moisture. Diodes D1203 and D1204 rectify the oscillator output and produce a negative signal to Q1211 when oscillation starts. This turns Q1211 off and Q1005 on. The resultant low at the collector of Q1005 is applied to pin 20 of Q1002, placing the VCR in automatic STOP.

### 3-5.6 Forward and Rewind Sensor Circuit

The forward and rewind sensors and related circuits are shown in Fig. 3-39. The cassette has nonmagnetic (aluminum) trailer tape on both ends of the magnetic tape, and the tape end is detected when the forward sensor (1) or rewind sensor (2) senses this trailer tape. When the tape end is sensed, the VCR is brought to an automatic STOP.

The rewind sensor circuit is shown in Fig. 3-39. The forward sensor circuit is similar. A Colpitts oscillator is formed by Q1214. The frequency of the oscillator (about 200 kHz) is determined by the rewind sensor coil and capacitors C1211 and C1212. Feedback to sustain oscillation is made at the junction of C1211 and C1212. The amplitude of oscillation (and thus the trigger point of the sensor circuit) is set by adjustment of VR1203.

When the tape reaches the tape end, the aluminum trailer approaches the rewind sensor, the Q of the sensor coil decreases, and the oscillation output (at TP1203) drops. D1208 provides temperature compensation for the oscillation circuit. The oscillator output is clipped by D1209 and rectified by D1210. When the oscillation signal amplitude is large (aluminum trailer not near the sensor), Q1215 is turned on. When the oscillation signal is decreased (aluminum near the sensor), Q1215 is turned off.

The automatic stop condition occurs when the collector of Q1215 goes high (Q1215 off), and Q1005 turns on, producing a low at the collector of Q1005. This low is applied to pin 10 of Q1002, placing the VCR in automatic STOP. Oscillation of the rewind sensor circuit also stops when the pinch roller solenoid signal is high, and the base voltage of Q1214 drops.

### 3-5.7 Reel Base Sensor Circuit

The take-up reel sensor and related circuits are shown in Fig. 3-40. When the take-up reel stops during a normal operating mode (PLAY, RECORD, etc.) because of a defect in the reel motor, belt rupture, dew condensation, and so on, the tape will become slack and may be damaged. This condition is prevented by the reel base sensor circuit.

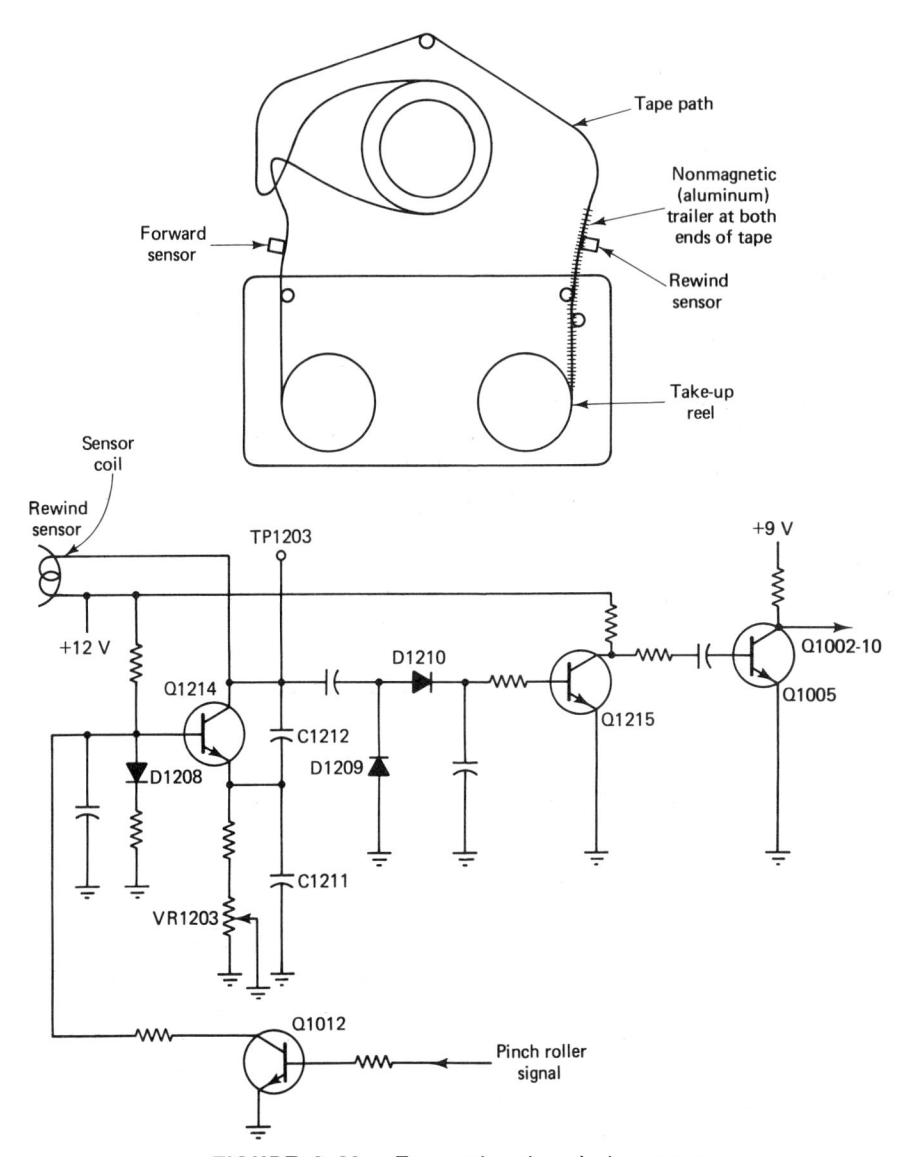

**FIGURE 3-39.** Forward and rewind sensors.

As shown in Fig. 3-40, the reel base sensor consists of a phototransistor and an LED, arranged under the take-up reel base. The phototransistor periodically receives intermittent light from the LED, passing through the slots at the bottom rim of the take-up reel base. When the take-up reel stops rotating, no intermittent light can reach the phototransistor, and the sensor circuit produces a signal that places the VCR in automatic STOP to protect the tape from damage.

150    Typical Beta VCR Circuits

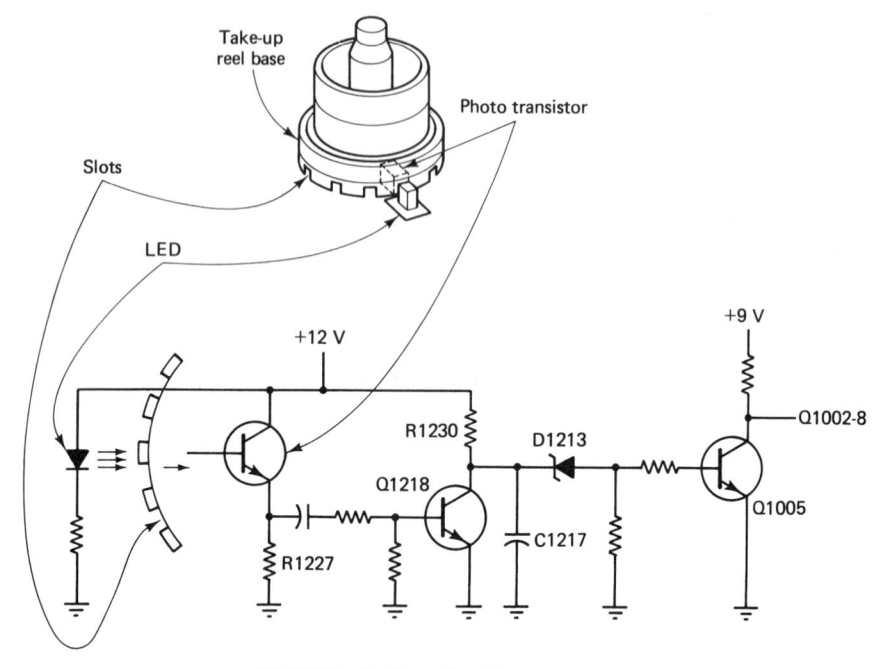

**FIGURE 3-40.**  Reel base sensor.

As shown in Fig. 3-40, the intermittent light (reel moving) produces pulses across R1227. These pulses switch Q1218 on and off. When Q1218 is off, C1217 charges through R1230. When Q1218 is on, C1217 discharges. If Q1218 remains off for any time (reel not moving), C1217 charges to the point where zener D1213 turns Q1005 on. The low at the collector of Q1005 is then supplied to pin 8 of Q1002, placing the VCR in automatic STOP.

In normal operation (reel moving), the pulsating light from the detector intermittently turns Q1218 on often enough to prevent C1217 from charging, and Q1005 remains off. The automatic STOP signal is delayed for about 2 s (because of circuit time constants) to prevent accidental STOP.

### 3-5.8  Timing Phase Circuit

The timing phase circuit is shown in Fig. 3-41. The relationships of the related pulses are shown in Fig. 3-42.

The timing phase circuit serves to record control pulses at equivalent intervals on the tape control track when the PAUSE mode is selected during RECORD. This prevents breakup of the picture during playback of recording interrupted by pause. The timing phase circuit receives three input signals: RF switching pulse, S-1, and MR-1.

The RF switching pulse is differentiated by C1218, R1232, and R1233 so that detection is executed only at the time of RF pulse rise. Each RF pulse produces an input to pins 2 and 9 of Q1202.

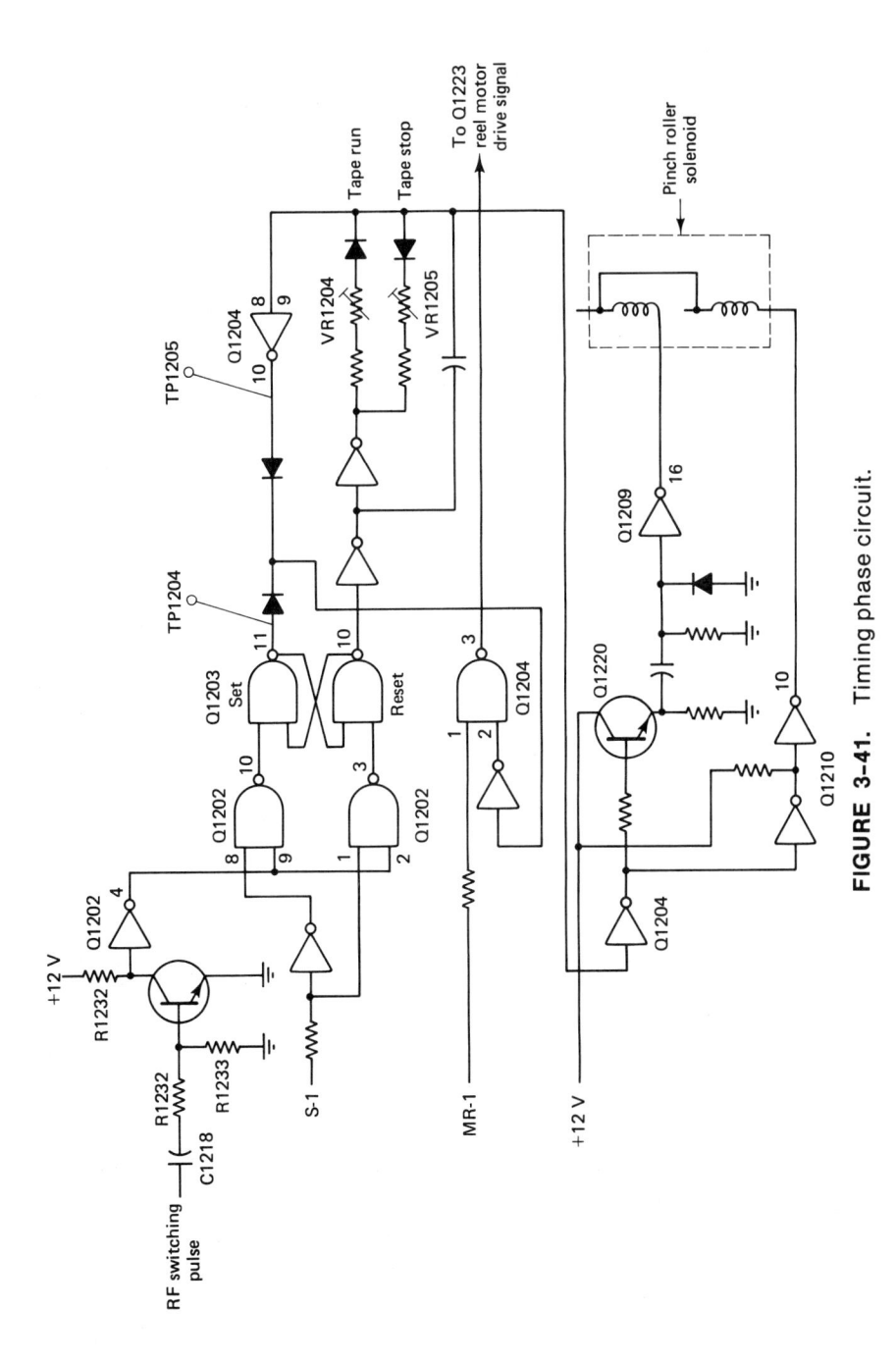

**FIGURE 3-41.** Timing phase circuit.

152  Typical Beta VCR Circuits

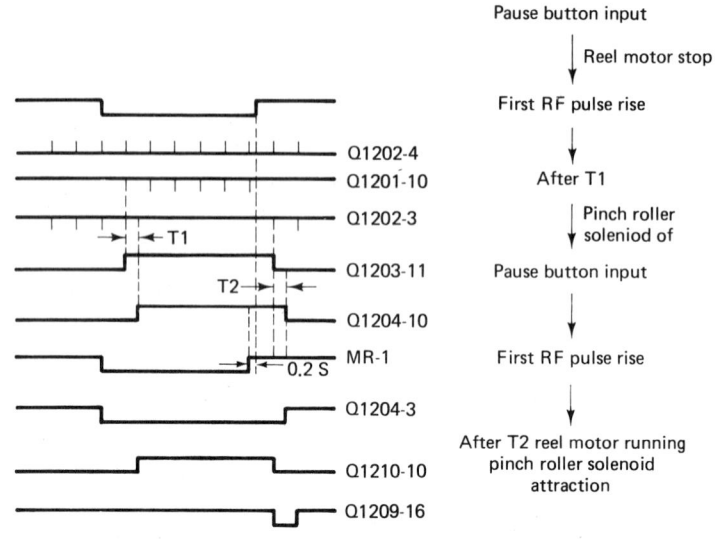

**FIGURE 3-42.** Timing phase pulse relationships.

Input S-1, taken from pin 32 of Q1003, becomes high at the time the pinch roller solenoid is attracted. When S-1 is applied as an input to pin 1 of Q1202 and the inverse of S-1 is applied to pin 8 of Q1202, the logic product appears at pins 3 and 10 of Q1202. This logic product is applied to the SET and RESET inputs of an RS flip-flop, composed of two NAND gates. The SET output of the RS flip-flop (synchronized with video head rotation) appears at pin 11 of Q1203. The RESET output (pin 10 of Q1203) is delayed by times T1 and T2, and appears at pin 10 of Q1204. Time T1 is adjusted by VR1205 and determines the timing at the time of tape stop. Time T2 is adjusted by VR1204 and determines the timing at the time of tape run.

Input MR-1 is applied to one input of a NAND gate. The other input of the NAND gate receives an input which represents the combined SET and RESET (delayed) outputs. The output of the NAND gate at pin 3 of Q1204 is the reel motor drive signal and is applied during the REC, PLAY, and AUDIO DUB modes. The RESET output is also used to operate the pinch roller solenoid (through Q1204, Q1209, Q1210, and Q1220). Operation of the pinch roller solenoid in the various modes is discussed in Chapter 5.

### 3-5.9 Reel Motor Drive

Figure 3-43 shows the reel motor drive circuit for all operating modes except F-SEARCH and R-SEARCH. During the REC, PLAY, and AUDIO DUB modes, the reel motor is operated by a signal from pin 3 of Q1204. When this signal goes low, Q1223 and Q1224 are turned on and the output is applied through D1215 and current-amplifying Darlington pair Q1233 and Q1234 to the motor. The amplitude of the drive motor voltage (during REC, PLAY, and

Beta System Control Circuit   153

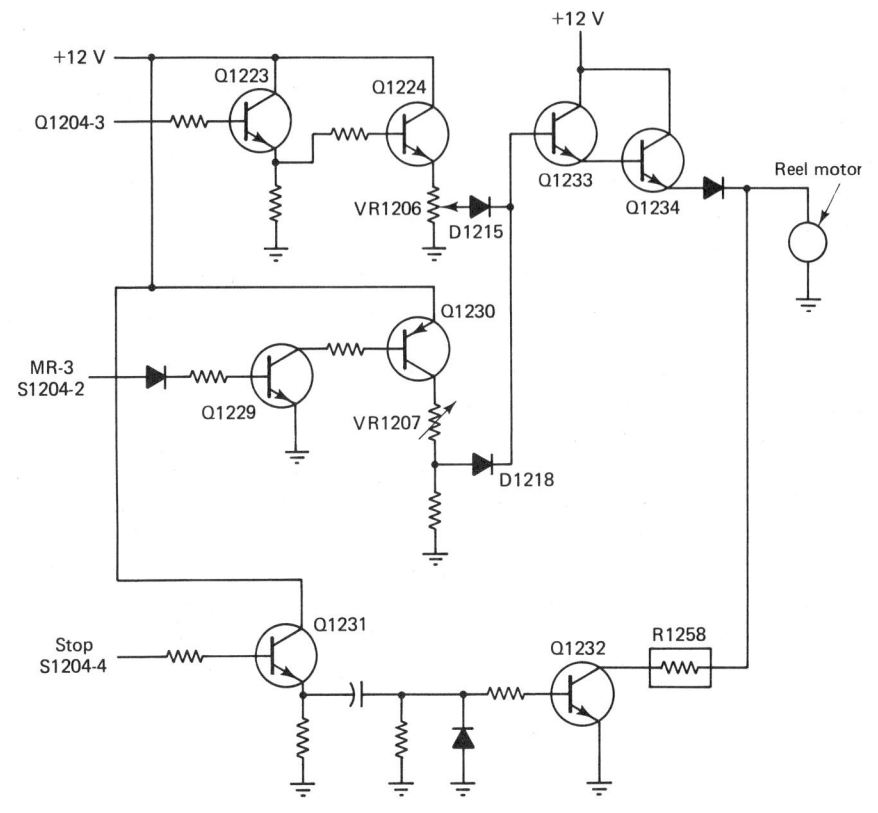

**FIGURE 3-43.** Reel motor drive circuit (except for F-SEARCH and R-SEARCH modes).

AUDIO DUB) is set by VR1206. During the F FWD, REW, and unloading modes, the reel motor is operated by a signal (MR-3) from pin 2 of S1204. When the MR-3 signal goes high, Q1229 and Q1230 are turned on and the output is applied through D1218 and Q1233-Q1234 to the motor. The amplitude of the drive motor voltage (during F FWD, REW, and unloading) is set by VR1207.

The signal at pin 4 of S1204 goes high in the STOP mode. This turns on Q1231 and Q1232, causing the motor to be grounded through R1258 and applying the brake to the reel motor. This serves to prevent abnormal tension on the tape at the time of reel motor stop.

### 3-5.10  Reel Motor Drive during F-SEARCH and R-SEARCH

Figure 3-44 shows the reel motor drive circuit used during F-SEARCH and R-SEARCH modes. The circuit is essentially an electronic governor for the reel motor (which operates at high speeds during search). During either forward or reverse search modes, Q1230 is turned on by a high signals at D1217 or D1231.

*154  Typical Beta VCR Circuits*

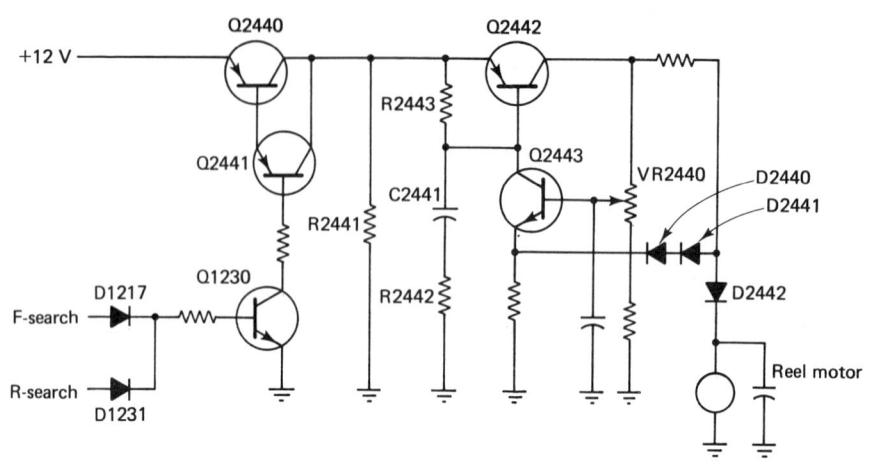

**FIGURE 3-44.**  Reel motor drive circuit for F-SEARCH and R-SEARCH modes.

When Q1230 turns on, switching transistors Q2440 and Q2441 turn on, and apply +12 V to the electronic governor transistors Q2442 and Q2443. With +12 V available, C2441 is charged through R2442 and R2443, thus turning on the governor circuit. In all other operating modes, Q1230 remains turned off, and R2441 discharges C2441, preventing operation of the electronic governor.

The electronic governor circuit controls reel motor speed, which is adjusted by VR2440. The governor is a typical current-type motor speed control circuit. Assume that motor speed increases, causing an increase in the counter electromotive force (CEMF) of the motor. An increase in CEMF causes a voltage increase at the motor terminals. This results in a voltage increase at the emitter of Q2443 (through D2440, D2441, D2442), and a current decrease at the collector of Q2442. The voltage applied to the reel motor decreases, and the reel motor speed decreases. An initial decrease in motor speed produces the opposite results (motor speed is increased to compensate for the initial decrease).

### 3-5.11  Solenoid Drive Circuit

The solenoids for the pinch roller, brake, and roller are driven by a Darlington array of Q1209 and Q1210 as shown in Fig. 3-37. A fuse resistance of 3.9 Ω is inserted in series with the solenoid winding (on the attraction side) to protect the attraction winding from burning out. The fuse resistance opens after about 1 minute when a continuous current flows through the attraction winding (due to a defect in the circuit).

### 3-5.12  Camera Pause Circuit

Figure 3-45 shows the camera pause circuit and related pulse timing. As shown, the pause circuit is composed of gates within Q1205 and Q1206. While the PAUSE signal from the VCR operating controls is of the push-and-release type,

**FIGURE 3-45.** Camera pause circuit.

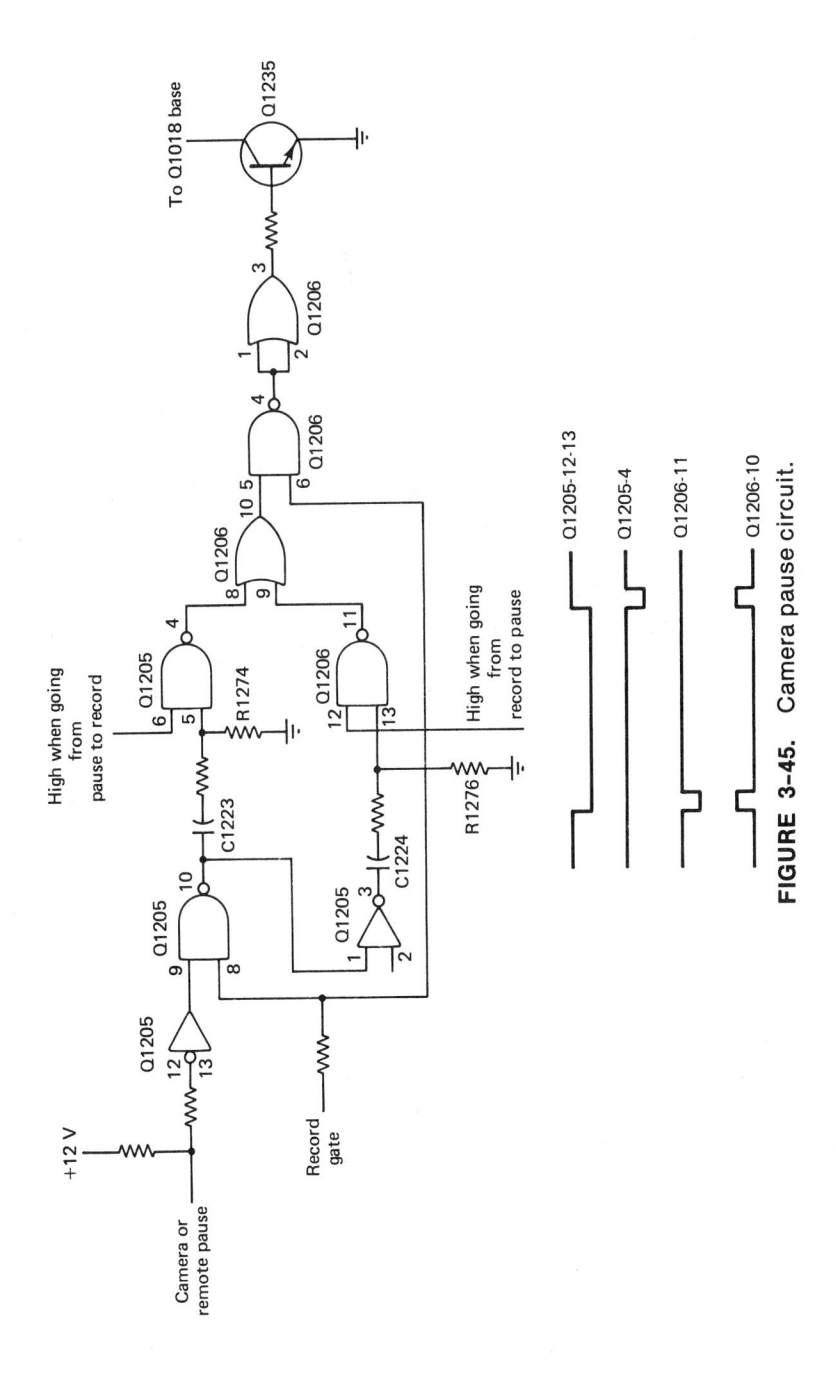

the camera PAUSE signal is of the push-lock type, so that the signal conversion provided by the circuit of Fig. 3-45 is required.

When the record gate signal is high, the NAND gate is enabled and the remote or camera pause signal input to pins 12 and 13 of Q1205 results in an output at pin 10 of Q1205. The trailing edge of this output is differentiated by C1223 and R1274. The differentiated output is available at pin 5 of Q1205. The output at pin 10 of Q1205 is also inverted, and the leading edge is differentiated by C1224 and R1276 to appear as an output at pin 13 of Q1206.

When switching from PAUSE to REC, the pinch roller solenoid is not attracted, pin 6 of Q1205 goes high, and an output is obtained at pin 4 of Q1205. When switching from REC to PAUSE, the pinch roller is attracted, pin 12 of Q1206 goes high, and an output is obtained at pin 11 of Q1206. These outputs are applied through an OR gate to an AND gate at pin 5 of Q1206. The AND gate is turned on by a signal at pin 6 of Q1206 when the record gate signal is high. An output is produced at pin 4 of Q1206 when both input signals are present. The AND function is controlled by the record gate signal to prevent output at pin 4 of Q1206 in any mode other than record.

The output at pin 4 of Q1206 is inverted and turns Q1235 on. This results in the equivalent of a PAUSE pushbutton input being applied to the I/O expander Q1002 of the microprocessor, and places the VCR in the PAUSE mode.

## 3-6 BETA AUDIO CIRCUIT

Figure 3-46 is an overall block diagram of the audio circuit. As shown, many of the audio circuits perform dual functions (during playback and record).

During record, there are three possible audio input signal sources: one from the TV tuner, one through the AUDIO IN connector, and line signals from other equipment (such as a microphone). These signals are selected (or automatically switched on or off) when a plug is inserted in the AUDIO IN connector or the microphone input. The selected audio input signal is applied to a preamplifier and then to a line amplifier. During record, the audio signal then enters the recording amplifier, where the level is increased and the highs are emphasized before recording on tape by the audio record/playback head. The oscillator circuit provides an erase current for both the full-width erase head and the audio erase head, as well as a bias current for the record/playback head. Note that these currents are applied to the record/playback head only during audio dubbing and record.

During playback, the audio playback signal from the audio record/playback head is applied to the preamplifier and equalizer where high-frequency components (around 7 kHz) are emphasized (to restore the original flat frequency response lost in record and playback). The low-level playback signal is then increased in amplitude to a suitable level by the line amplifier and applied to the audio line through a relay controlled by the muting circuit.

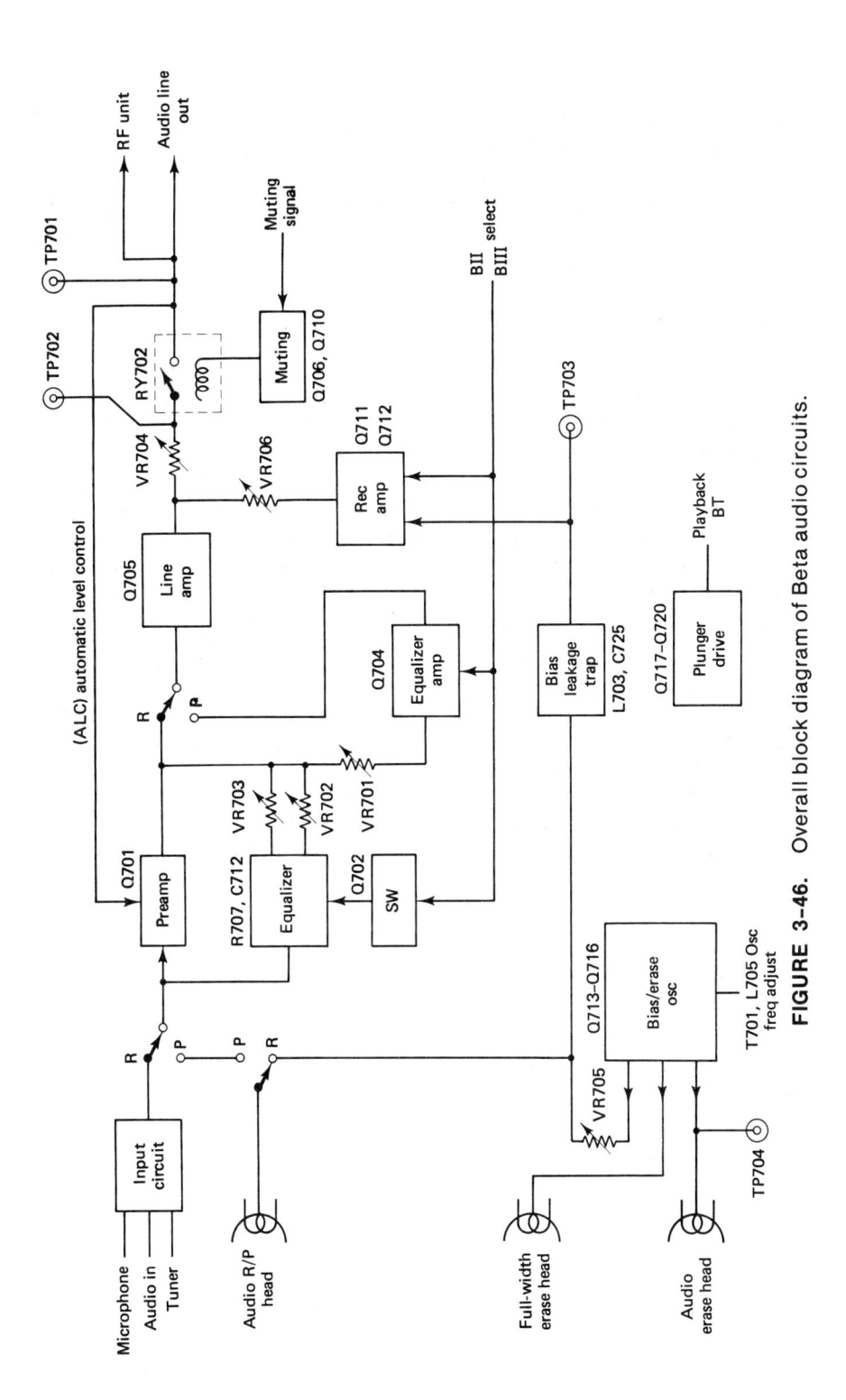

**FIGURE 3-46.** Overall block diagram of Beta audio circuits.

The muting circuit operates during playback and functions to disconnect the audio line from the line amplifier under the following conditions: for a few seconds after the PLAY button is pressed, when no signal is recorded on the tape, for 1 or 2 s following switch-on of the POWER switch, and during F-SEARCH, R-SEARCH, and STILL operation. From this, it can be seen that the muting circuit operates only whenever sound reproduction is not required.

The plunger driving circuit also operates only during playback. The B+ supply is applied to the plunger driving circuit (during playback) to switch the various record/playback switches (shown in Fig. 3-46) into the playback position.

### 3-6.1 Input Circuit

The audio input signals from the TV tuner and from the AUDIO IN connector are of the same signal level, and are reduced to a lower level by resistance division before application to the preamplifier.

### 3-6.2 Preamplifier and Equalizer Amplifier

The preamplifier Q701 receives an input from the input circuit during record, or from the record/playback head during playback. A feedback loop provides a flat frequency response, whereas an equalizer network (adjusted by VR701, VR702, and VR703) provides emphasis of certain frequencies. The equalizer characteristics are selected by a relay (RY701 driven by Q703), depending on tape speed (BII or BIII).

Equalizer amplifier Q704 emphasizes the higher-frequency components of the playback signal. The LC-resonance circuits at the emitter of Q704 provide emphasis at about 7 kHz for BII and 5 kHz for BIII.

### 3-6.3 Line Amplifier

The line amplifier Q705 operates as a flat-response amplifier for both record and playback modes. Three outputs are taken from the line amplifier: the audio line output, an output applied to the RF unit (for audio modulation of the signal sent to the TV set), and an ALC (automatic level control) output. This ALC output is rectified and then applied to the preamplifier Q701 as a control signal. The control signal keeps the recording level constant.

### 3-6.4 Recording Amplifier

The recording amplifier Q711-Q712 is used to increase the line amplifier output to a level suitable for recording on tape by the record/playback head. The recording level is set by VR706. LC resonances in the emitter of Q712 emphasize higher frequencies of the recording signal. The Q711 circuit is used to change the recording amplifier characteristics, depending on the tape speed selected (BII or BIII). The output of the recording amplifier is applied to the record/playback head through a bias leakage trap L703-C725. Coil L703 is adjustable to provide maximum bias current rejection.

### 3-6.5 Oscillator Circuit

The output signal generated by the oscillator Q713-Q716 is used for audio track erasure, full-width track erasure, and record bias. The bias output level to the record head is adjusted by VR705. The oscillator frequency (about 70 kHz) is adjusted by T701. Since the full-width erase head is not used during audio dubbing, L705 is inserted as a dummy load for balancing the oscillator circuit. L705 also adjusts oscillator frequency during the audio dub mode.

The bias output and full-width erase output are used during normal record. This erases all three tracks (audio, video, control) on tape and places the tape in a condition to record new information. The audio track output is used during audio dub mode. This preserves the video and control tracks, but erases the audio track so that new audio can be recorded. The oscillator circuit is disabled during other operating modes (playback, etc.) by the muting circuit.

The 70-kHz Hartley oscillator Q713 is normally turned off except during record and audio dub. Start-stop of oscillation is controlled by Q715, which operates as a gate having two inputs (from D709 during audio dub and D710 during record). When Q715 is turned on, Q714 begins to conduct and oscillation starts. Q716 is used to delay the oscillation startup. Thus, any noise at the time the REC button is pressed is not recorded, and no recording is made until the cassette tape gains a steady operating speed.

### 3-6.6 Plunger Driving Circuit

The plunger driving circuit operates the various record/playback changeover switches in the audio system. The switches are operated by a plunger which can be attracted to either of two positions (record or playback) by one of two corresponding coils or solenoids. When the POWER switch is first turned on, C738 charges through R757, turning on Q719-Q718 and causing the record coil to be energized. This places the record/playback switches in REC (record). When the PLAY button is pressed, a playback B+ is applied, Q720 and Q721 turn on, and the plunger is attracted in the opposite direction by the playback coil. This places the record/playback switches in PLAY (playback).

When the playback mode is selected, Q717 is turned on, allowing C738 to discharge. This turns Q719-Q718 off, removing power to the record coil. When the STOP button is pressed, the power supply is cut off, Q717 turns off, and C738 again charges through R757, turning on Q719-Q718 and causing the record coil to be energized. This places the record/playback switches in REC.

### 3-6.7 Muting Circuit

Output of the line amplifier is controlled by relay RY702, which, in turn, is operated by Q706. When Q706 is on, the line is connected to the line amplifier output. The line is disconnected when Q706 is off. When B+ is first applied by the POWER switch, Q706 does not turn on until capacitor C723 is sufficiently charged. This keeps the line amplifier output from reaching the audio line for a few seconds and thus eliminates noise at the instant of POWER turn-on. When the POWER switch is turned off, the relay opens immediately, disconnecting the line amplifier output and eliminating noise at the instant of POWER turn-off. Q709 shorts and discharges C723 at the time of POWER turn-off, making ready for the next POWER turn-on operation.

Q710 receives an input when the PLAY mode is selected. This turns Q710 on and keeps Q706 from turning on immediately. However, the muting control signal from system control is applied to the base of Q706 through D705. This turns Q706 on (and operates RY702 to connect the line amplifier output to the line) a few seconds after the PLAY button is pressed. Capacitor C722 is charged by the signal at the base of Q706 and serves to hold Q706 and RY702 for 1 or 2 s after the control signal is removed.

## 3-7 BETA TIMER CIRCUIT

Figure 3-47 is an overall block diagram of the timer circuit. As shown, the timer is composed of an LSI (large-scale integration), a time-displaying digitron, an IC, and six discrete transistors. Q1507 is an LSI designed exclusively for digital clock and timer operation. Q1507 uses the frequency of commercial power source as the timer base and has a counter function to display the time of day. Q1507 also has counter circuits to control starting time and shutoff time for the timer.

Three types of display (time of day, start time, and stop time) are selected by means of switches in a power supply gate and control circuit. Q1507 has a circuit which compares the timer-start time and timer-stop time with the time of day, and produces turn-on and turn-off signals at the coincidence of the two times compared. Q1507 produces these turn-on and turn-off signals at the same time of day, each day. To prevent repetition, a timer signal processing circuit prevents the timer signal from appearing more than once unless the timer-start and timer-stop are set again.

### 3-7.1 Waveforming and Power Supply Circuit

The waveforming and power supply circuit is shown in Fig. 3-48. As shown, a clock pulse from the commercial power source (60 Hz) is stepped down and applied to the base of Q1506. The shaped output pulses from Q1506 are applied to

Beta Timer Circuit 161

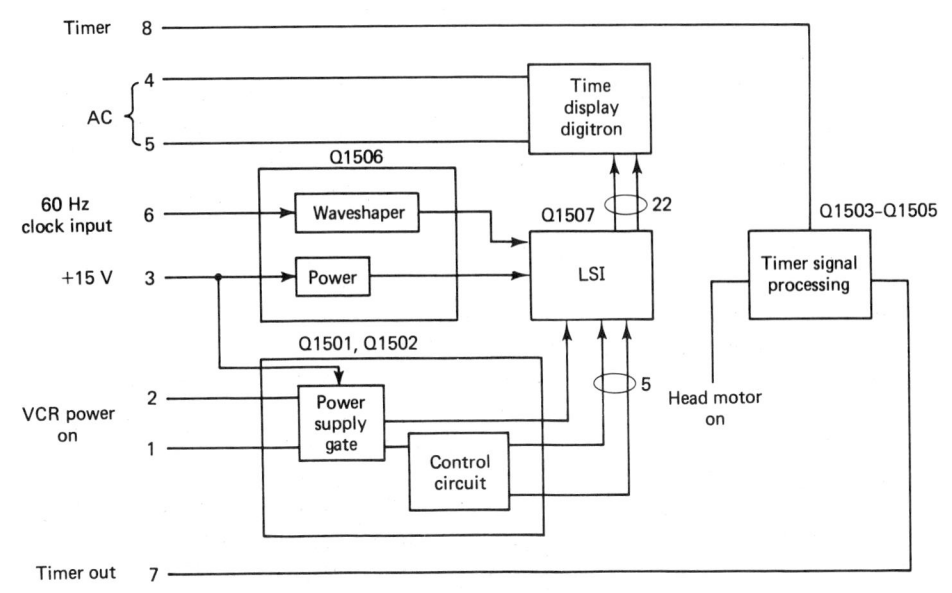

**FIGURE 3-47.** Overall block diagram of Beta timer circuits.

Q1507. The +15 V at D1501 is applied to the $V_{SS}$ terminal of Q1507 through D1502 and to zener diode D1503. A regulated +12.5 V is taken from the junction of D1503 and R1507 and is applied to the $V_{DD}$ terminal of Q1507. Diodes D1501 and D1502 prevent C1501 from discharging immediately should a power failure occur. This keeps the timer circuit operating for about 1 s after power is removed, and prevents timer failure due to momentary power interruption.

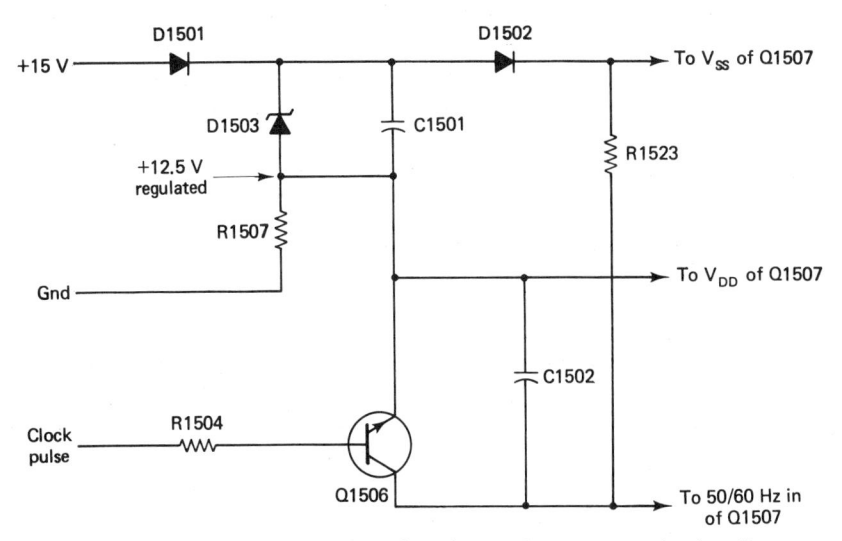

**FIGURE 3-48.** Timer waveforming and power supply circuit.

## 3-7.2 LSI Circuit

The functions of the LSI are shown in Fig. 3-49. The 60-Hz input from the waveforming circuit (Fig. 3-48) is shaped by a Schmitt trigger and divided down to 1-Hz pulses by a 1/60 divider. The 1-s pulses are applied to a real-time circuit which counts off seconds, minutes, and hours. In addition, there are two other timer circuits (record start and record stop) used to preset timer-start and timer-stop times. These two preset times and the real time (time of day) are applied to a comparator. When the times coincide, the comparator produces timer-start and timer-stop signals. Note that the timer-start signal is forcibly cut off by the timer-stop signal. The three types of times (real, start, and stop) are also applied to the seven-segment digitron readout through a segment decoder and segment driver.

## 3-7.3 Power Supply Gate Circuit and Control Circuit

The power supply gate and control circuits are shown in Fig. 3-50. Transistor Q1501 operates as a gate circuit for the +15-V power supply and is turned on by ON or VCR POWER signals applied to the base. When this occurs, +15 V is available to the control circuit. An ON signal occurs when the VCR POWER switch is turned on, and a VCR POWER signal occurs when the VCR is made ready to operate by pressing the control buttons.

As long as +15 V is supplied through Q1501, a high-level signal developed through D1511 at the timer-off input terminal of Q1507 (Fig. 3-49) forcibly cuts

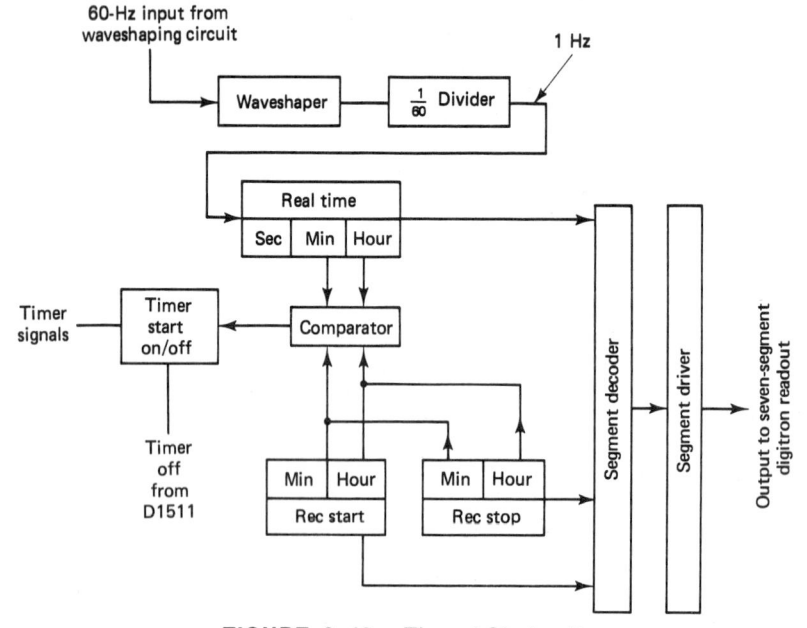

**FIGURE 3-49.** Timer LSI circuit.

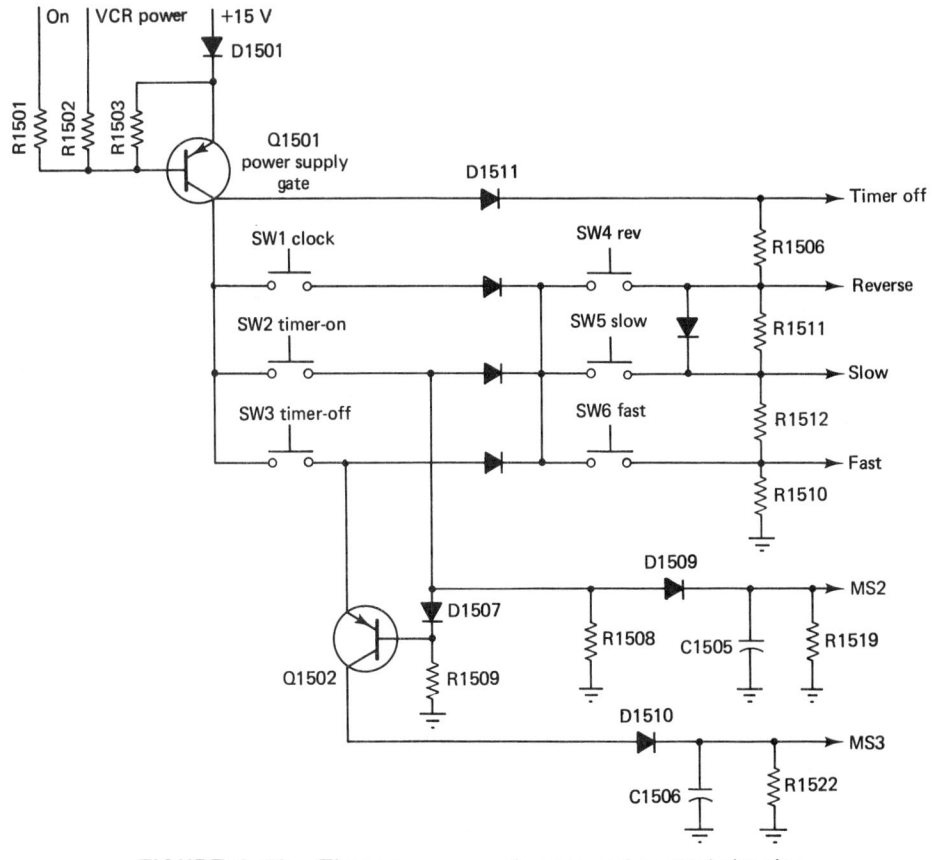

**FIGURE 3-50.** Timer power supply gate and control circuits.

off the timer-start signal. In this state, the time of day, and a timer-start and a timer-stop time, can be set or reset by means of switches SW1 through SW6. The purpose of Q1502 is to give priority to SW2 when SW2 and SW3 are operated simultaneously. Resistors R1506, R1510, R1511, and R1512 pull down the respective terminal to a low level. C1505, C1506, R1519, and R1522 are included to prevent erroneous performance on timer setting.

### 3-7.4 Timer Signal Processing Circuit

The timer signal processing circuit and related pulses are shown in Fig. 3-51. The circuit operates only when the POWER switch is set to TIMER. The timer signal (A) coming from pin 24 of Q1507 is inverted by Q1503 and applied to the CP terminal of the timer output gate Q1508. The timer signal (A) is also applied to Q1504 together with a signal (C) from the Q terminal of Q1508. The resultant output signal (D) is amplified by Q1505 to become the timer output signal.

164  Typical Beta VCR Circuits

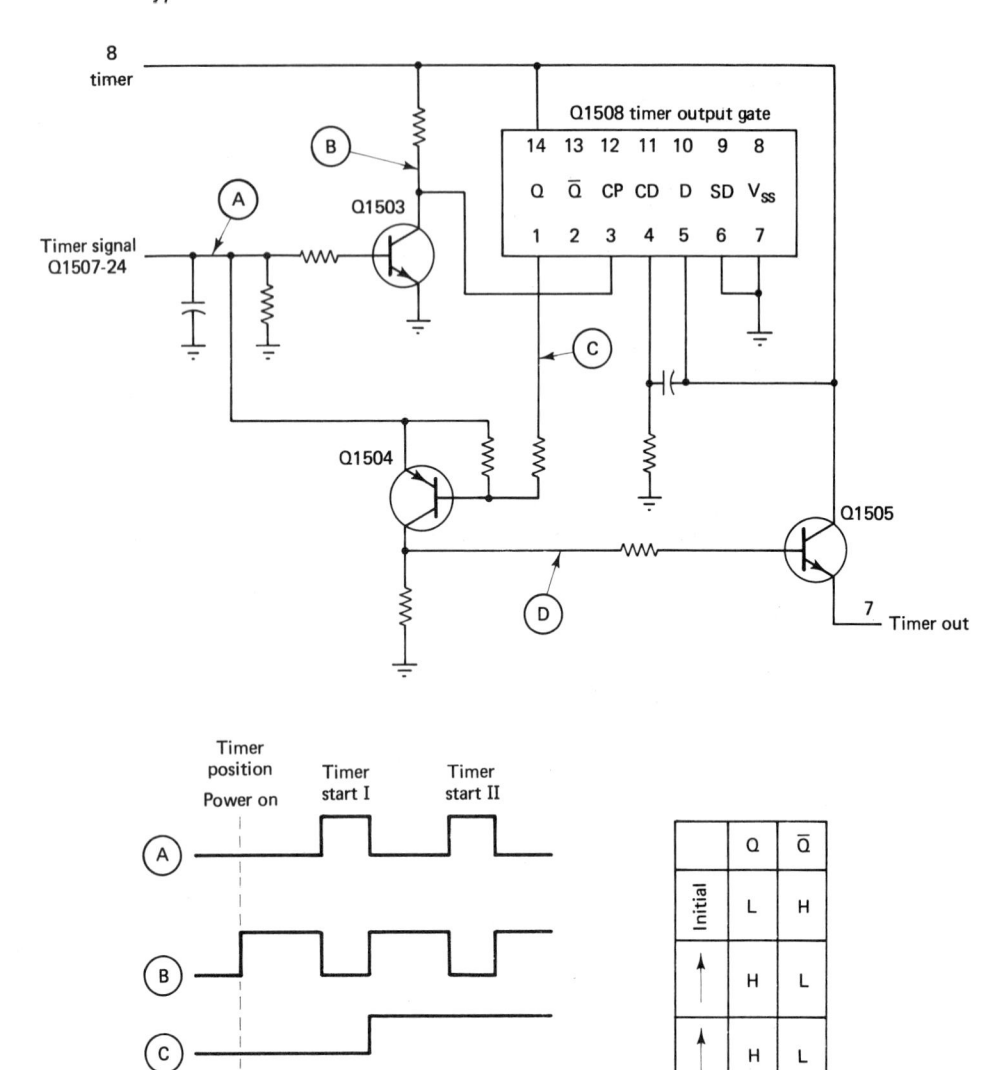

**FIGURE 3-51.** Timer signal processing circuit and related pulses.

The input signal (B) to the CP terminal of Q1508 becomes effective when the first timer pulse starts to fall. Thereafter, the Q output signal (C) is always high in spite of any successive pulse application to the CP terminal. Also, as shown in the timing diagram of Fig. 3-51, output pulse (D) appears only when the first timer pulse is present, and does not reappear on successive timer pulse signals. Once the POWER switch is released from the TIMER position, and then returned to the TIMER position, Q1508 is reset to the initial state, and the next timer start signal can set the timer signal output.

## 3-8 BETA TUNER CIRCUIT

Figure 3-52 is an overall block diagram of the tuner circuit. As shown, the circuit consists of a VCR-TV selector switch and distributor located on the W14 PC board, a UHF tuner, a VHF tuner, a U/V splitter, and circuits on the W5 PC board.

### *3-8.1 W14 PC Board Circuits*

The W14 PC board has RF signal input and output terminals labeled ANTENNA IN and TELEVISION OUT, a VCR-TV selector switch and an RF distributor. TV station signals (usually from a combination UHF/VHF antenna) supplied through a coaxial cable to the ANTENNA IN terminal are divided by the RF distributor into two branches: the VCR-TV selector switch and the U/V splitter. The VCR-TV selector switch connects the TELEVISION OUT terminal to either the RF converter (in VCR) or to the RF distributor (in TV). The U/V splitter separates the UHF and VHF signals and applies them to the corresponding UHF or VHF tuner.

### *3-8.2 VHF Tuner Circuit*

The VHF tuner selects the desired VHF channel signal and converts it into an IF signal for application to the IF amplifier on the W5 PC board. The VHF tuner is quite similar to those used in TV receivers (RF amplifier, frequency converter, local oscillator). The VHF tuner receives AGC (automatic gain control) and AFT (automatic fine tuning) signals from circuits on the W5 PC board.

### *3-8.3 UHF Tuner Circuit*

The UHF tuner selects the desired UHF channel signal and converts it into an IF signal. The frequency converter in the VHF tuner operates as an amplifier for the IF signal from the UHF tuner. The UHF tuner is also quite similar to those used in TV receivers (tuning stage, frequency converter using a diode, and a local oscillator).

### *3-8.4 W5 PC Board Circuits*

Figure 3-53 shows the W5 PC board circuits in more detail. Note that most of the functions such as VIF (video IF), SIF (sound IF) detector, RF and IF AGC, AFT, video detector, and video amplifier are contained within an IC Q901. The following paragraphs describe operation of these various circuits.

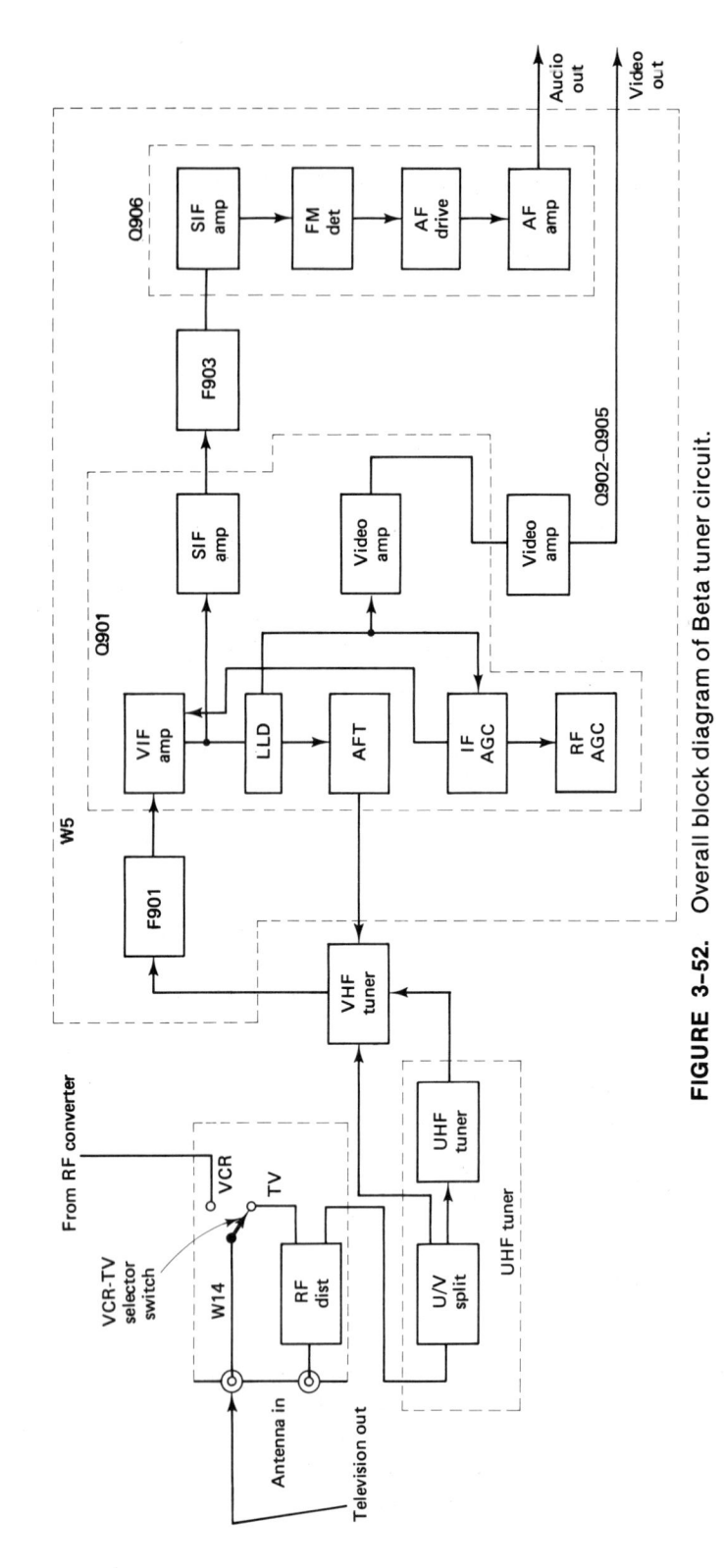

**FIGURE 3-52.** Overall block diagram of Beta tuner circuit.

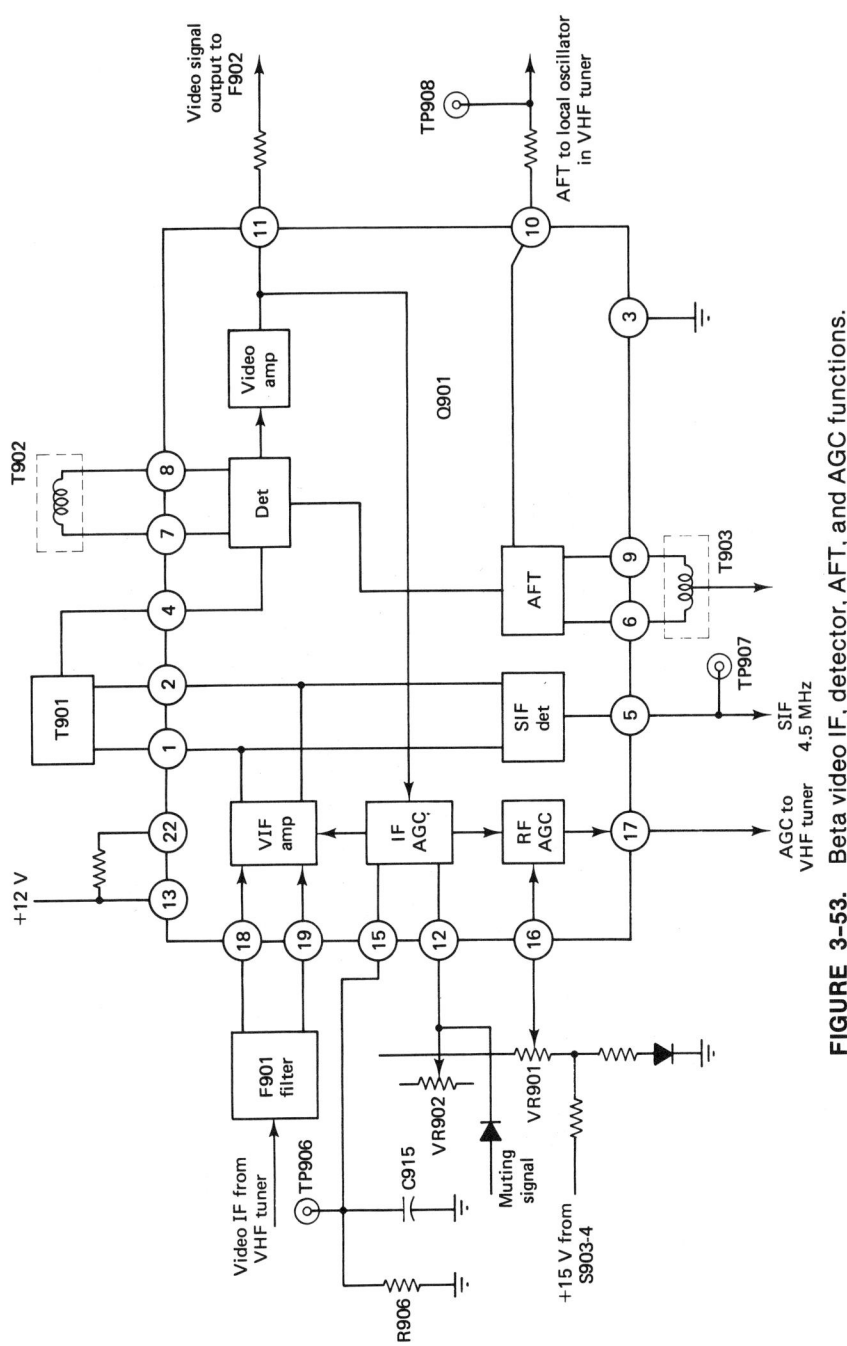

**FIGURE 3-53.** Beta video IF, detector, AFT, and AGC functions.

**Video IF Amplifier.** The video IF signal taken from the VHF tuner is applied through filter F901 to pins 18 and 19 of Q901. The video IF signal is amplified in Q901 and is applied to T901 through pins 1 and 2. A sound trap associated with T901 eliminates the 41.25-MHz sound carrier. A part of the amplified IF enters a sound IF detector. The detected 4.5-MHz sound IF signal is taken from pin 5.

**Video Detector.** The VIF signal is fed to a detector through pin 4 of Q901. The detector circuit (tuned by T902) amplifies and detects the IF signal which then appears as the video signal output at pin 11.

**AFT.** The AFT circuit (in association with a tuned circuit T903) is essentially a phase shifter which produces a d-c voltage corresponding to the frequency deviation of the input signal from the VIF carrier signal. This d-c voltage is applied to the local oscillator in the VHF tuner and shifts the oscillator frequency as necessary to correct the undesired frequency deviation. When pin 9 of Q901 is grounded by the AFT switch, operation of the AFT circuit is disabled, resulting in no output at pin 10.

**AGC.** The AGC circuit keeps the level of the detected video signal constant, in spite of changes in input signal level. The AGC circuit is a peak AGC system where the peak level of the sync signal in the video is detected, and fed back to keep the sync peak level constant. A filter, R906 and C915, is connected to pin 15. This filter smooths out the AGC voltage by eliminating ripple. The AGC voltage is applied to the IF amplifier in Q901, and is adjustable with VR902 in order to set the signal level at the video output (pin 11).

The AGC voltage for the RF amplifier stage is obtained from pin 17 and supplied to AGC terminal of the VHF tuner as a reverse AGC. The RF AGC does not function when the incoming signal is low in amplitude. The point at which the RF AGC takes hold is adjusted by VR901. The initiation point of RF AGC for the UHF signal must be set at a higher level than that of the VHF signal, because the amplification gain of the UHF tuner is greater than that of the VHF tuner (by several dB). The changeover in the RF AGC initiation point is accomplished by +15 V is applied from terminal 4 of S903.

### 3-8.5 Video Amplifier Circuit

Figure 3-54 shows the video amplifier circuit (external to Q901). The purpose of this circuit is to provide a 1-V (peak-to-peak) video signal with a flat frequency characteristics and a 75-$\Omega$ termination. The sound IF signal contained in the video signal at pin 11 of Q901 is filtered out by a ceramic trap F902. The video signal is amplified by Q902 and applied to Q903. The video signal detected by Q901 is somewhat reduced at high frequencies. A resonant circuit L905-C931 at the emitter of Q903 has a peak resonance at about the color subcarrier frequency. Since the circuit is series resonance, the color subcarrier frequency is bypassed to

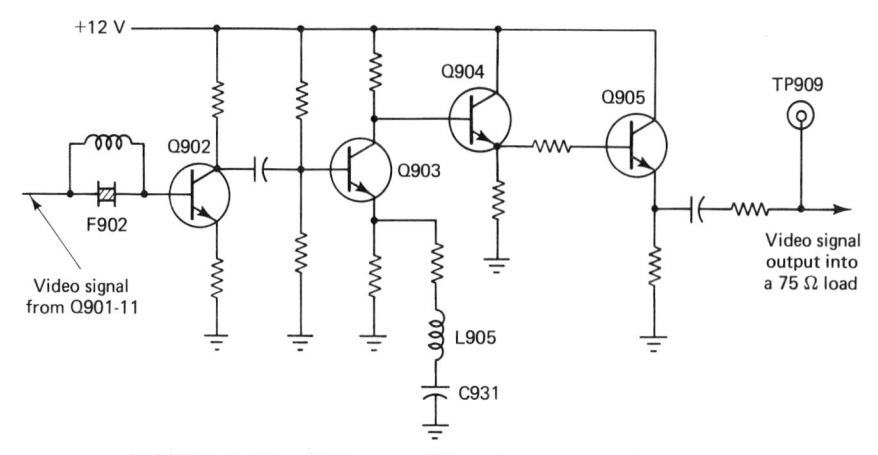

**FIGURE 3-54.** Video amplifier circuit (external to Q901).

ground, and thus reduced in relation to the luminance frequency signals. This results in a flat frequency response for the video signals. Q904 and Q905 form a power amplifier that delivers a current sufficient for a video signal output into a 75-$\Omega$ load.

### 3-8.6 SIF Amplifier

Figure 3-55 shows the SIF amplifier circuit (external to Q901). Note that most of the SIF functions are performed by IC Q906. The circuit picks up the 4.5-MHz sound IF signal from pin 5 of Q901, and then amplifies, limits, and detects the

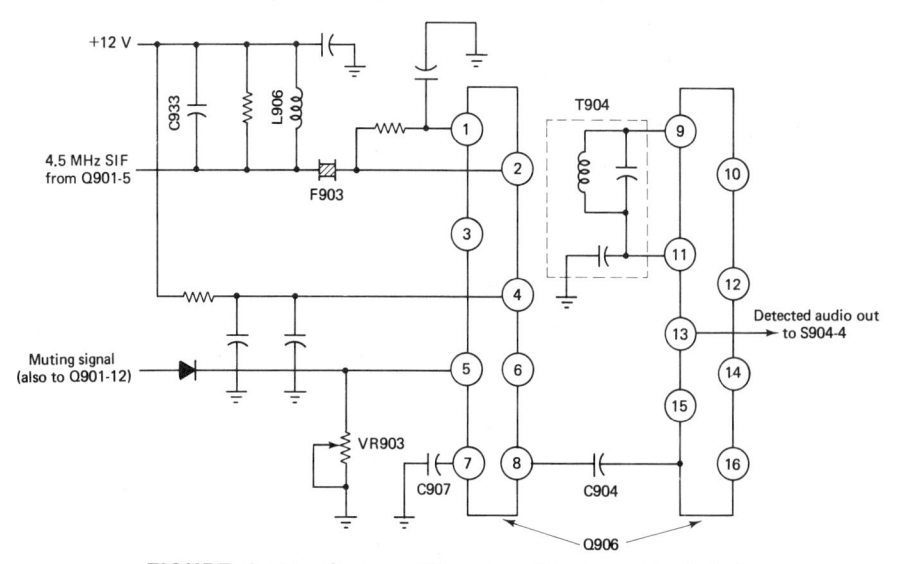

**FIGURE 3-55.** SIF amplifier circuit (external to Q901).

FM audio signal. The detected audio signal is then applied to the VCR audio system for recording on the tape audio track.

A tuned circuit composed of L906, C933, and ceramic filter F903 are included to eliminate the video signal component, and take out only the sound IF signal, which is then amplified and amplitude limited in Q906 to eliminate any AM component. A phase discriminator within Q906, tuned by T904 external to Q906, demodulates the FM signal. Capacitor C907 is part of a deemphasis circuit (within Q906) which is used to compensate for the enhanced high-frequency response, and to restore the original flat response. The detected audio signal is taken from pin 8 of Q906 and applied through C940 to pin 16, where the audio is amplified further. The audio output at pin 13 appears as audio from an emitter follower.

### 3-8.7 Muting Circuit

During playback, both the video and audio signals from the tuner circuit are muted to prevent interference between the playback signal and the demodulated tuner signal. Muting is done by means of a voltage applied to pin 12 of Q901 (Fig. 3-53) and a voltage applied to pin 5 of Q906 (Fig. 3-55). The voltage at pin 12 of Q901 increases the AGC so as to cut off the video signal at pin 11. The voltage at pin 5 of Q906 reduces the audio output level (from pin 13 of Q906) to zero. The level of the muting signal can be adjusted by VR903 (Fig. 3-55).

# 4

# Typical VHS VCR Circuits

This chapter describes the theory of operation for a number of VHS VCR circuits. (Similar coverage for a cross section of Beta circuits is provided in Chapter 3.) The VHS circuits described here include video signal processing, servo, and system control. Note that the RF unit and tuner circuits are not covered. As discussed in Chapter 3, this is typical for most VCR service literature. In the case of the RF unit (and in many cases the tuner), if the unit fails, it must be replaced as a unit, preferably at a factory service facility. Since the RF unit is essentially a miniature TV broadcast station or transmitter, any improper adjustment can produce interference with normal television broadcasting.

By studying the circuits found in this chapter, you should have no difficulty in understanding the schematic and block diagrams of similar VHS VCRs. This understanding is essential for logical troubleshooting and service, no matter what type of electronic equipment is involved. No attempt has been made to duplicate the full schematics for all circuits. Such schematics are found in the service literature for the particular VCR. Instead of a full schematic, the circuit descriptions are supplemented with partial schematics and block diagrams that show such important areas as signal flow paths, input/output, adjustment controls, test points, and power source connections. These are the areas most important in service and troubleshooting. By reducing the schematics to these areas, you will find the circuit easier to understand, and you will be able to relate circuit operation to the corresponding circuit of the VCR you are servicing.

Note that, as shown in the illustrations of this chapter, many circuit parts are contained within integrated circuits which carry the reference designation of

IC. Individual transistors outside the ICs carry the designation Q. Both the ICs and transistors are mounted on printed circuit boards, carrying the reference designations PCB.

The majority of the video circuits described in this chapter are part of the Mitsubishi HS-300U. This VCR uses the conventional VHS color-under recording system described in Sec. 1-7 and the VHS high-density PI (phase inversion) described in Sec. 1-9. Note that the overall block diagrams showing video luminance and color (chroma) operation in both record and playback are given in Chapter 1 and are not duplicated here. However, the following paragraphs (which describe circuit details) make direct reference to these block diagrams in Chapter 1.

## 4-1 VHS LUMINANCE SIGNAL CIRCUITS DURING RECORD

Figure 1-41 is the overall block diagram of the luminance circuits used during record. Overall functions of these circuits are described in Sec. 1-9.1. The following paragraphs describe details and characteristics for some of the circuits.

### 4-1.1 Low-Pass Filter and AGC Circuits

The video signal coming from the VIDEO IN terminal, during a color broadcast, is first fed to the low-pass filter LPF2F4 to remove the burst signal inserted in the back porch of the horizontal sync signal. The filter characteristics are shown in Fig. 4-1. The filter impedance is 1 kΩ and the filter delay is about 0.35 μsec.

The signal passing through LPF2F4 is added to the AGC circuit, which uses a peak AGC system. The video signal from LPF2F4 passes through the AGC (Q201, Q202), and video amplifier (Q203, Q204) to LPF2F5. The video signal of the video amplifier is passed through the sync separator Q206 to extract only the signal at the AGC detector (which controls the AGC function). The peak

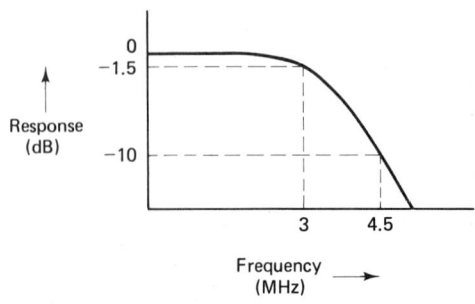

**FIGURE 4-1.** LPF2F4 characteristics.

AGC system uses the peak level of the sync signal to keep the amplitude of the video signal constant.

### 4-1.2  Low-Pass Filter and Video Amplifier Circuits

The video signal from the AGC circuit is amplified by Q203 and Q204, and applied to LPF2F5 where the 3.58-MHz color subcarrier is reduced in amplitude to prevent beat interference due to mixing of the 3.58-MHz subcarrier and the FM modulation frequency. The characteristics of LPF2F5 are shown in Fig. 4-2. The filter impedance is 1 kΩ, and the filter delay is about 0.35 $\mu$s. The signal passing through LPF2F5 is amplified by the video amplifier (of IC2F2) and applied to the nonlinear emphasis circuit.

### 4-1.3  Nonlinear Emphasis Circuit

Since the FM signal usually deteriorates in signal-to-noise (S/N) ratio at the high frequencies, the signal is preemphasized during record to compensate for this deterioration. However, in this case, the narrower the video track thickness, the worse the S/N ratio becomes if the signal level decreases. To correct this condition, the high-frequency component of the input signal is recorded with nonlinear emphasis (more emphasis on low-level signals than on high-level signals). On playback, the signal is passed through a circuit with the reverse characteristics. This overall process improves the S/N ratio by about 3 dB.

Figure 4-3 shows the nonlinear emphasis circuit. On a 2-hour (standard) recording, the base of Q2H5 goes low, turning Q2H5 off, and Q2H1 on. The video signal from pin 22 of IC2F2 is supplied to pin 19 of IC2F2 through Q2H1, R244, R245, VR2G2, and C2N5, resulting in no emphasis. VR2G2 sets the level of FM deviation during record. On 6-hour (extended) recording, a high is applied to the base of Q2H5, turning Q2H5 on and Q2H1 off. Q2H4 turns on because of the high applied to the base. The video signal from pin 22 of IC2F2 is amplified by Q2H2 and Q2H3 and is taken from the emitter of Q2H3. Under these condi-

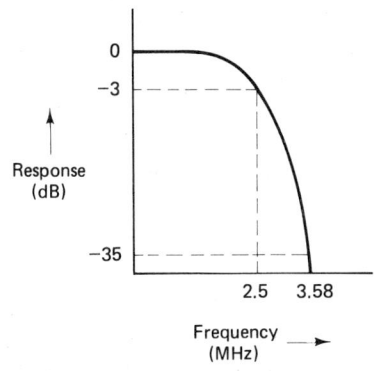

**FIGURE 4-2.**  LPF2F5 characteristics.

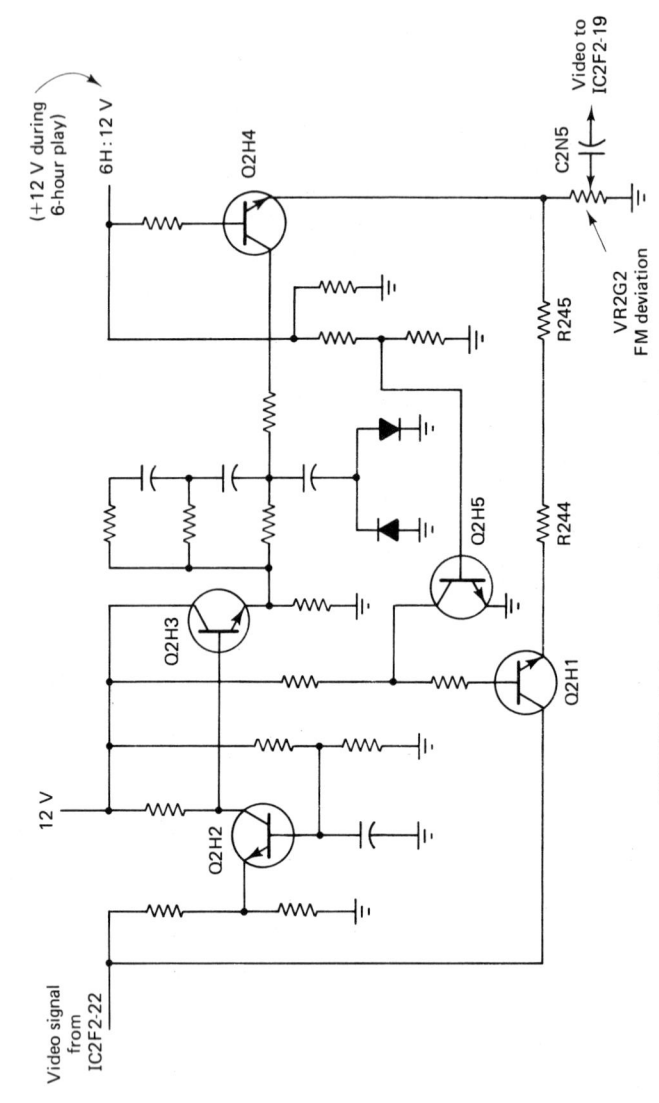

**FIGURE 4-3.** Nonlinear emphasis circuit.

tions, the video signal supplied to pin 19 of IC2F2 passes through the nonlinear emphasis circuit of Q2H4, VR2G2, and C2N5. As a result, though the rate of preemphasis is the same as a conventional circuit when signal level is high, the high-frequency region is further increased by about 8 dB when the signal level is small. The rate of preemphasis varies in this low-signal-level region.

### 4-1.4 Clamp, Preemphasis, White and Dark Clip Circuits

The video signal from the nonlinear emphasis circuit (set to the correct level for proper FM deviation by VR2G2) is amplified by the video amplifiers of IC2F2 and applied to the clamp. The signal amplified by the video amplifier is applied to the clamp through pin 16 to keep the sync tip accurately at a constant potential so that the sync tip remains at 3.4 MHz during FM modulation.

The video signal is then applied to preemphasis, where the high-frequency spectrum is emphasis to improve the S/N ratio at demodulation. Generally, the FM signal is subject not only to a change in amplitude by noise, but also to phase modulation. The higher the frequency, the more the FM signal is influenced by noise. In some literature, such noise is called "triangle noise" because of the characteristics as plotted in Fig. 4-4, which shows the noise at FM signal demodulation. Preemphasis is determined by the frequency response of the feedback loop from the collector of Q2F8 to pin 16. Figure 4-5 shows the typical preemphasis characteristic.

The video signal passing through the preemphasis circuit is applied to the white clip and dark clip circuits. Since a sharp overshoot and undershoot are caused at the rise and fall of the signal by preemphasis and can cause overmodulation when FM modulation is applied, the white level exceeding a specified level is clipped by the white clip. Similarly, the dark clip circuit clips the black level that goes below a specified point. VR2F3 controls white level, while VR2F9 controls the black level. The signal thus processed is applied to the FM modulator.

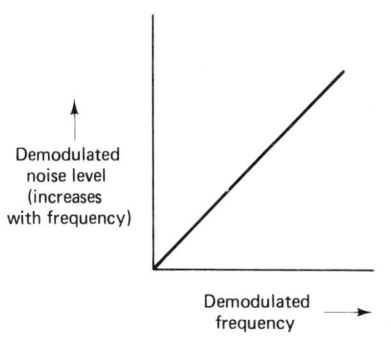

**FIGURE 4-4.** Noise at FM signal demodulation.

176  Typical VHS VCR Circuits

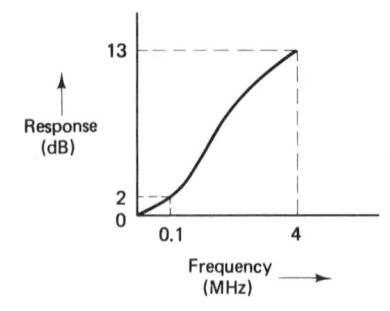

**FIGURE 4-5.** Typical preemphasis characteristics.

### 4-1.5  FM Modulator, Carrier Interleave, Amplifier, and High-Pass Filter

The video signal supplied from the white and dark clip circuits is fed to the FM modulator, which is adjusted by VC2F2. The FM modulator uses an emitter-coupled unstable multivibrator which is designed to operate from 3.4 MHz (at sync tip) to 4.4 MHz (at white peak). The FM-modulated signal, after being amplified by Q2G1 and converted into a lower impedance by emitter-follower Q2G2, is applied to the high-pass filter.

As shown in Fig. 4-6, during a color broadcast, the chroma signal is down-converted (heterodyned) to 629 ± 500 kHz to be superimposed on the FM signal. There is a part of the lower sideband of the FM signal which can overlap the color signal. This lower sideband is attenuated by the high-pass filter (HPF).

A 30-Hz pulse is applied to the FM modulator from the FM carrier interleave circuit to advance the phase of the channel 2 (head B) track by $\frac{1}{2}$ fH. Figure 4-7 shows the carrier interleave circuit. On 6-hour play, a high is fed from the collector of Q408 in the 2H/6H discriminator circuit to the cathode of D2G4. This turns D2G4 off and permits the 30-Hz pulses from the drum or cylinder servo to be applied at the base of Q2H0 through R2S5, D2F8, and D2F9. The drum pulse is inverted at the collector of Q2H0. During record, the magnitude of the video signal to be recorded is adjusted by VR2G6 and is fed to the FM modulator circuit. Since the drum pulse changes from high to low with each field, the carrier interleaving circuit interleaves the FM carrier by $\frac{1}{2}$ fH for each

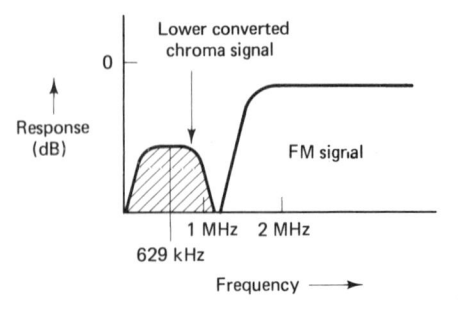

**FIGURE 4-6.** Relationship of chroma signal and FM signal during record.

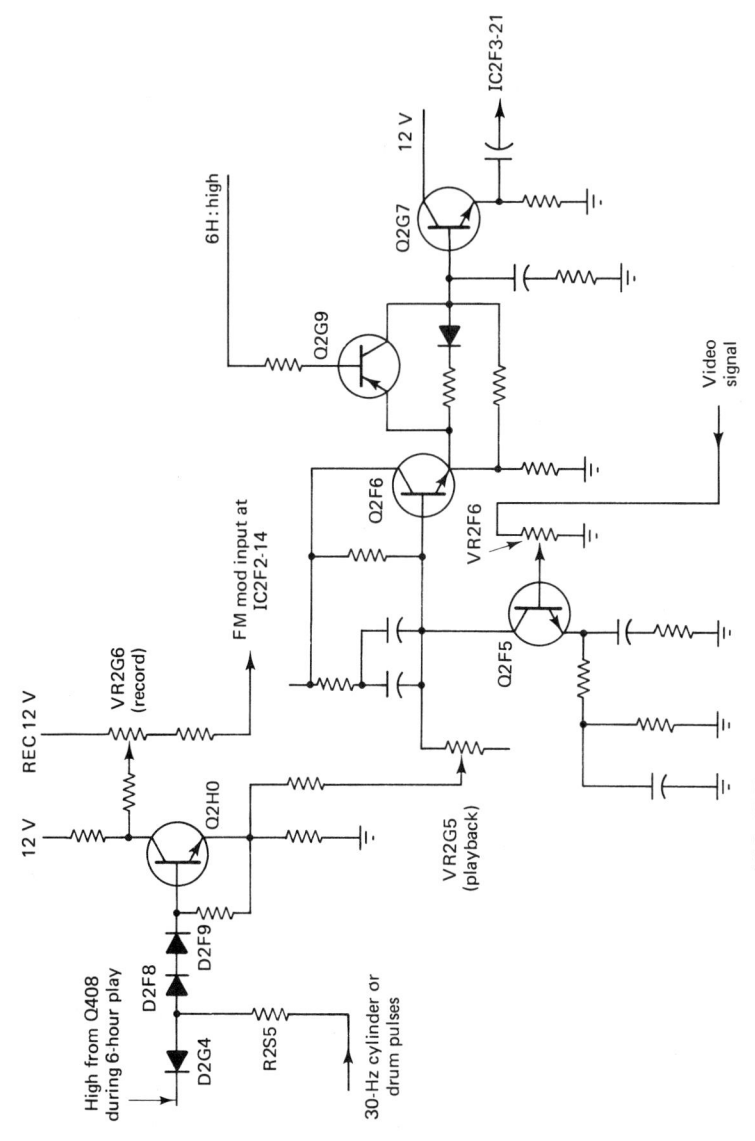

**FIGURE 4-7.** Carrier interleave and squelch circuit.

178  Typical VHS VCR Circuits

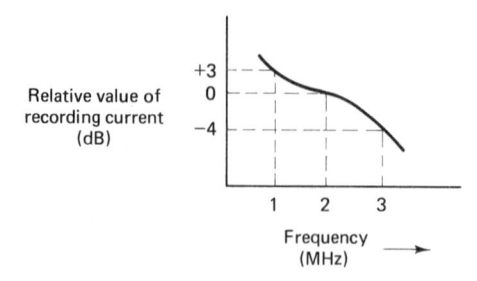

**FIGURE 4-8.** General characteristics of optimum recording current.

video track. This makes noise invisible on the picture (due to the integrating effect of the human eye). Note that the video signal is applied to the base of Q2F5 through VR2F6, which sets the level of the video signal.

### 4-1.6  Squelch and Record Amplifier Circuit

The signal from emitter of Q2H0 is supplied to the squelch circuit through VR2G5, as shown in Fig. 4-7. The squelch circuit functions to prevent signals from being fed to the record amplifier for about 1.5 seconds after completion of loading. This prevents the recorded signal from being erased if the tape runs near the drum during the loading process (if the tape is incorrectly threaded).

The signal passing through the squelch circuit is applied to the record amplifier IC2A1, which produces the optimum recording current for each signal frequency to be recorded. Figure 4-8 shows the general characteristics of optimum recording current. The record amplifier uses a complementary SEPP [single-ended (push-pull) circuit] which minimizes crossover and switching distortion while still providing a low load resistance.

## 4-2  VHS LUMINANCE SIGNAL CIRCUITS DURING PLAYBACK

Figure 1-42 is the overall block diagram of the luminance circuits used during playback. Overall functions of these circuits are described in Sec. 1-9.2. The following paragraphs describe details and characteristics for some of the circuits.

### 4-2.1  Preamplifier, Switch and Mixer Amplifier

Since the playback signal from the video head is in the order of a few millivolts, the playback signal is amplified in the first, second, and third preamplifiers to a level of about 0.3 V (peak to peak). The preamplifier circuits are provided with adjustment controls to produce an overall flat response as shown in Fig. 4-9. This is done by adjusting both the resonance frequency and the Q of the video

### VHS Luminance Signal Circuits during Playback 179

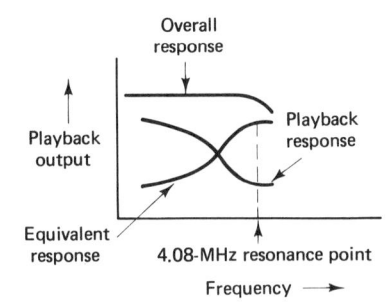

**FIGURE 4-9.** Reproduction characteristics to produce overall flat response during playback.

heads. VC2A1 and VC2A0 are adjusted to set the resonance of the heads at about 4.08 MHz. VR2A1 and VR2A0 are adjusted to set the video head Q so that the overall response characteristics of Fig. 4–9 are obtained at playback.

The signals amplified by the preamplifiers are mixed by the switching circuit and mixer amplifier to produce a continuous noise-free signal as shown in Fig. 4–10. Note that the overlap of the channel 1 and channel 2 signals at the heads is eliminated by the drum FF pulses, which switch the channel 1 and channel 2 output so that channel 1 is off at the instant channel 2 is on (and vice versa). The continuous signal obtained from the switching circuit is adjusted for proper balance between the channels by VR2A2. This signal is amplified by the mixer

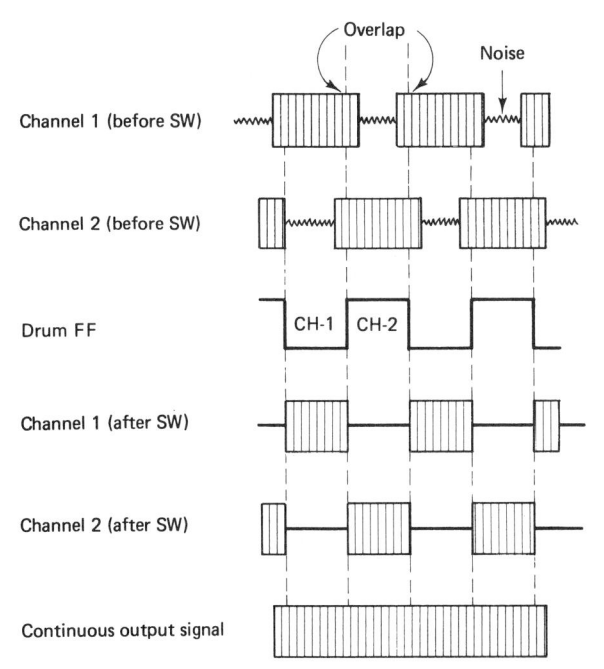

**FIGURE 4-10.** Switching and mixing process to produce a continuous signal from the video heads.

180  *Typical VHS VCR Circuits*

amplifier, adjusted to the correct FM level by VR2F4, and applied to the video amplifier circuit.

### 4-2.2  Video Amplifier and High-Pass Filter

The signal from VR2F4 is amplified by Q2F0 and passed through high-pass filter HPF2F0 to Q2F1. The high-pass filter removes the 629-kHz color signal. Emitter follower Q2F1 converts the output to a low impedance.

### 4-2.3  Dropout Compensation Circuit

Figure 4-11 shows the dropout compensation (DOC) circuit in block form. The DOC circuit is disabled during black-and-white operation by a color killer voltage applied to the changeover switch. During black-and-white operation, the switch connects the mixer to the FM input through pin 9, so that the FM signal passes to pin 10 and is not affected by the DOC circuit. During a color broadcast, the changeover switch connects the FM input from pin 7 to the mixer, and then to the circuit output at pin 10. This same output is applied through the 1H delay line and amplifier to the gate circuit. The gate remains closed when there is no dropout. The FM input at pin 7 is amplified by the DOC amplifier and applied to the DOC detector. If dropout is present, the detector produces a pulse that is shaped by the Schmitt trigger and applied to the gate. The pulse opens the gate

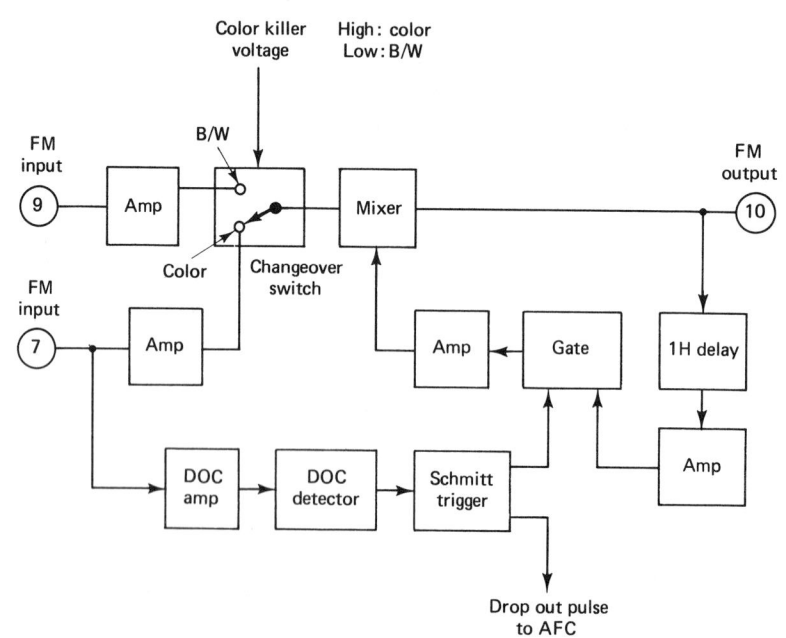

**FIGURE 4-11.**  Basic dropout compensation circuit.

and permits signal passing through the 1H delay line to be amplified and mixed with the signal from pin 7, thus eliminating dropout. As discussed in Sec. 1-9.6, the dropout pulse is also applied to the AFC circuit.

### 4-2.4 Double Limiter Circuit

Figure 4–12 shows the double limiter circuit in block form. A double limiter is used to offset the effects of preemphasis of the signal at various stages. Such preemphasis can result in sharp overshoot, as shown in Fig. 4–12. If a signal with ex-

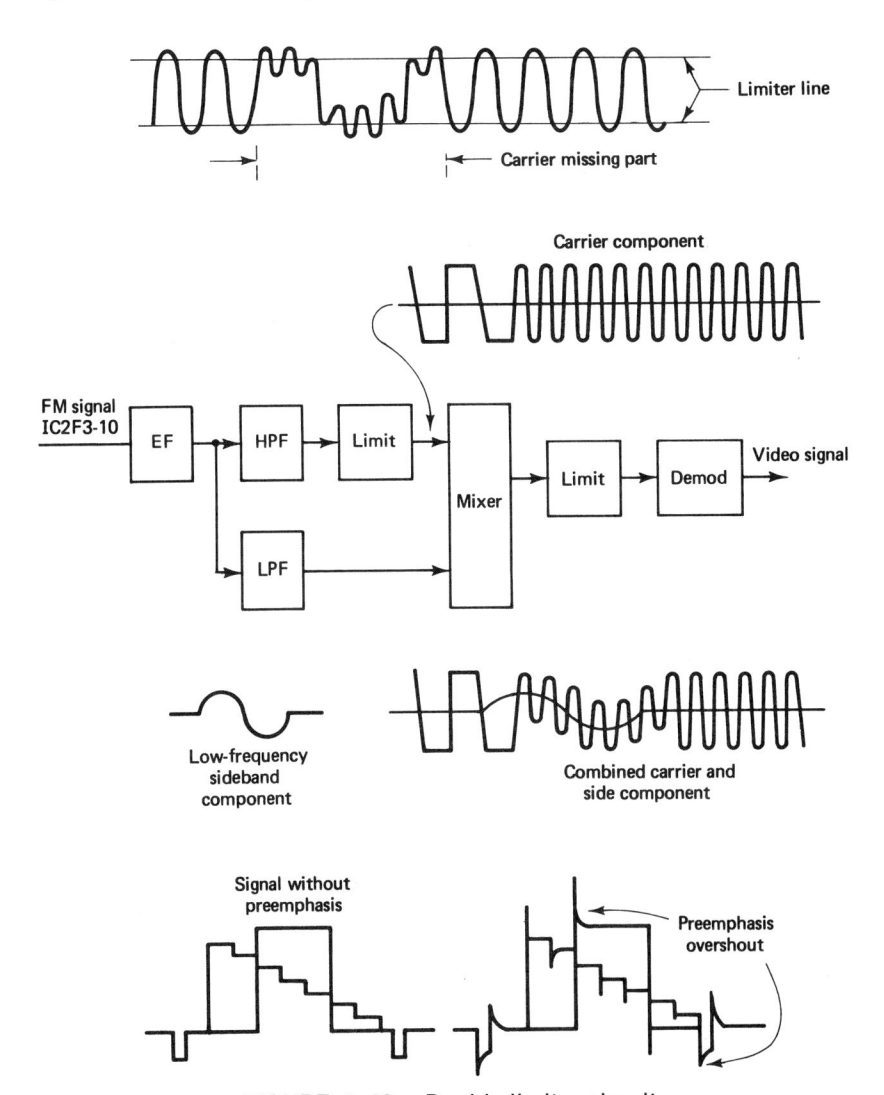

**FIGURE 4-12.** Double limiter circuit.

cessive overshoot is FM modulated, overmodulation occurs and the low-frequency sidebands can be increased in amplitude. This results in possible AM modulation and can cause picture reversal or negative picture (white parts turn black, and vice versa). The signal-to-noise ratio also deteriorates. Although the white and dark clip circuits serve to reduce overshoot, any AM modulation must be eliminated by limiting (as in the case of any FM system). However, conventional limiting can result in loss of low-frequency sidebands (which, in turn, results in loss of high-frequency signals after demodulation, and picture deterioration).

In the double limiter of Fig. 4–12, the FM signal is passed through a high-pass filter and a low-pass filter to separate the signal into a carrier component and a low-frequency sideband component. The carrier is further limited and mixed with the sidebands. The combined signal is again limited and demodulated. This double limiter system compensates for the ratio of the carrier and sidebands, without amplifying noise, and thus ensures no loss of carrier.

### 4–2.5  FM Demodulator Circuit

Figure 4–13 shows the FM demodulator circuit and the associated waveforms. The FM signal entering pin 14 of IC2F6 is limited by the limiter circuit and is fed to the mixing circuit through an internal IC2F6 circuit and through an external circuit. This external circuit is adjusted to provide the proper time constants and balance by VR2G8 and VR2G9 (limiter and carrier balance, respectively). The time period $T_0 = 1/FC$ of the FM waveform coming from pin 6 of IC2F6 is delayed by $1/4\ T_d$. The resultant signal is then mixed, detected, and rectified into waveform D of Fig. 4–13. The video portion of this signal is applied to the deemphasis and feedback circuits through an emitter follower Q2F4 and a low-pass filter LPF2F2.

### 4–2.6  Deemphasis and Feedback Circuits

The deemphasis circuit is provided to reduce high-frequency noise during playback by attenuating the high-frequency components using a characteristic almost exactly opposite that of the preemphasis circuits during record. The level of the playback luminance signal entering the deemphasis circuit is set by VR2F6. The deemphasized signal is supplied to the edge noise canceler.

### 4–2.7  Edge-Noise Canceler Circuit

Figure 4–14 shows the edge noise canceler circuit and the associated waveforms. The video signal from the emitter of Q2F5 is differentiated by C2Q6, R248, and R247 and is fed to the base of Q2H7. The signal is amplified and inverted by Q2H7. After the noise component is eliminated by D2G2 and D2G3, the signal is applied to the base of Q2H8, where the signal is amplified and superimposed on the video signal at the collector of Q2F5 to cancel noise.

# VHS Luminance Signal Circuits during Playback 183

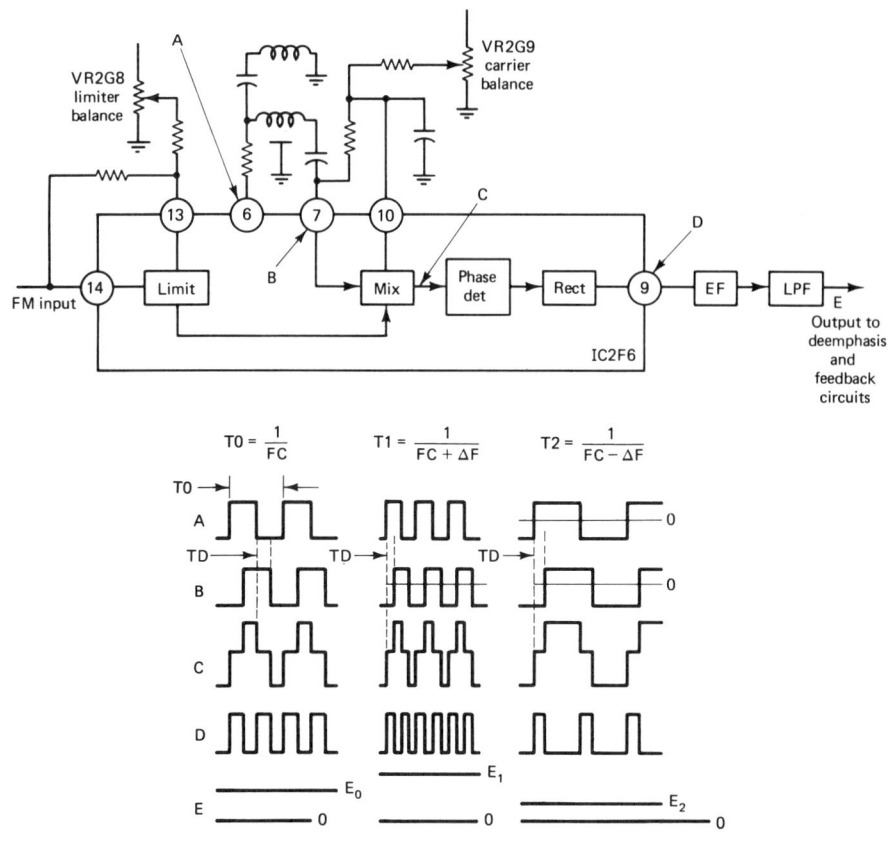

**FIGURE 4-13.** FM demodulator and associated waveforms.

$$T0 = \frac{1}{FC} \qquad T1 = \frac{1}{FC + \Delta F} \qquad T2 = \frac{1}{FC - \Delta F}$$

Since, at recording, the channel 2 signal is advanced by $\frac{1}{2}$fH by the carrier interleave circuit (Sec. 4-1.5 and Fig. 4-7), a small variation in the d-c level occurs during playback. Therefore, the 30-Hz pulse derived from the emitter of Q2H0 is fed to the base of Q2F6 through VR2G5 to compensate for the d-c variation.

## 4-2.8 Nonlinear Deemphasis Circuit

Figure 4-15 shows details of the nonlinear deemphasis circuit used during playback. During 2-hour (standard) playback, the base of Q2G9 goes low to turn on Q2G9 and short circuit the filter network (R2S1, R2D3, D2G6). Similarly, the base of Q2G8 goes low, turning Q2G8 off and disabling the filter network (R2D4, C2R0, D2F5). During 6-hour playback [called expanded playback (EP)] a high is supplied to the base of Q2G9. This turns Q2G9 off and connects the filter (R2S1, R2D3, D2G6) into the circuit. A high is also supplied to the base of Q2G8. This turns Q2G8 on and connects the filter R2D4, C2R0, D2F5) into the circuit.

## 184 Typical VHS VCR Circuits

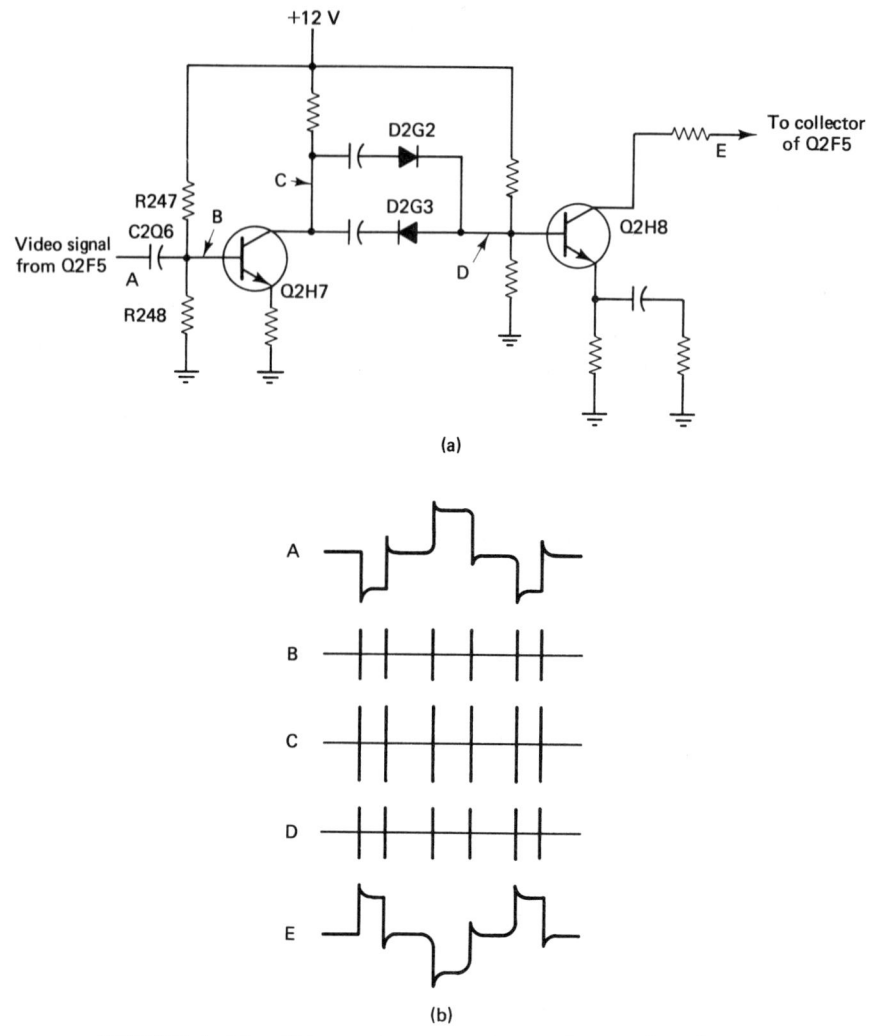

**FIGURE 4-14.** Edge noise canceler and associated waveforms.

When processing normal signals during playback, the high-frequency component of the video signal at the emitter of Q2F6 is attenuated by the filter of R2S1, C2R1, and R2G2. In the case of a signal where the black level changes rapidly into a white level, undershoot and noise occur. During 6-hour play, D2G6 is inserted and turned on by the undershoot. The turn-on of D2G6 connects R2S1 and R2D3 in parallel, reducing the combined resistance and increasing the filter effect to reduce the undershoot.

If the high-frequency component of the video signal becomes larger (during 6-hour play), D2F4 and D2F5 are turned on, attenuating the signal at the base

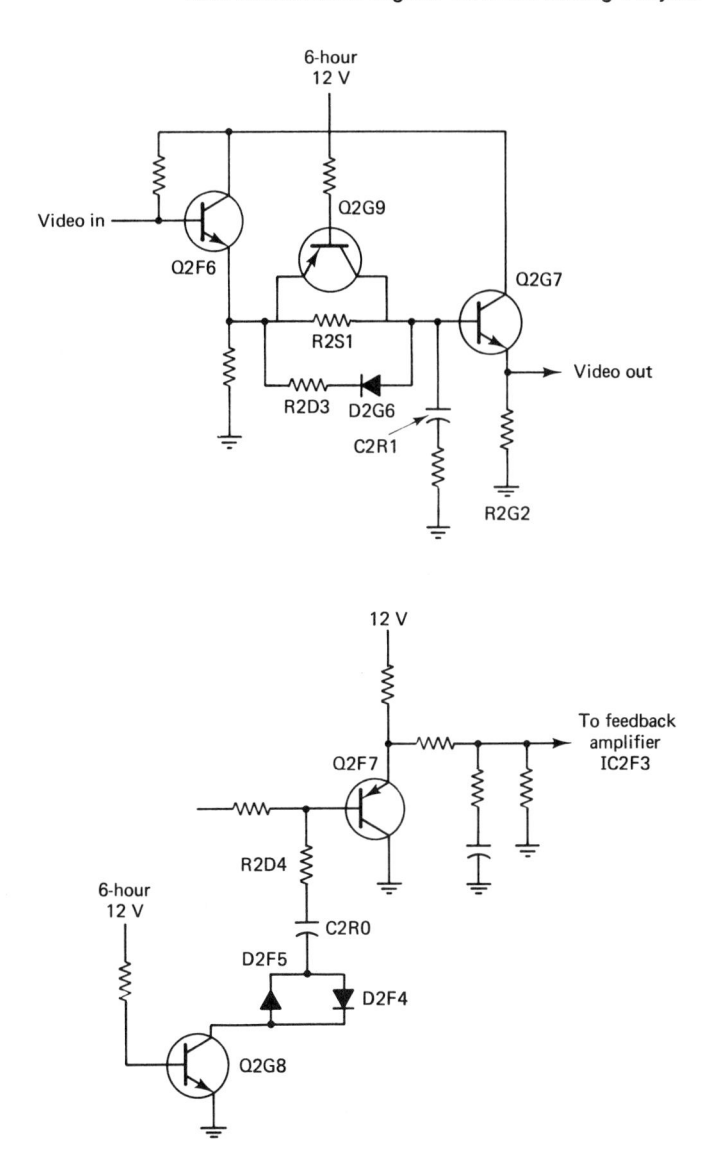

**FIGURE 4-15.** Playback nonlinear deemphasis circuit.

of Q2F7 (through R2D4, C2R0, D2F4, D2F5, and Q2G8). The larger the high-frequency component, the more the Q2F7 base level is attenuated. The output of Q2F7 is applied to feedback amplifier IC2F3, where the high-frequency component of the video signal is boosted. In this case, the larger the signal, the more the high-frequency component is boosted.

To summarize, the nonlinear deemphasis circuit is designed so that the high-frequency component can be attenuated by the filter, and boosted by the feedback amplifier, to correct for the nonlinear emphasis at recording.

## 4-2.9 Color/BW Switch

Figure 4-16 shows details of the color/BW switch network. The switch is operated by a color killer voltage and connects the low-pass filter LPF2F3 and video amplifier into the signal path when the color killer voltage is high. LPF2F3 ensures that the point where the video signal overlaps the demodulated chroma signal is free of noise, and the video amplifier compensates for any attenuation by LPF2F3. During a black-and-white playback, the color killer voltage goes low, and the switch operates to bypass both LPF2F3 and the video amplifier. This is necessary because black-and-white picture resolution can deteriorate if the signal is passed through the low-pass filter.

## 4-2.10 Noise Canceler, Clamp, Mute, EE/VV Switch

As shown in Fig. 1-42, the signal from the color/BW switch is processed in the high-pass amplifier (HPA) to extract only the high-frequency component to be amplified. The signal is then limited and applied to the noise canceler. The direct signal from the color/B&W switch is also added to the noise canceler so that only the high-frequency component is canceled. This type of noise canceler circuit, using a low-frequency boost circuit, improves picture quality.

The color signal is superimposed on the video signal in the Y/C mixer and is fed to the clamp/mute circuit, which receives a quasi-sync signal at pin 26. This quasi-sync signal is added to the video signal to replace or supplement the normal sync during the still, frame transfer, speed search, or slow playback modes (which may be disturbed by noise during these modes). The E-E/V-V switch is provided to automatically switch video playback and monitor (E-E) signals. Both the playback video signal from the clamp/mute circuit and the monitor (E-E) signals from the E-E amplifier are applied to the E-E/V-V switch. During record, the E-E signals are passed through the E-E/V-V switch to the RF unit so that the program being recorded can be viewed on the TV set. During playback, the E-E/V-V switch disconnects the E-E input and connects the playback video from the clamp/mute circuit to the RF unit.

## 4-2.11 Quasi-Sync Signal Generator

Figure 4-17 shows the quasi-sync signal generator and the associated waveforms. The drum FF pulse is applied to pin 8 of IC501, and a high from the speed search discriminator is applied to pin 9 during still, frame transfer, speed search, or slow playback. During normal playback, pin 9 goes low and pin 10 remains high, even though the drum FF pulse is being applied to pin 8. As a result, no quasi-sync signal is produced.

When a high is applied to pin 9, the drum FF pulse passes and appears at pin 10 (in inverted form). The pulse is then applied to another NAND gate (at pins 12, 13) and appears at pin 11 (again inverted). The drum FF pulse at pin 10 is dif-

## VHS Luminance Signal Circuits during Playback 187

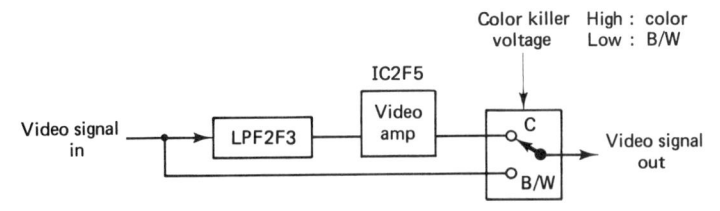

**FIGURE 4-16.** Color/BW switch network.

ferentiated by C502 and R506 and is applied to pin 1. Similarly, the pulse at pin 11 is differentiated by C503 and R501 and is applied to pin 2. This results in a 60-Hz pulse at pin 3. The 60-Hz pulse is differentiated by C504 and R502 and is applied to pins 5 and 6. The output at pin 4, a positive 60-Hz pulse of about 0.3-ms width, is inverted by Q501 and fed to the clamp/mute circuit as the quasi-sync signal.

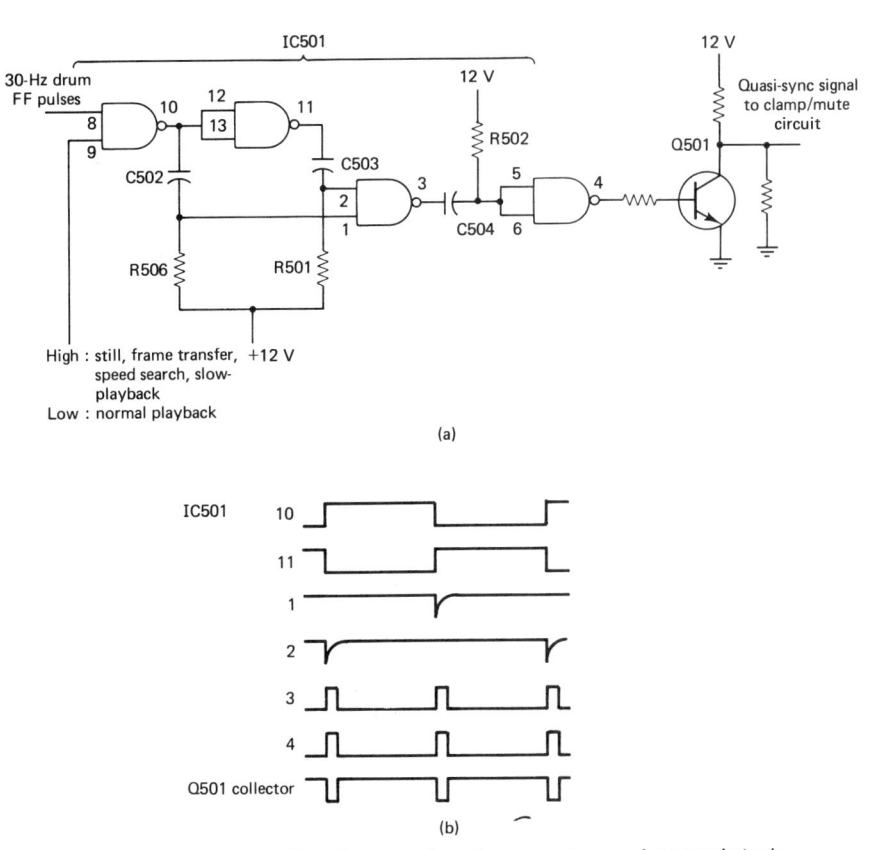

**FIGURE 4-17.** Quasi-sync signal generator and associated waveforms.

## 4-3 VHS COLOR SIGNAL CIRCUITS DURING RECORD

Figure 1-43 is the overall block diagram of the color circuits used during record. Overall functions of these circuits are described in Sec. 1-9.3. The following paragraphs decribe details and characteristics for some of the circuits.

### 4-3.1 Bandpass Filter, Amplifier, Expander, and Emitter Follower

As shown in Fig. 4-18, the video signal from the tuner is applied to BPF6F0, where only the 3.58-MHz color burst signal is passed. The response of BPF6F0 is also shown in Fig. 4-18. The signal passing through BPF6F0 is supplied to the emitter of Q6F0, which is connected in a grounded-base configuration. The amplified signal is applied through emitter follower Q6F1 to the ACC circuit. The horizontal pulse from pin 15 of IC6F4 is passed through a low-pass filter

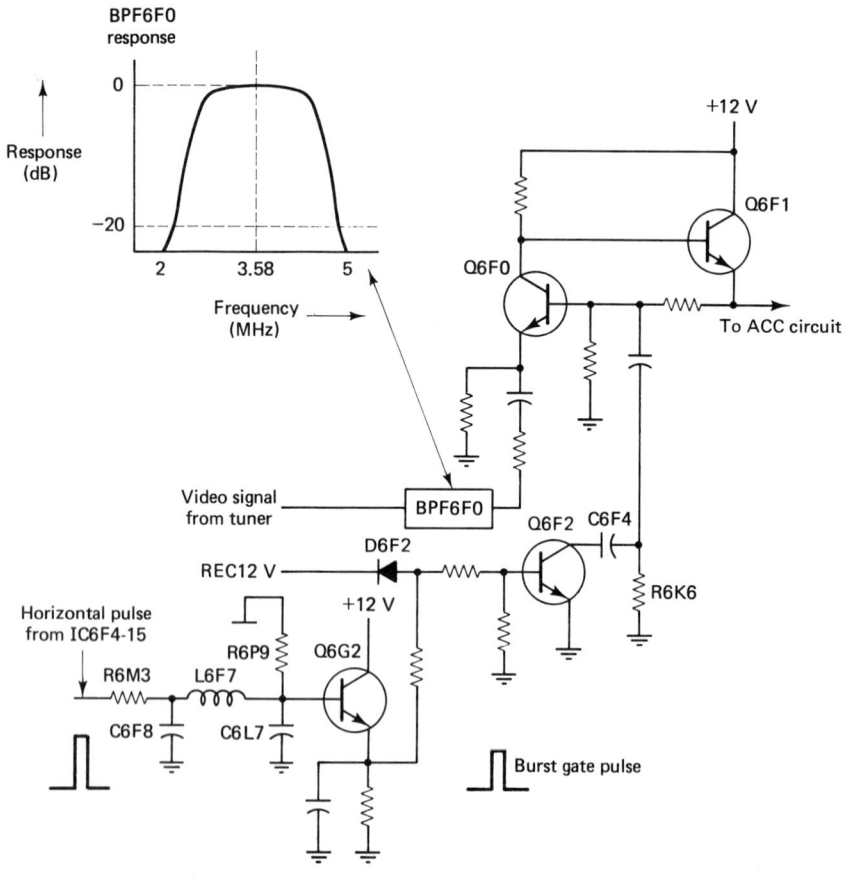

**FIGURE 4-18.** Bandpass filter, amplifier, expander, and emitter-follower circuits.

(R6M3, C6F8, L6F7, C6L7, R6P9), and an emitter follower Q6G2 to the base of Q6F2. This turns Q6F2 on and connects C6F4 in parallel with R6K6 to increase the gain of Q6F0 by about 6 dB during the horizontal pulse. Thus, the 3.58-MHz color signal is passed to the ACC circuit during the horizontal pulse. The low-pass filter delays the horizontal pulse so that the pulse output from the emitter of Q6G2 corresponds to the color burst portion. During playback, D6F2 is turned on so that Q6F2 remains turned off and keeps the gain of Q6F0 low during playback. This prevents tuner noise from passing to the ACC circuit during playback.

### 4-3.2 ACC and Detector Circuit

As shown in Fig. 1-43, the output of the ACC (automatic color control) at pin 11 is passed to the main converter and through the Q6F3 switch (which is turned on by the REC12V voltage) to the IC6F0 burst gate. The REC12V voltage is also applied to the burst gate and to a switch Q6F5, turning on both stages. During record, Q6F5 connects C6H1 in parallel with C6G8, R6H1, and C6G9 to change the time constants as necessary to alter the detector output. Within the detector, the peak value of the 3.58-MHz color signal is converted to a d-c voltage used as a bias or control voltage by the ACC (at pin 5). This control voltage keeps the 3.58-MHz color signal constant during record, and keeps the 629-kHz color signal constant during playback. VR6F0 sets the ACC output level during both record and playback.

### 4-3.3 Main Converter, Low-Pass Filter, and Emitter Follower

As shown in Fig. 1-43, the 3.58-MHz output from the ACC is applied to the main converter IC6F0 at pin 3. IC6F0 also receives a 4.2-MHz signal from the APC circuit (Sec. 4-5) at pin 5. This produces both sum and difference frequencies at the main converter output (pin 7). As described in Secs. 1-9.5 and 1-9.6, the 4.2-MHz signal is rotated in phase by +90° each 1H period for channel 1 and −90° for channel 2. As a result, the difference frequency at pin 7 is also rotated in phase by ±90°. This difference frequency of 629 kHz ± 90° is passed by LPF6F0, where the sum frequency is rejected. The difference-frequency signal is passed through Q6F6 and VR2G4 to the record amplifier. VR2G4 sets the level of the color signal during record. During black-and-white programs, Q6F6 is turned off to prevent noise in the color circuits from being recorded.

## 4-4 VHS COLOR SIGNAL CIRCUITS DURING PLAYBACK

Figure 1-44 is the overall block diagram of the color circuits used during playback. Overall functions of these circuits are described in Sec. 1-9.4. The following paragraphs describe details and characteristics for some of the circuits.

## 4-4.1 Amplifiers, Low-Pass Filter, and Emitter Follower

As shown in Fig. 1-44, the playback signal (video FM signal plus 629-kHz color signal) from pin 10 of IC2F0 is applied through the FM level adjustment VR2F4 to be amplified by Q2F3. A trap (C2H1-L2F5) is inserted at the base of Q2F3 to attenuate the high-frequency sound bias recorded on the tape. The amplified signal from Q2F3 is applied through LPF2F0, where the 629-kHz color signal is passed. The color signal is then amplified by Q6F0 and applied through emitter follower Q6F1 to the ACC circuit of IC6F0.

## 4-4.2 ACC Circuit

As shown in Fig. 1-44, the output from Q6F1 is applied through the ACC circuits of IC6F0 to the main converter of IC6F0, where the 629-kHz signal is mixed with the 4.2-MHz signal to restore the 3.58-MHz color signal. The 3.58-MHz signal is applied through BPF6F1, Q6F7, DL6F0, and IC6F1 to the burst gate IC6F0. The gate opens only during the burst period, and supplies only the burst signal to the detector. Within the detector, the peak value of the 3.58-MHz color signal is converted to a d-c voltage used as a bias or control voltage by the ACC. This control voltage keeps the 629-kHz color signal passing through the ACC constant during playback. VR6F0 sets the ACC output level during both playback and record.

## 4-4.3 Compressor Circuit

As shown in Fig. 4-19, the 3.58-MHz signal from BPF6F1 is passed through Q6F7, which is controlled by compressor Q6F8. During record, the burst signal is increased by the expander circuit (Sec. 4-3.1, Fig. 4-18). During playback, the

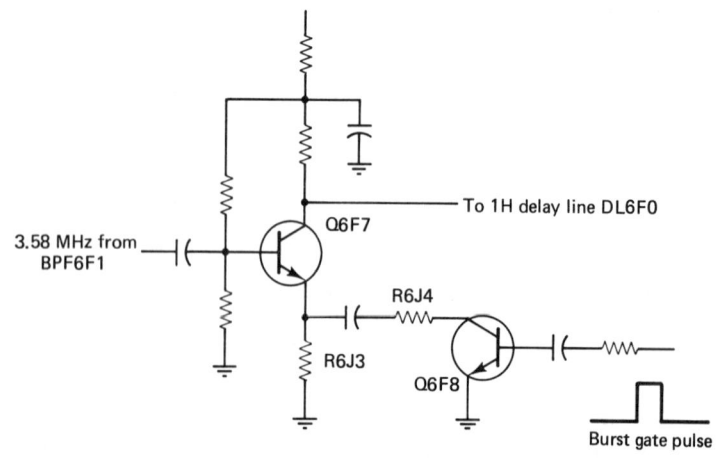

**FIGURE 4-19.** Burst compressor circuit.

compressor circuit of Fig. 4-19 reduces the burst signal by 6 dB. When there is no burst gate pulse, Q6F8 is turned on, placing R6J4 in parallel with R6J3, reducing the emitter resistance and increasing the gain of Q6F7. When the burst gate pulse is applied, Q6F8 is turned off, leaving only R6J3 in the emitter of Q6F7. This increases the emitter resistance and decreases the gain of Q6F7 (by about 6 dB).

## 4-4.4 1H Delay

As shown in Fig. 1-44, the amplified signal from Q6F7 is passed through 1H delay line DL6F0 to amplifier IC6F1. Figure 4-20a shows the equivalent circuit of DL6F0, which is designed to pass only the 3.58-MHz signal. As discussed in Sec. 1-9, with the VHS color recording system, the channel 1 track is advanced by 90°, while the channel 2 track is delayed by 90° for each 1H period. This produces a recording pattern similar to that of Fig. 4-20b. Figure 4-20c shows what can occur if the channel 1 head traces slightly on the channel 2 track during playback (producing crosstalk).

When both the major signal and crosstalk are passed through a 1H delay line, the major signal becomes equivalent to the component vector (about double), and the crosstalk is canceled. The 1H delay line circuit improves the signal-to-noise ratio by about 3 dB.

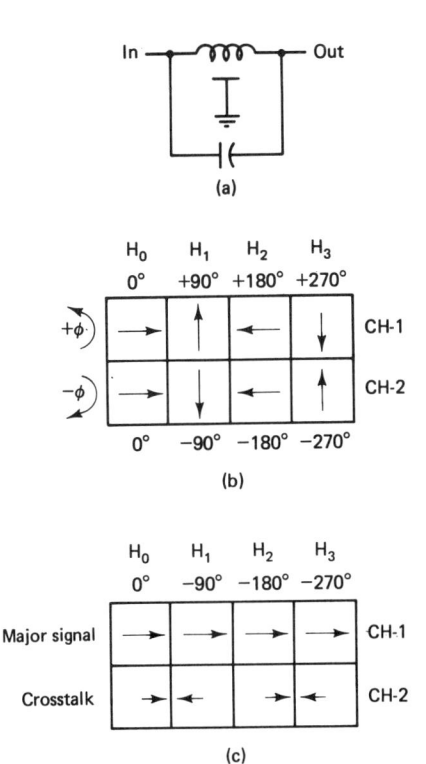

**FIGURE 4-20.** 1H delay functions.

### 4-4.5 Amplifier and Killer Amplifier

As shown in Fig. 1-44, the color signal from which the crosstalk components have been removed by DL6F0 is amplified by IC6F1, killer amplifier IC6F0, and is applied to IC2F3 (Fig. 1-42) through color output adjust VR6F1. The color signal is mixed with the luminance (Y) signal in IC2F3. Killer amplifier IC6F0 also receives a control signal from the color killer circuit applied at pin 6. When pin 6 is high, the color signal is passed from pin 10 of IC6F0. During a black-and-white program, the color killer control signal is low at pin 6, and no output is obtained from pin 10 of IC6F0. This prevents noise in the color circuit from passing during black-and-white operation.

## 4-5 VHS SERVO SYSTEM

The majority of the servo circuits described in this chapter are part of the Hitachi VT-8500A. Overall functions of the servo system for this VCR are described in Sec. 1-9.7. The following paragraphs describe details and characteristics for some of the circuits.

### 4-5.1 Cylinder Phase Control Circuit

Figure 4-21 is the overall block diagram of the cylinder phase control circuit. Figure 4-22 shows the related waveforms.

As shown in Fig. 4-21, the 30-Hz cylinder tach pulse from the pulse generator installed in the lower part of the cylinder shaft is supplied to the comparison signal circuit for cylinder phase control in IC501 through pin 19. This cylinder tach pulse is used in both playback and recording. The pulse at pin 19 is applied to the positive pulse amplifier and negative pulse amplifier to be detected separately as positive and negative pulses. The pulse applied to the positive amplifier is detected as the positive pulse and amplified to a level suitable to trigger monostable multivibrator MM 1, which acts as an adjustable delay circuit to determine the video head switching phase. The delay can be changed by the time constant of C522 and R541 connected to pin 18. Channel 1 switching phase is adjusted by R505. The pulse applied to the negative amplifier is delayed by MM 2, as adjusted by R504.

The output from MM 1 and MM 2 is supplied to a flip-flop FF. MM 1 is connected to the reset input of the FF, with MM 2 connected to the set input. The pulse from the FF (waveform F of Fig. 4-22) is the video head switch pulse, called SW30 and supplied to the video luminance and color circuits. Pulse SW30 is adjusted by R504 and R505 to become a 30-Hz rectangular wave with a 50% duty cycle. The output of the FF is converted to a trapezoidal waveform which is applied to the sample-and-hold circuit as a comparison signal. The sample-and-hold circuit also receives a reference signal which is compared with the

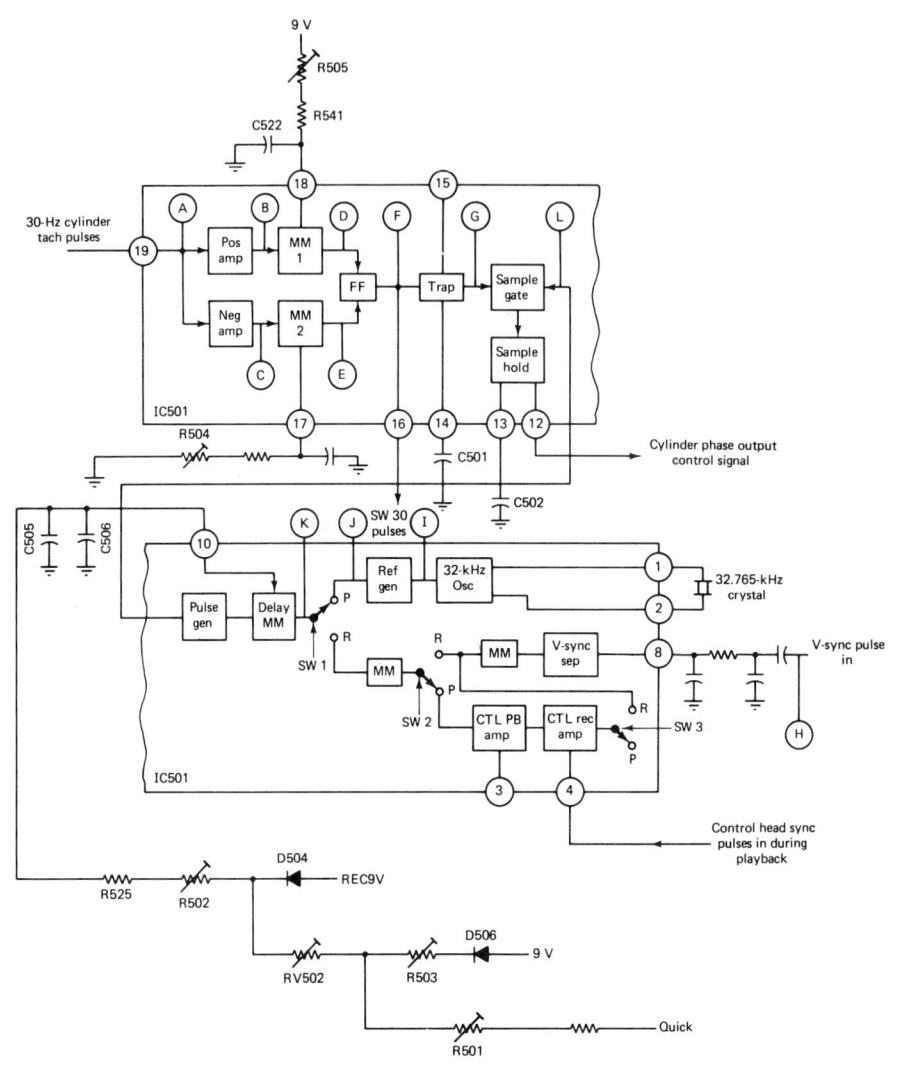

**FIGURE 4-21.** Cylinder phase control circuit.

trapezoidal signal to produce the desired cylinder phase output control signal at pin 12.

During record, switches SW 1, SW 2, and SW 3 in IC501 are electrically connected to the record position, and the broadcast V-sync signal is used as the reference signal. The V-sync component of the composite sync signal supplied by the luminance circuit is applied to pin 8. The V-sync signal is separated by the V-sync separator and shaped by the monostable multivibrator to produce a 30-Hz rectangular pulse. This pulse is amplified through the control pulse recording amplifier and supplied to the control head (through pin 4), where the

*194 Typical VHS VCR Circuits*

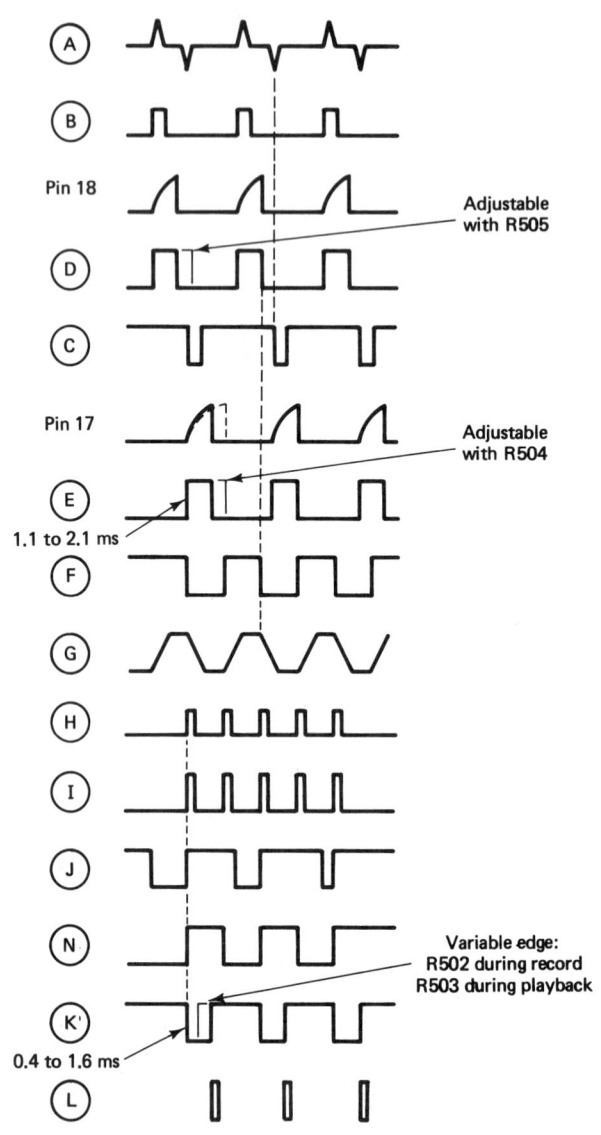

**FIGURE 4-22.** Waveforms associated with cylinder phase control circuit.

pulse is recorded on tape to become the capstan control signal during playback.

During playback, switches SW 1, SW 2, and SW 3 are electrically connected to the playback position, and the internal crystal-controlled signal is used as the reference signal. The playback reference signal is obtained by dividing the 32.765-MHz signal (developed by the crystal oscillator using an external crystal connected at pins 1 and 2) by 1093. This 1093 division occurs in the reference generator circuit.

During either playback or record, the delay MM circuit receives a 30-Hz reference signal through switch SW 1. During playback, the time constant of the delay MM is increased to provide correct tracking. Diode D504 is off during playback, and the delay MM time constant is approximately equal to the total of C505, C506, R502, R503, R525, and RV502. Diode D504 is turned on during record by a 9V control voltage, removing RV502 and R503 from the circuit, and decreasing the time constant of the delay MM to the equivalent of C505, C506, R502, and R525. Variable resistor R502 is the record timing control used to set the recorded V-sync signal at the correct position. R501 and R503 provide a similar function during playback (R501 for the quick mode, and R503 for other modes). Refer to Sec. 4-6 for a discussion of the various operating modes.

The pulse generator differentiates the pulse obtained through the delay MM to produce the reference sampling pulse applied to the sample-and-hold (S/H) circuit. When the trapezoidal wave (comparison signal) and the sampling pulse (reference signal) are applied to the S/H circuit, and error voltage corresponding to the phase difference is generated and applied to the capstan speed control circuit (Sec. 4-5.3) through pin 12. Capacitor C502 holds the error voltage between samplings.

If the speed and phase of the cylinder are correct, the trapezoidal and sampling pulses line up as shown in Fig. 4-23. Notice that the sampling pulse lines up in the center of the trailing edge of the trapezoid. When the trapezoid waveform and the sampling pulse are supplied to the S/H stage, the voltage level of the trapezoid ramp is sampled (at the point where the sampling pulse intersects the ramp). This is given as X-volts in Fig. 4-23.

If the cylinder motor speed increases, the trapezoid wave phase leads with respect to the sampling pulse as shown in Fig. 4-24. The sampling position moves lower on the ramp, and the error voltage decreases, making the cylinder motor rotate at a lower speed. If the cylinder motor speed decreases, the trapezoidal wave lags behind the sampling pulse as shown in Fig. 4-25, and the sampling

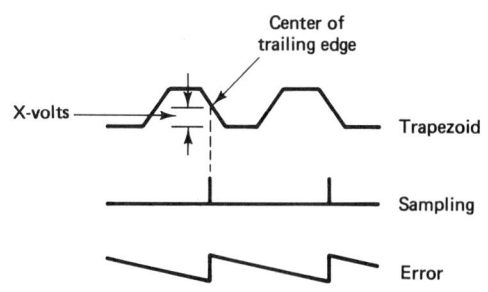

FIGURE 4-23. Relationship of pulses when cylinder speed is normal.

## 196 Typical VHS VCR Circuits

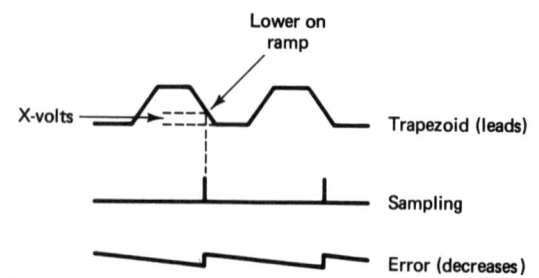

**FIGURE 4-24.** Relationship of pulses when cylinder speed increases.

position moves higher on the ramp. As a result, the error voltage increases, making the cylinder rotate faster.

In summary, when the sampling pulse is used to sample the trapezoid ramp, a variable voltage results that is in direct relationship to the relative position of the sampling pulse on the ramp. Because this voltage represents video head position, the voltage can be used to control the cylinder motor phase. However, the phase control voltage developed by this circuit is limited to that which can be detected on the slope of the ramp, and is used only as a vernier speed control (or phasing control) voltage. The actual speed of the cylinder motor is controlled by the speed control circuit as discussed in Sec. 4-5.3.

### 4-5.2 Capstan Phase Control Circuit

Figure 4-26 is the overall block diagram of the capstan phase control circuit. Figure 4-27 shows the related waveforms.

As shown in Fig. 4-26, during record, the comparison signal of the capstan phase control is obtained from the frequency generator (FG) built into the capstan motor. The record comparison signal (capstan FG pulse) frequency is 720 Hz in SP, 360 Hz in LP, and 240 Hz in EP operating modes. Refer to Sec. 4-6

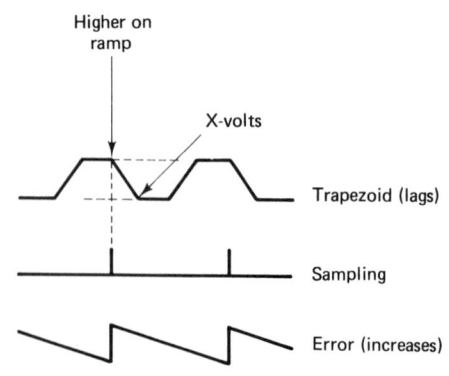

**FIGURE 4-25.** Relationship of pulses when cylinder speed decreases.

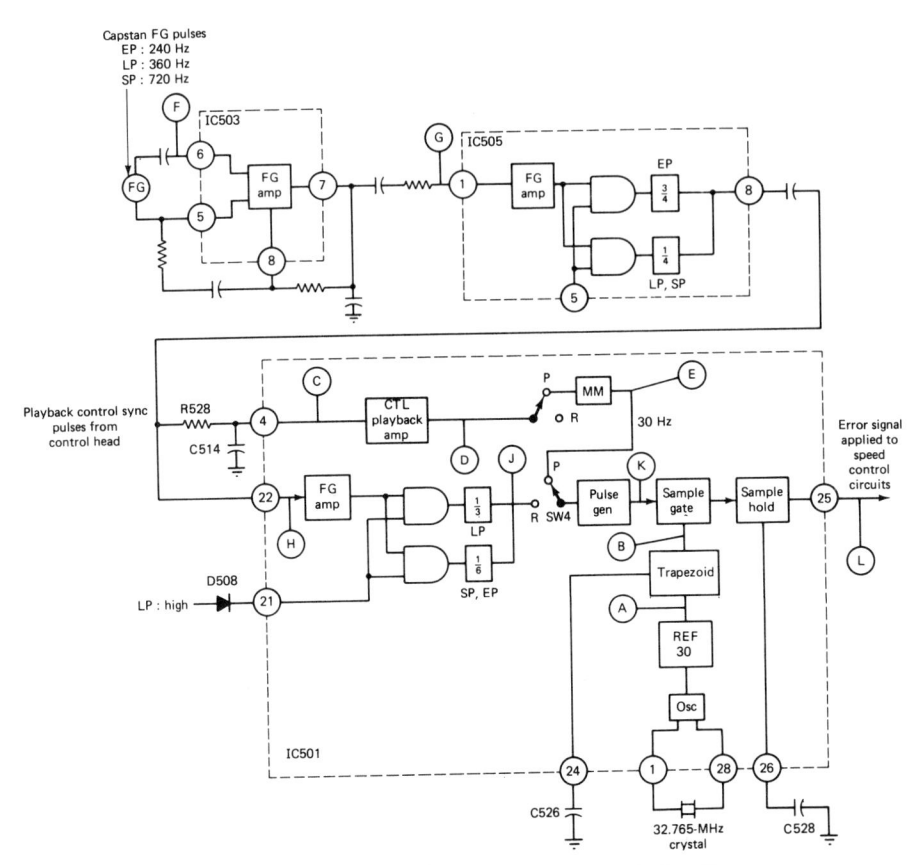

**FIGURE 4-26.** Capstan phase control circuit.

for a discussion of the operating modes. During playback, the control pulse recorded on the control track during record is detected by the control head and used as the comparison signal.

During record, the capstan FG pulses are amplified by the FG amplifier in IC503. The output from pin 7 of IC503 is applied through pin 1 of IC505 and control gates to a divider circuit. The FG pulses are divided by $\frac{1}{4}$ in the LP and SP modes and by $\frac{3}{4}$ in the EP mode. The output from pin 8 of IC505 is applied through pin 22 of IC501, another FG amplifier, and control gates, to another divider circuit. The FG pulses are further divided by $\frac{1}{6}$ in the SP and EP modes and by $\frac{1}{3}$ in the LP mode.

During playback, the control pulse picked up by the control head is applied to a monostable multivibrator through pin 4 of IC501, low-pass filter C514 and R528, and a control playback amplifier. The 30-Hz playback pulse triggers the multivibrator, which produces a 30-Hz rectangular pulse.

During either playback or record, the pulse generator receives a 30-Hz reference signal through switch SW4. The rectangular pulse supplied to the pulse

*198 Typical VHS VCR Circuits*

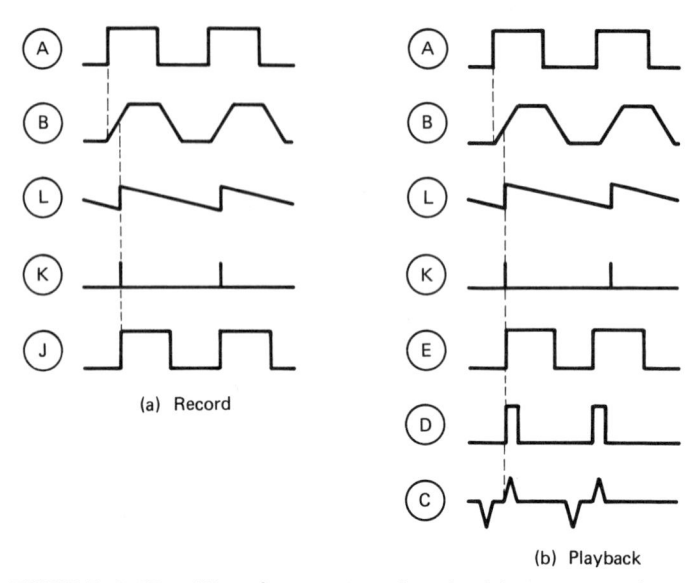

**FIGURE 4-27.** Waveforms associated with capstan phase control circuit.

generator is converted to a sampling pulse and applied to the S/H circuit as the comparison signal. The S/H circuit also receives a reference signal in the form of a trapezoidal wave. This trapezoidal waveform is obtained by dividing the 32.765-MHz signal (developed by the crystal oscillator) by 1093. The divided signal (in the form of a 29.98-Hz rectangular wave) is converted to a trapezoidal waveform (waveform B in Fig. 4-27).

The trapezoidal wave is sampled in the S/H circuit by the 30-Hz signal produced by the control pulse generator. This produces an error voltage corresponding to the phase difference. The error voltage is applied to the speed control circuits through pin 25. Capacitor C528 holds the error voltage between samplings. Capacitor C526 determines the time constant of the trapezoidal wave.

Basically, the sampling action of the capstan phase control system is similar to that of the cylinder phase control system (Sec. 4-5.1). The difference is, in the capstan servo, the reference signal is the trapezoidal wave and the comparison signal is the sampling pulse. As a result, the sampling position is on the leading edge of the trapezoidal wave (whereas the position is on the trailing edge for the cylinder phase control). When the capstan motor is at the correct speed, the sampling position is at the middle of the ramp as shown in Fig. 4-28. When the capstan motor speeds up slightly, the phase of the comparison signal advances with respect to the reference signal as shown in Fig. 4-29. As a result, the sampling position moves lower on the ramp, and the error voltage is reduced, decreasing motor speed. If the capstan motor slows down, the phase of the comparison signal lags behind the reference signal as shown in Fig. 4-30. As a result, the sampling position moves up the ramp and the error voltage increases, increasing motor speed.

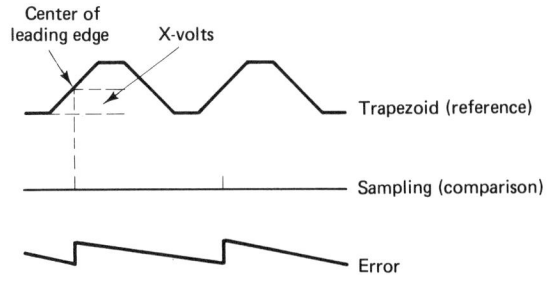

**FIGURE 4-28.** Relationship of pulses when capstan speed is normal.

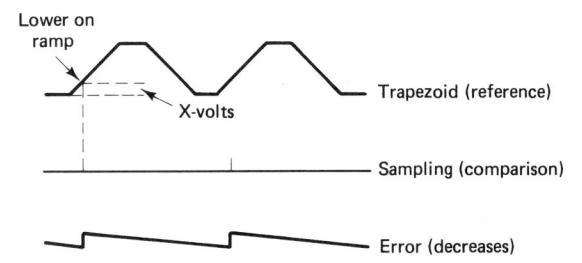

**FIGURE 4-29.** Relationship of pulses when capstan speed increases.

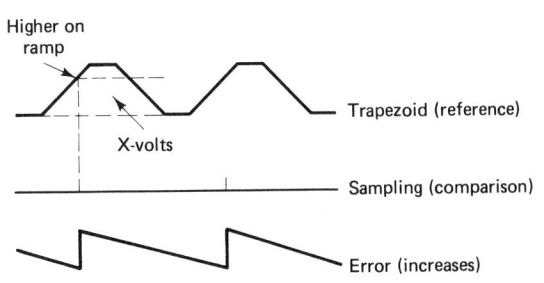

**FIGURE 4-30.** Relationship of pulses when capstan speed decreases.

*199*

As in the case of cylinder phase control, the error voltage developed by the capstan phase control system is limited to that which can be detected on the slope of the ramp and is used only as a vernier speed control (or phasing control) voltage. The actual speed of the capstan motor is controlled by the speed control circuit as discussed in Sec. 4–5.3.

### 4–5.3 Speed Control Circuit

Figure 4–31 is the overall block diagram of the cylinder speed control circuit. Figure 4–32 shows the related waveforms. A similar circuit is used for capstan speed control.

As shown in Fig. 4–31, the 120-Hz signal output from the cylinder FG is applied to the FG amplifier through pin 6 of IC502 and C543. The FG amplifier has a differential input at terminals 5 and 6 and a feedback output at pin 8. A reference voltage applied to pin 5 is adjusted by R506. The feedback signal is applied through R593 and C549. The output of the FG amplifier is applied to a delay circuit, and to a S/H circuit, through a doubler and former circuit. The signal is shaped by the former and converted to double frequency by the doubler. The delay circuit produces an approximate 20$\mu$s delay.

The output from the delay circuit is used as the trigger for the sawtooth generator, which produces a sawtooth wave. The slope of the sawtooth waveform is determined by the time constant of the "C" and "R" connected to pins 11 and 9, respectively. Note that the resistance is made adjustable so that the speed can be set. As the sawtooth wave slope becomes sharper, speed increases.

The output of the sawtooth generator is applied to the S/H circuit, which also receives a delayed sampling pulse from the delay circuit. Capacitor C536 holds the voltage between samplings. The two inputs to the S/H circuit produce an error voltage output at pin 13 (waveform G in Fig. 4–32). The error voltage is applied to the cylinder motor drive (Sec. 4–5.4) through an adder, comparator, amplifier, and motor switch.

The sampling action is shown in Figs. 4–33, 4–34, and 4–35. As shown in Fig. 4–33, the sawtooth ramp is sampled at a fixed level determined by the delayed sample pulse amplitude. Since the lag is constant, when motor speed increases, the sawtooth ramp is steeper, the sampling position moves lower on the ramp, and the error voltage is reduced, as shown in Fig. 4–34. This reduces motor speed. When motor speed is reduced, the sawtooth ramp is less steep, the sampling position moves higher on the ramp, and the error voltage is increased, as shown in Fig. 4–35. This increases motor speed.

The error signal at pin 13 is passed through a low-pass filter composed of R560 and C535 (to remove the sampling frequency component) and applied to the adder through pin 14. The adder also receives an error signal from the phase control circuit (Sec. 4–5.1). The two error signals are added and applied to the motor drive at pin 3 as the speed control signal (Sec. 4–5.4).

The comparator compares the sum of the speed control, phase control, and

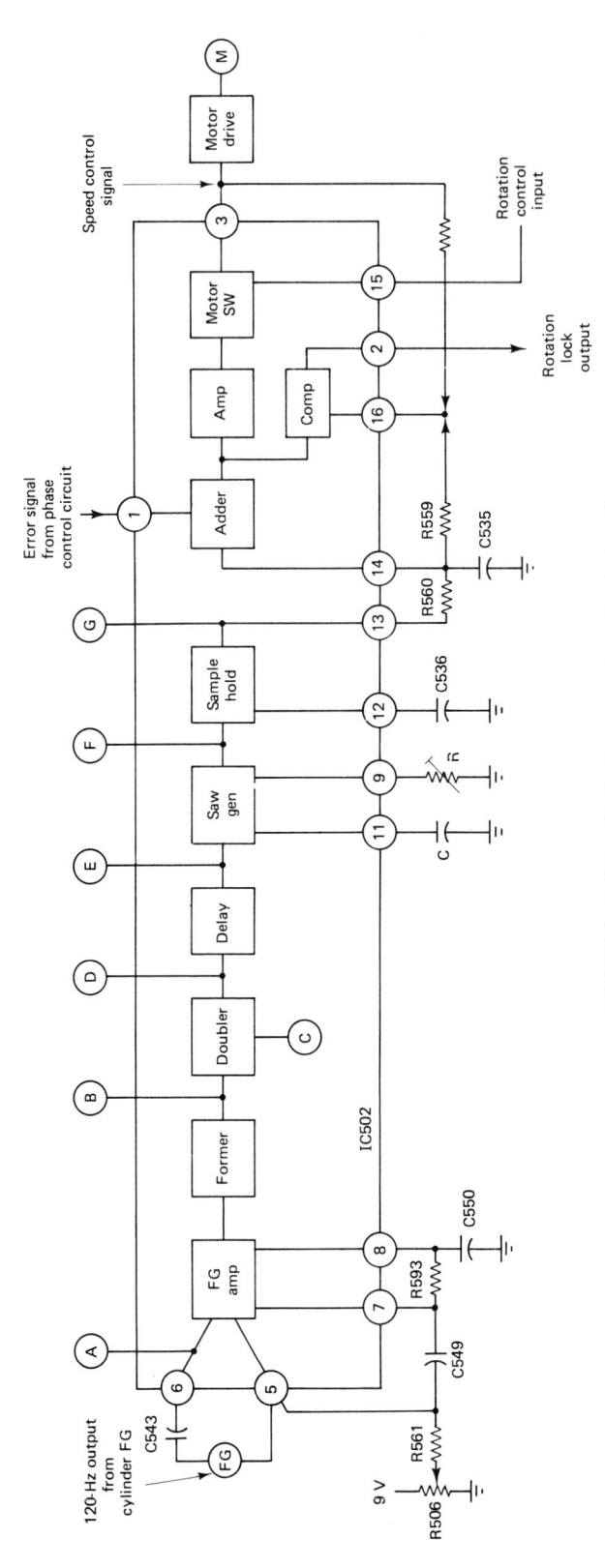

**FIGURE 4-31.** Cylinder speed control circuit.

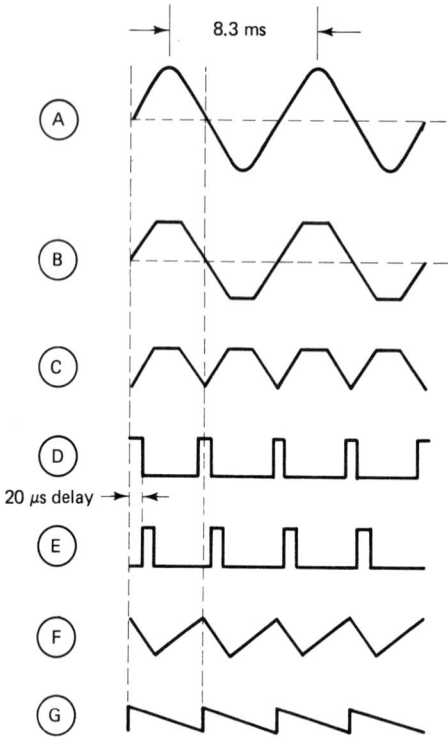

**FIGURE 4-32.** Waveforms associated with cylinder speed control circuit.

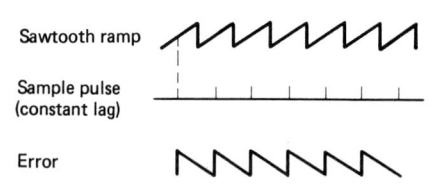

**FIGURE 4-33.** Sampling action when cylinder speed is normal.

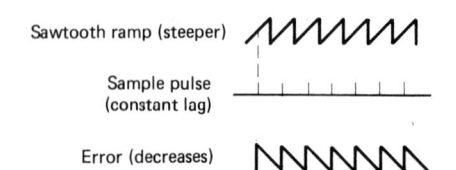

**FIGURE 4-34.** Sampling action when cylinder speed increases.

202

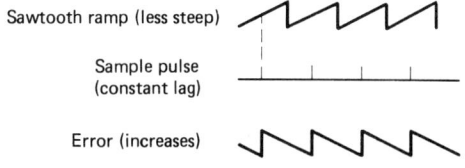

**FIGURE 4-35.** Sampling action when cylinder speed decreases.

reference voltage outputs. If the motor slows down for any reason (to a point where the servo cannot control the speed properly), pin 2 goes high and the motor stops (as discussed in Sec. 4-6). Pin 2 is normally low.

### 4-5.4 Motor Drive Circuit

Both the cylinder motor and capstan motor are three-phase motors. Figure 4-36 shows the positions of the rotor, stator coils, and Hall element devices of the capstan motor, while Fig. 4-37 shows the same elements for the cylinder motor. (Hall elements operate on the Hall effect principle, and produce a current that is controlled by the magnetic field surrounding the element. When the magnetic field is alternating, the output current from the Hall element alternates.)

Figure 4-38 shows the basic motor drive circuit. Figure 4-39 shows the

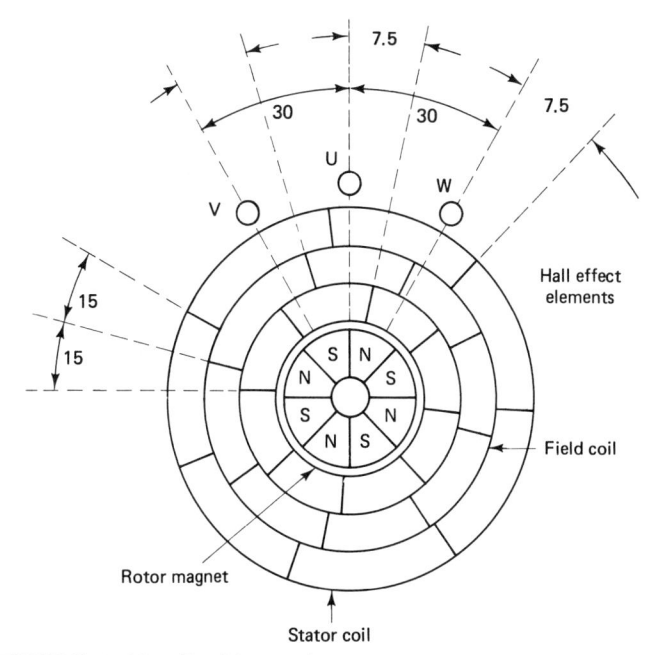

**FIGURE 4-36.** Positions of rotor, stator coils, and Hall element devices of capstan motor.

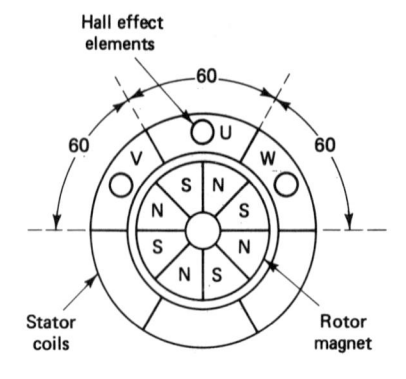

**FIGURE 4-37.** Positions of rotor, stator coils, and Hall element devices of cylinder motor.

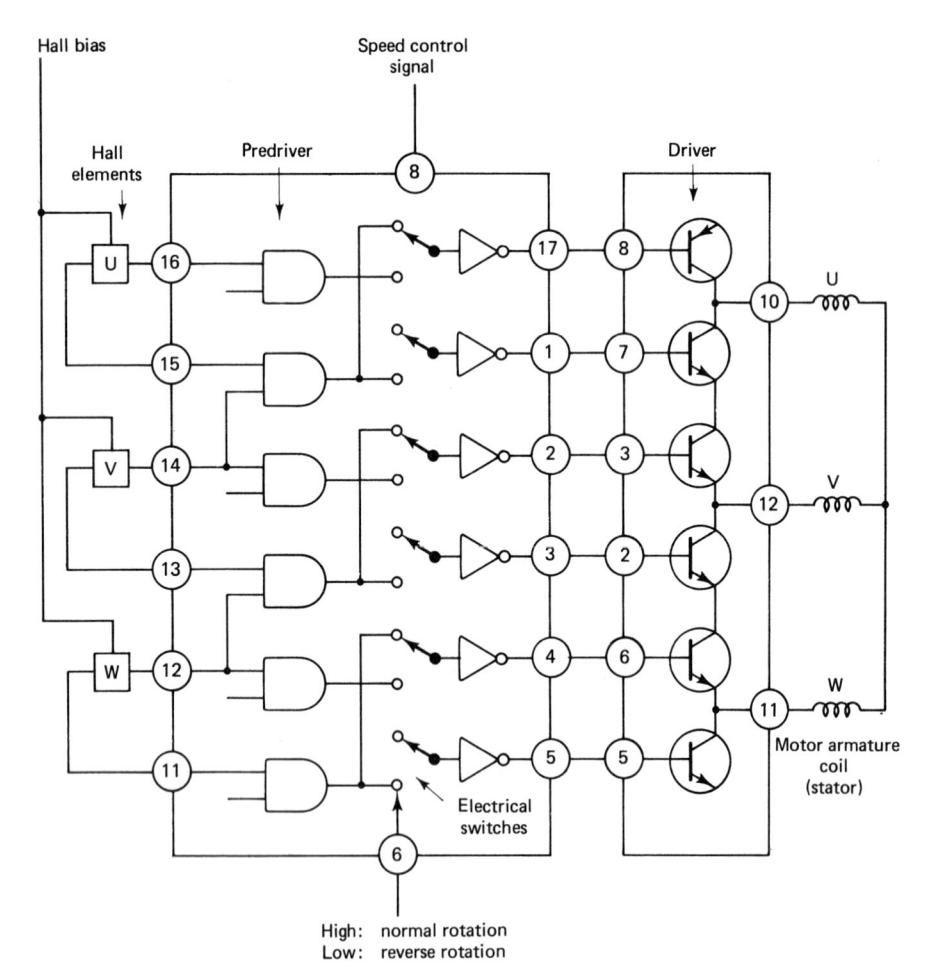

**FIGURE 4-38.** Basic motor drive circuit.

204

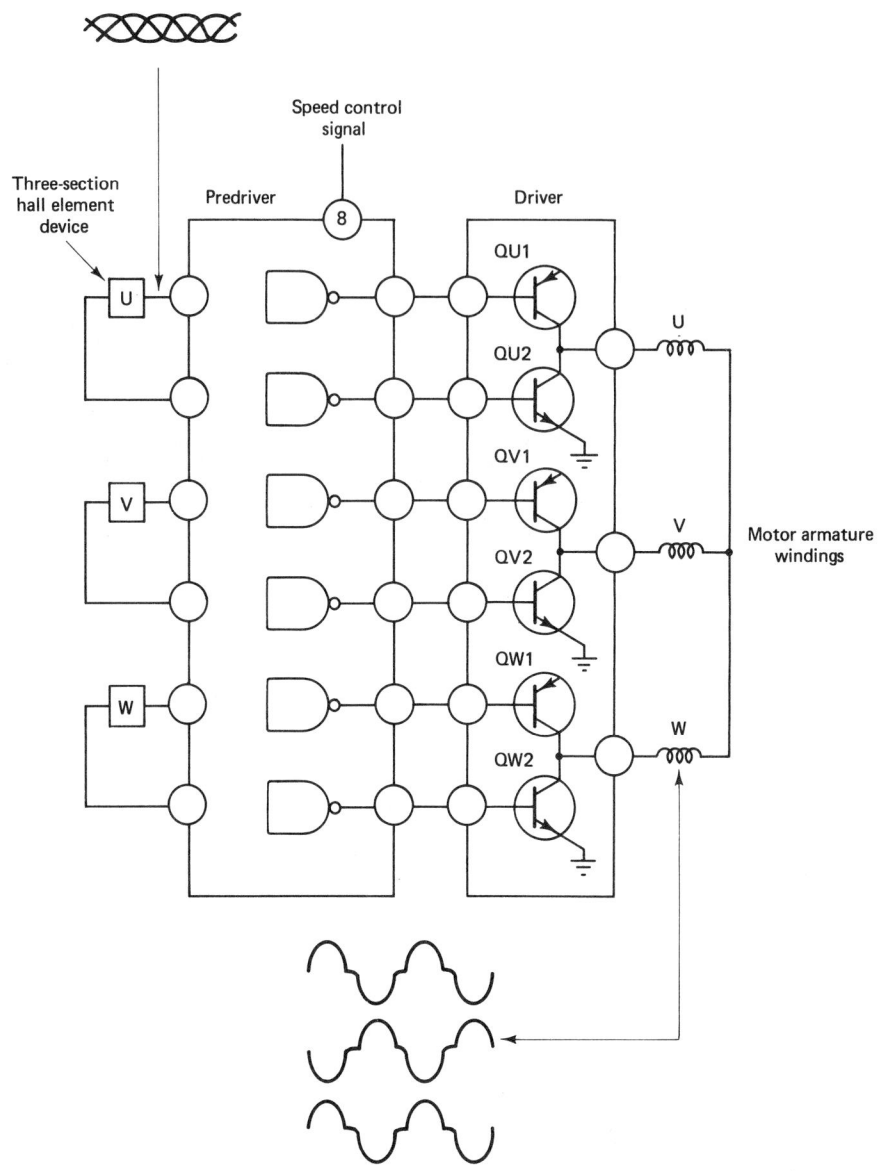

**FIGURE 4-39.** Cylinder motor drive circuit and related waveforms.

drive circuit for the cylinder motor and related waveforms. As shown, the motor drive system is composed of a Hall element device, a motor predriver, and a driver. When the motor turns, the polarity of the magnetic field applied to the Hall element alternates (because of rotation by the rotor magnet). This produces a three-phase sine wave with phases U, V, and W (separated in phase by $\frac{2}{3}\pi$) obtained from the Hall element device. The three-phase sine wave is selected and amplified by the predriver, and the resultant current is applied to the motor armature by the driver.

The drive systems of the capstan motor and cylinder motor are essentially the same. However, the cylinder motor does not have a motor reversal provision (at pin 6 as shown in Fig. 4–38).

As shown in Fig. 4–39, the cylinder motor has a three-section Hall element device. The signal obtained from this device is actually a six-phase sine wave. The predriver selects three pairs of Hall element outputs and converts them to three pairs of controlled outputs which are applied to transistors within the driver. Transistors QU1, QV1, and QW1 are active-low and control the current applied to the corresponding armature windings of the motor. Transistors QU2, QV2, and QW2 are active-high, and ground the corresponding windings.

Motor speed depends on the current supplied to the windings. In turn, the amount of current depends on the amplitude of the speed control signal (Sec. 4-5.3) applied to pin 8. The direction of rotation for the cylinder motor is always the same, and only motor speed is subject to control. In the case of the capstan motor, both speed and direction of rotation are controlled. Pin 6 of the predriver is normally high. However, when the capstan must be reversed, pin 6 goes low and the electrical switches within the predriver go to the low position. This reverses the polarity of all three phases of the motor drive signals.

## 4-6 VHS SYSTEM CONTROL CIRCUIT

Operation of the system control circuit is determined primarily by microprocessors. The use of such microprocessors for system control is typical for many present-day VCRs. The microprocessors accept logic control signals from the VCR operating controls (called operation keys) and from various tape sensors. In turn, the microprocessor sends control signals to the various circuits, as well as drive signals to solenoids and motors.

We will not go into operation of microprocessors here since such information is beyond the scope of this book (and each VCR has its own particular microprocessor applications). However, we do discuss the circuit inputs and outputs to and from the microprocessor, since such circuits are typical for many VCRs. If you want a thorough discussion of microprocessors, your attention is invited to the author's best-selling *Handbook of Microprocessors, Microcomputers, and Minicomputers* (Englewood Cliffs, N.J.: Prentice-Hall, Inc., 1979).

Note that many system control functions are closely related to mechanical operation of the VCR, as described in Chapter 5. Therefore, it is essential that you study the related sections of Chapter 5 when reviewing system control operation.

## 4-6.1 Overall System Control Circuit Operations

Figure 4-40 is an overall block diagram of the system control circuits. As shown, system control includes microprocessor IC901, which reads in and decodes data from interfaces (from both sensors and operation keys), and microprocessor IC902, which receives decoded sensor and operation key information from IC901. Microprocessor IC901 sends control signals to the main brake and take-

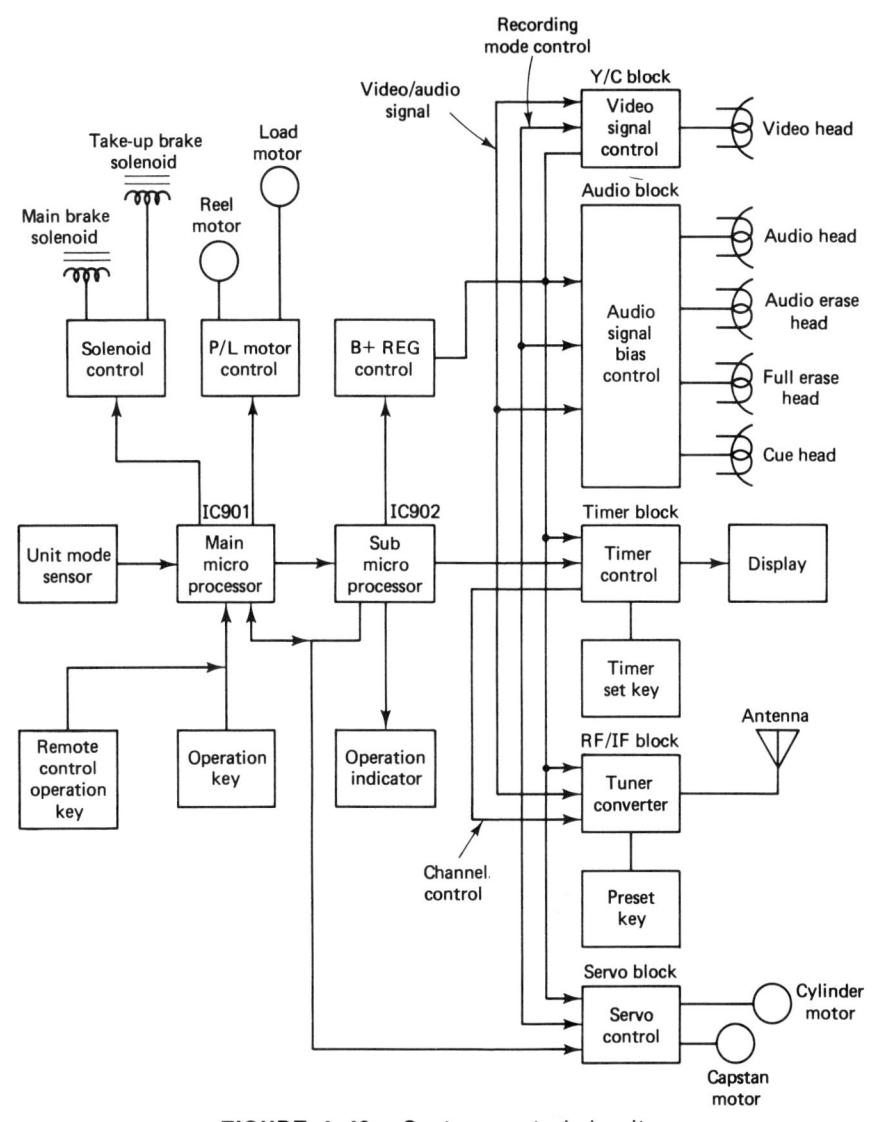

**FIGURE 4-40.** System control circuits.

## 208 Typical VHS VCR Circuits

up brake solenoids, as well as the reel and load motors. All the remaining control signal outputs come from IC902, which also provides corresponding outputs to the operating indicators (PLAY indicator lamp, REC indicator lamp, etc.).

### 4-6.2 Operating Key Input

Figure 4-41 shows the basic operation key input circuits. To reduce the number of operation key input lines to IC901, the key operation for each of the 14 modes is converted into a 14-step voltage (designated as V1). This voltage V1 is compared with a voltage V2 produced by a 4-bit D/A (digital-to-analog) converter. IC901 pins 36 to 39 serve as the D/A input. The emitter resistance of Q2 is

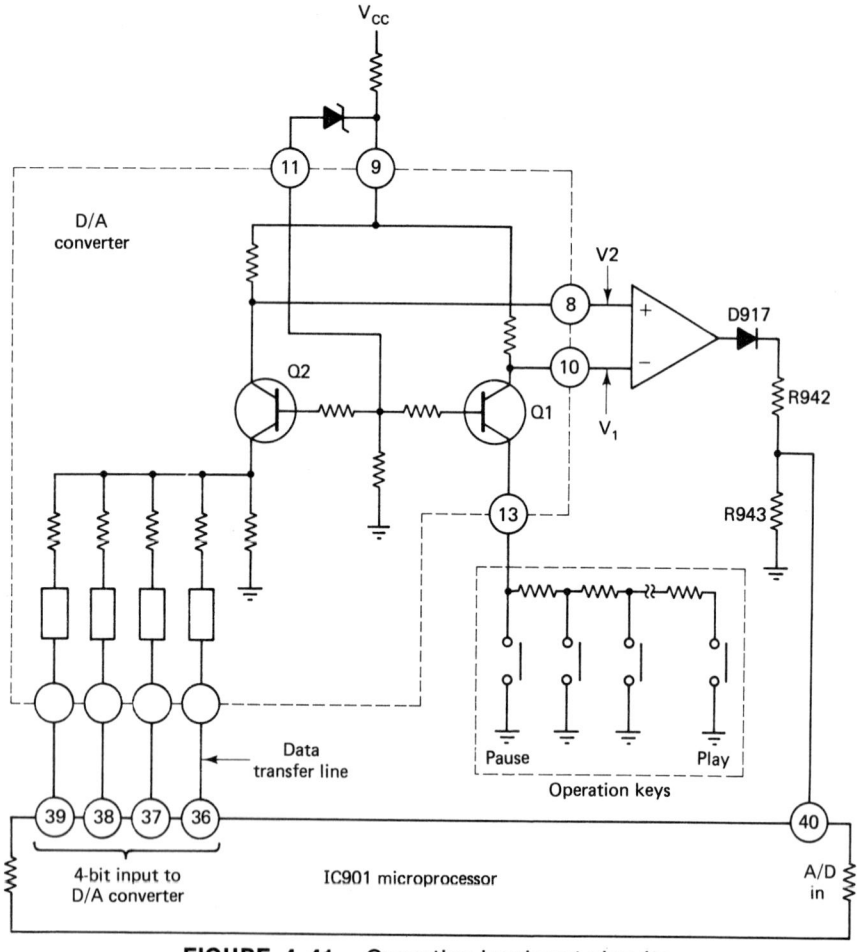

**FIGURE 4-41.** Operation key input circuits.

changed in 16 steps by means of this input from IC901, thus generating a 16-step voltage V2. The 14-step voltage V1, which corresponds to the selected operation key, is generated by varying the emitter resistance of Q1.

Pin 40 is the input to IC901 for comparison of V1 and V2. Each time the D/A converter counts one step, IC901 checks the comparison at pin 40 to determine the point where V2 is greater than V1 (when the 16-step voltage has reached the selected operation key voltage). When this occurs, IC901 sends the necessary control and indicator signals to IC902. Note that pins 36 to 39 of IC901 form the data transfer line to IC902, as discussed in Sec. 4-6.3.

If you are not familiar with operation of D/A converters, or A/D converters, your attention is invited to the author's best-selling *Handbook of Digital Electronics* (Englewood Cliffs, N.J.: Prentice-Hall, Inc., 1981).

### 4-6.3  Data Transfer and Read-In

Figure 4-42 shows the relationship between microprocessors IC901 and IC902. The outputs from IC902 are applied to the mode and indicator control circuits. The control signals from IC901 to IC902 are passed (using time sharing) through the 4-bit transfer line, and a 1-bit timing control line (pin 1 of IC901 and pin 8 of IC902). This 5-bit system allows the number of output lines to be reduced.

The 1-bit timing control pulse (pin 1 of IC901) is known as the TX pulse and is used to synchronize operating modes of both ICs. Data bits are stored in IC902

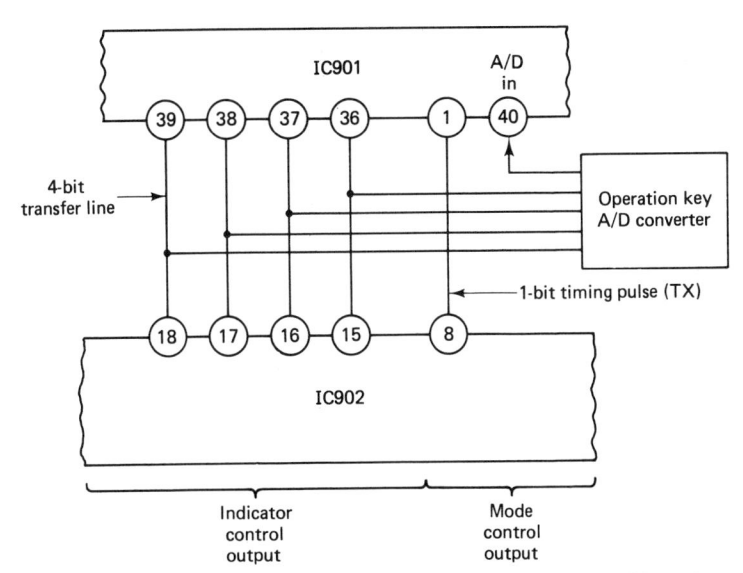

**FIGURE 4-42.** Relationship between IC901 and IC902 in system control.

210  Typical VHS VCR Circuits

until the next data byte is transferred. Data transfer is performed periodically to prevent the information from being altered by noise, and so on. The data bits are renewed repeatedly so that information from IC901 coincides with that of IC902.

### 4-6.4  Operation Mode Control

As shown in Fig. 4-40, when the operation keys are pressed, signals to indicate the circuit composition and mode (operation indicator) are produced, simultaneously with the mechanism drive outputs for loading, unloading, and so on. The outputs (both indicator and control) continue until a different operation key is pushed. The following is a summary of the 14 operating modes.

In the STOP mode, unloading is completed; the cylinder motor, capstan motor, and reel motor stop; the main brake is applied; and the STOP indicator is lit by the output of IC902 pin 1. When entering the STOP mode from the REC (record) and PLAY modes which require unloading, unloading continues until all the stop functions are performed, but the STOP indicator lights before the start of unloading. IC902 is reset when the POWER switch is turned off and on. However, when the POWER switch is not yet turned on after IC901 is reset, the STOP indicator does not light, but the mechanism condition is the same as in the STOP mode. When the power plug is pulled out, or a power failure occurs during a certain operation, IC901 is reset and the mechanism enters the same mode as the STOP mode (after unloading is completed).

When the PAUSE operation is performed in the STOP mode, the PAUSE indicator is lit by the output of IC902 pin 2. (PAUSE is called FREEZE or FREEZE FRAME in some VCRs.) However, the mechanism keeps in the STOP mode. When going to PAUSE from PLAY, REC, or DUB operation, the PAUSE is latched by IC902 and the VCR enters the corresponding PAUSE mode (a freeze frame picture appears on the TV).

When the PLAY operation is performed, the PLAY indicator is lit by the output of IC902 pin 7, a PB 9V is generated by the output of pin 19, and the loading operation is performed. The PB 9V is applied to audio/video/servo circuits as necessary to perform the playback operation.

The following operations are possible after the PLAY operation has been selected:

With the PAUSE button pressed during playback, the PAUSE indicator is lit and the VCR is in the STILL mode.

With the VCR in the STILL mode, the frame advances each time the FRAME ADVANCE button is pressed.

When the SLOW button is pressed during playback, the capstan servo system is controlled to play at one-half speed.

When the QUICK button is pressed, the capstan servo system is controlled to play at three times normal speed.

When the VIDEO SCANNING ►► mode button is pressed during

playback, the capstan servo system is controlled to play at 10 times the EP (extended play) speed. (This mode is called SEARCH or VIDEO SEARCH in some VCRs.)

When the VIDEO SCANNING ◄◄ mode button is pressed during playback, the capstan servo system is controlled to play in the reverse direction at 10 times the EP speed. Reverse play is performed by turning the capstan motor in reverse (Sec. 4-5.4). The tape is taken up by driving the tape supply reel disk with the reel motor. If the VCR is kept in the reverse video scanning mode for more than about 5 minutes, the operation changes over to normal playback. This is to protect the reel motor. Operation of the reel motors is discussed further in Sec. 4-6.7.

The following modes require that the PLAY button be pressed (either simultaneously with, or after, the desired operating mode button is pressed):

When the RECORD button is pressed and the cassette safety tab is not broken, a REC 9V is generated by the output of pin 20, and the audio/video/servo circuits are changed over to the record mode. If the cassette safety tab is broken (Sec. 1-10.2), the record mode is disabled.

When the PAUSE button is pressed in the RECORD mode, the capstan motor turns slightly in reverse and rewinds tape to improve phase matching. Then the loading motor turns in reverse to release the pressure roller from the capstan shaft. The capstan motor continues turning. The PAUSE indicator is also lit. Figure 4-43 shows this sequence.

When the DUB button is pressed and the cassette safety tab is not broken, a PB 9V is generated, the video/servo systems enter the PLAY mode, and the audio system enters the RECORD mode. Under these conditions, audio (from an external microphone or other audio source) is recorded on the audio track of the tape.

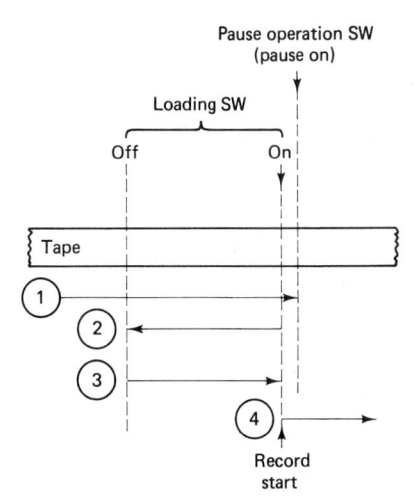

**FIGURE 4-43.** Sequence when the PAUSE button is pressed in the RECORD mode.

*212  Typical VHS VCR Circuits*

When the PAUSE button is pressed after the DUB button is pressed, the VCR enters the DUB PAUSE mode. The condition of the mechanism is the same as in the STILL mode. The PLAY/DUB/PAUSE indicators are lit.

The following modes do not require that the PLAY button be pressed:

When the F FWD (fast forward) button is pressed, the loading motor rotates normally and a supply brake pressure is applied to the supply reel (to provide back-tension). The reel motor then turns, compressing the FF/REW idler against the take-up reel disk, and the tape is taken up. The F FWD indicator is lit.

When the REWIND button is pressed, the reel motor drives the supply reel to take up the tape.

## 4-6.5  Stop Control

The VCR is placed in the STOP mode when the STOP button is pressed, when the tape runs to either end (forward and reverse), and when there is mechanical trouble. Figure 4-44 shows the circuits involved. The following summarizes operation of the stop control circuits.

**End Sensor Circuit.**  Both ends of VHS tape are clear (transparent) plastic. The cassette tape passes between an end sensor lamp (also known as a cassette lamp in some VCRs), and two end sensor photo transistors Q81 and Q82. The magnetic portion of the tape prevents the cassette light from reaching the photo transistors. When the tape reaches either end (supply or take-up), the cassette light passes through the transparent portion of the tape onto one end of the end sensor photo transistors. When either Q81 or Q82 receive light, they produce a signal (of about 5 V) to pin 24 (supply) or 25 (take-up) of IC901. This stops and unloads the VCR.

**End Sensor Lamp Failure Detector.**  Should the end sensor lamp burn out or fail for any reason, the VCR is placed in the STOP mode. If this feature were not included, the tape could be broken when run to either end. The end sensor lamp is connected in parallel across ZD904 and R928. If the lamp burns out, the cathode of DZ904 increases. This increase is applied to pin 33 of IC901 through the OR gate and places the VCR in STOP.

**Dew Sensor Circuit.**  The dew sensor output is also applied to pin 33 of IC901 through the OR gate. When humidity is less than about 80%, the resistance of the dew sensor is about 100 MΩ. When humidity increases over 80%, the resistance is less than 3 MΩ and the voltage at the junction of dew sensor R82 and R935 increases. This increase is applied to pin 33 of IC901 and places the VCR in STOP. The STOP indicator flashes to indicate that condensation is present.

**Reel Lock Circuit.**  This circuit detects when the reel motor has stopped rotating, except during modes when the tape should not be running at the normal speed (UNLOADING, LOADING, PAUSE, STOP, SLOW PLAYBACK). NAND gate Q905 is active when the reel disk is rotating, or when operating mode

VHS System Control Circuit 213

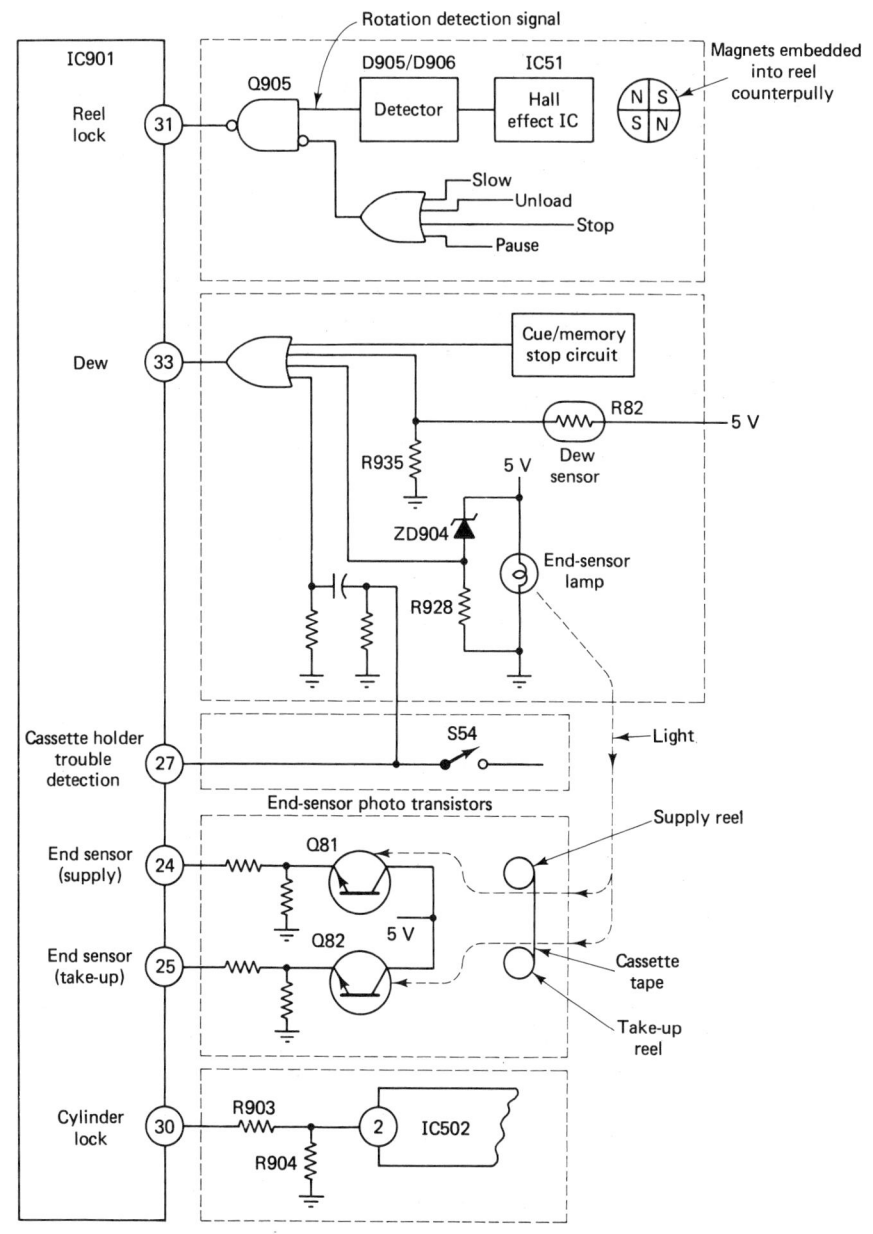

**FIGURE 4-44.** Stop control circuit.

signals are applied to the OR gate. When reel rotation stops, Q905 is cut off, and IC901 produces unloading and stop after about 8 to 10 s. This is prevented by override mode signals applied to the OR gate. These mode signals take the place of the rotation detection signal.

The rotation detection signal is developed by detectors D905/D906 (which act as a voltage doubler and rectifier), a Hall effect element IC51, and magnets embedded into the reel counterpully. When the reel is rotating, the magnetic field also rotates and causes the Hall IC to produce an alternating current. This alternating current is rectified and doubled by D905/D906 to become the rotation detection signal. If rotation stops, the alternating current stops, as does the detection signal.

**Cylinder Lock Circuit.** This circuit detects when the cylinder motor stops rotating, or when rotation slows down to a point where the servo can no longer provide proper control. As discussed in Sec. 4–5.3, pin 2 of the cylinder motor speed control IC502 is normally low. When cylinder motor speed drops below a certain point, pin 2 goes high. This high is applied to pin 30 of IC901, through R903/R904, and IC901 instructs unloading and stop.

**Cassette Holder Trouble Detection.** This circuit detects if the cassette holder is in the eject condition. Switch S54 is operated by the holder. If the eject button has been pushed, the VCR is placed in the STOP mode by switch S54.

## 4–6.6 Cue/Memory Counter Circuit

Figure 4–45 shows the cue/memory counter circuits in simplified form. The VCR has a tape counter known as the memory counter. This counter displays a number that increases as the tape moves forward and decreases as the tape moves backward. If you make a note of the number corresponding to a particular point on the tape, you can use the counter to help you locate that place. The VCR also has a cueing circuit which records a signal on tape at the start of a program (to detect the start of the program during playback).

**Memory Counter Circuit.** The counter has a memory capability. To use the memory, you stop the tape at the point you want to return to later, and set the counter reading to "000." To detect that position after running the tape, set switch S2001 to M (memory) and perform rewind. When the VCR rewinds to "000" (actually "999," or one count before "000" in order to avoid missing the starting point), the VCR goes into STOP. This is done when the counter switch closes, applying a 9-V signal to pin 33 of IC901 through Q2003 and D2001. Pin 33 is the same stop control input to IC901 used by the dew sensor, lamp failure detector, and so on, described in Sec. 4–6.5.

**Cueing Circuit.** The cueing circuit involves both record and playback functions. The cueing signal recorded on tape is obtained from the SW30 signal. This 30-Hz signal is superimposed on the full-erase head bias signal through

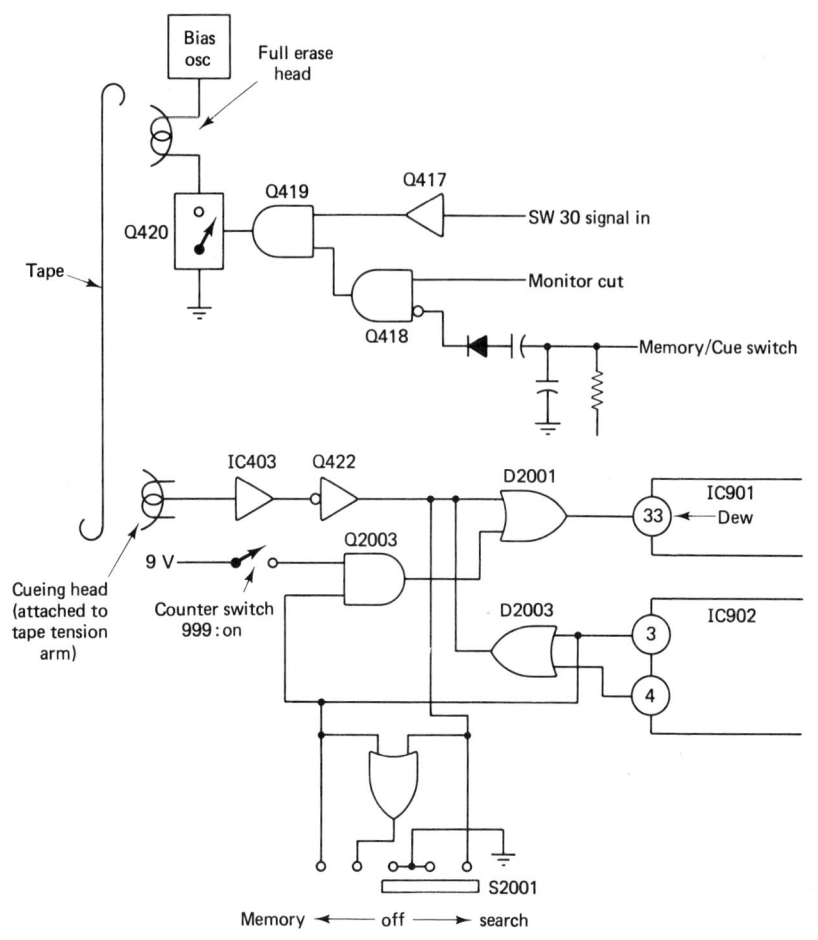

**FIGURE 4-45.** Cue/memory counter circuit.

Q417, Q419, and Q420. The recording period of the cueing signal is regulated by the time constant circuit connected to the base of Q418 during the trailing edge of the monitor signal. This monitor signal occurs (each time the RECORD button is pressed) during the loading period (at the time the tape is loading for a recording). The SW30 signal is transmitted to the full-erase head at that time, provided that the memory/cue switch is set to cue.

The cueing signal is picked up by the cueing head attached to the tape tension arm, when the memory/cue switch is set to cue, and the servo is operated in the fast forward or rewind modes. Under these conditions, the trailing edge of the recorded SW30 signal is detected by the cueing head is applied to pin 33 of IC901, causing the VCR to go into STOP mode. If several programs have been recorded on tape, with the RECORD button pressed at the beginning of each program, the VCR will stop at each program (during fast forward or rewind).

### 4-6.7 Loading Motor/Reel Motor Control

Figure 4-46 shows the loading and reel motor control circuits. As shown, the loading and reel motors are controlled by signals from microprocessor IC901 using negative logic (a low output from the IC produces the necessary control action). The microprocessor provides six specific instructions. The reel motor receives a fast-forward control (pin 7, IC901), a rewind control (pin 8), and slow-speed forward control (pin 9). The loading motor receives a loading control (pin 5), an unloading control (pin 6), and a brake control (pin 10).

A tristate signal is applied to the terminals of both the loading and reel motors as necessary. This tristate signal is (1) about 10 V, (2) 0 V or ground, and (3) high-impedance or open. Switches S1 through S6 control these signals. The switches are controlled by logic circuits within IC905 and IC906, which, in turn, are controlled by the six instructions from IC901. When no instruction is produced by IC901, both terminals are set to the high-impedance or open condition. During braking, both terminals are set to 0 V or ground, thus short circuiting the motor. During other instructions, one terminal is connected to 10 V with the opposite terminal connected to ground, causing the motor to rotate in the desired

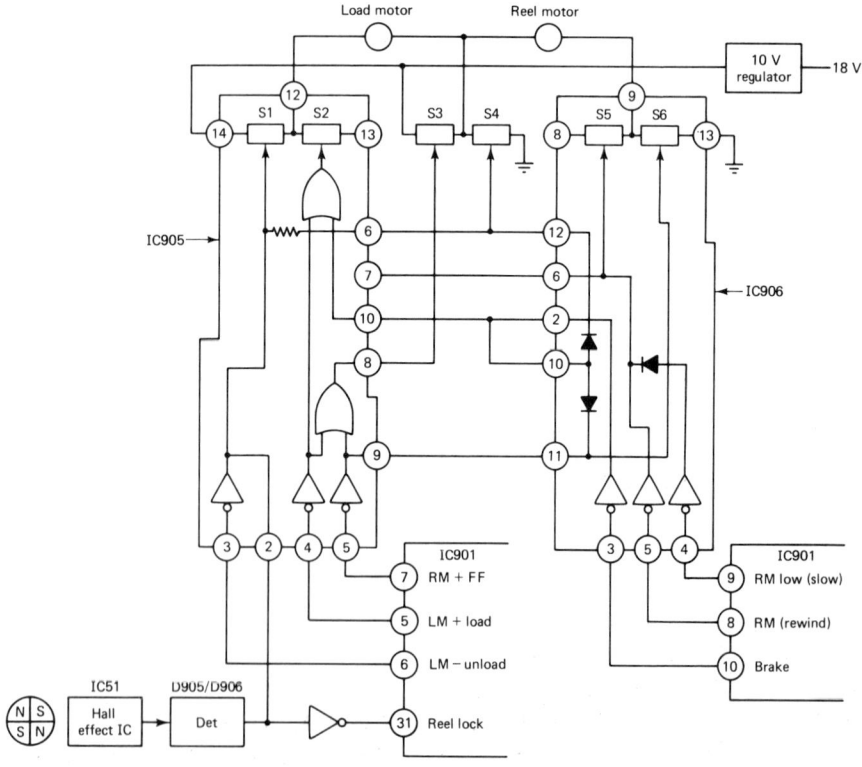

**FIGURE 4-46.** Loading and reel motor control circuits.

direction. When the slow-speed forward control is applied to the reel motor, one terminal is connected to 5 V with the opposite terminal connected to ground. This causes the reel motor to rotate in the forward direction, but at a reduced speed.

### 4-6.8  Main Brake Solenoid Control

Figure 4-47 shows the main brake solenoid control circuit and associated waveforms. The main brake solenoid applies braking to both reels in the STOP mode or in a mode transition period. This prevents tape slack. The main brake solenoid releases braking when conducting and its shaft is actuated. This conducting state is latched electrically. IC901 produces 50-ms pulses at pins 12 and 13. These pulses trigger the latching action.

For example, when the brake is to be turned off, the 50-ms pulse at pin 12 is inverted by Q53 and causes Q54 to close for 50 ms. This connects the control side of the main brake solenoid to ground, full current is applied, the brake shaft is actuated, and the brake is released. The brake is held in this condition by the latching action. When Q54 closes, the input to Q51 is connected to ground through R55. This produces a 50-ms low which is inverted to a high by Q51. The high at the output of Q51 closes Q52 and connects the control side of the braking solenoid to ground through R55. This provides sufficient current to hold the solenoid in the brake release condition.

When the brake is to be applied, the negative 50-ms pulse at pin 13 is applied to Q52, causing Q52 to open and disconnecting the brake solenoid from the ground.

### 4-6.9  Take-Up Brake Solenoid Control

Figure 4-48 shows the take-up brake solenoid control circuit and associated waveforms. Braking is applied to the take-up reel to prevent the tape from being pulled out of the take-up reel during loading and unloading (to provide some back tension on the tape). Braking is also applied during reverse scanning (or reverse visual search) to provide back tension on the tape.

The take-up brake solenoid is controlled by negative pulses from pin 11 of IC901 (except during reverse scanning when the pulses are taken from pins 3 and 4). The take-up brake instruction from pin 11 is inverted by Q907 and applied through D927/D904, and a differentiating network consisting of C51/R64, to Q56. The signal from D927/D924 is also applied directly to Q55. The differentiated signal causes Q56 to close, full current is applied to the solenoid, and the brake is applied momentarily. The brake is held in this condition temporarily by latching action. When Q55 closes, the control side of the braking solenoid is connected to ground through R65. This provides sufficient current to hold the solenoid in the braking condition. During all modes except reverse scanning, the brake is applied long enough to complete the loading and unloading operation.

*218  Typical VHS VCR Circuits*

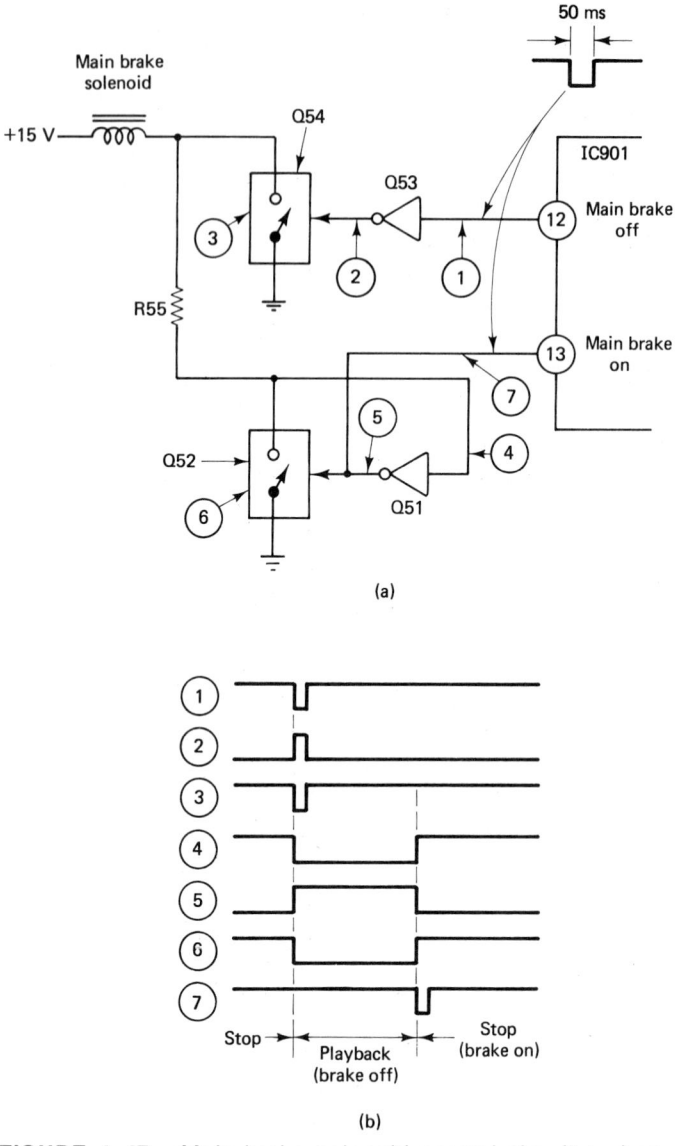

**FIGURE 4-47.** Main brake solenoid control circuit and associated waveforms.

During reverse scanning, continuous low signals appear at pins 3 and 4, producing a high at the output of Q903. This high causes Q55 and Q56 to close, full current is applied, and the brake is applied. Q56 opens once the differentiated pulse passes, but Q55 remains closed since Q55 is connected directly to the output of D927/D904. With Q55 closed, the solenoid is connected to ground

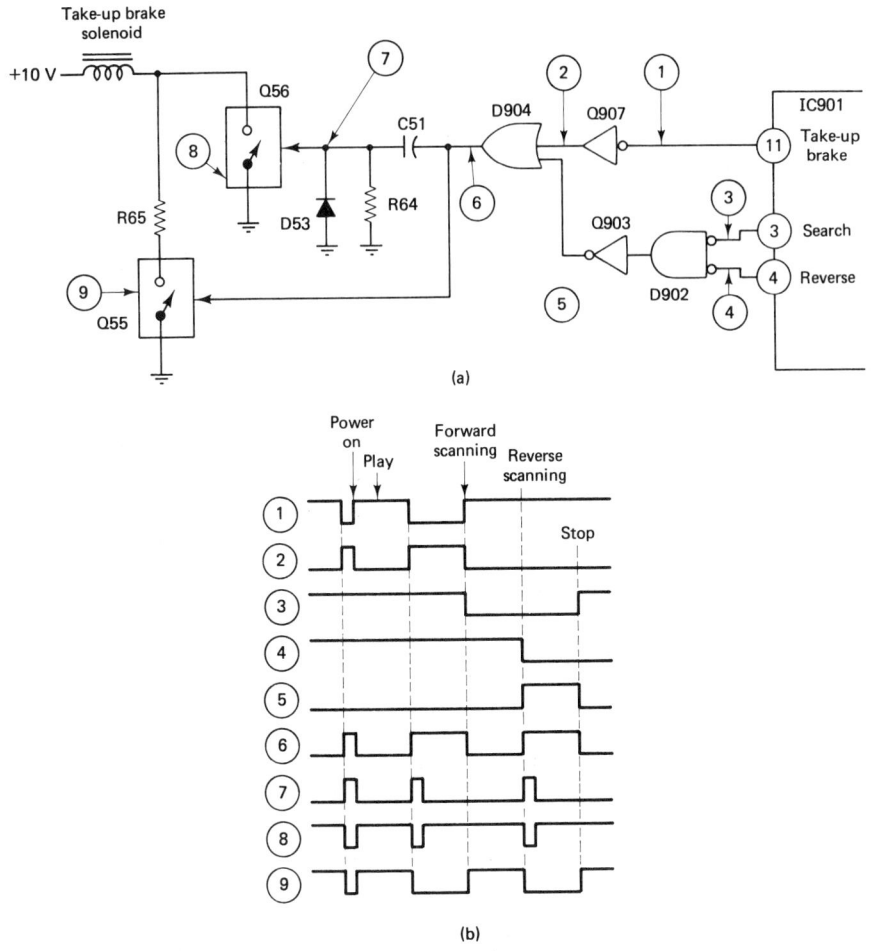

**FIGURE 4-48.** Take-up brake solenoid control circuit and associated waveforms.

through R65. This provides sufficient current to hold the solenoid in the braking condition, as long as the signals at pins 3 and 4 of IC901 are applied during reverse scanning.

### 4-6.10 Tape Protection Control

Figure 4-49 shows the tape protection control circuit and associated waveforms. The purpose of this circuit is to prevent the tape from breaking when a reverse instruction is given from a fast-forward condition. This can be accomplished if the unloading instruction occurs simultaneously with the reverse instruction. When the stop instruction is given, braking is applied to the take-up reel first and, after

**220** Typical VHS VCR Circuits

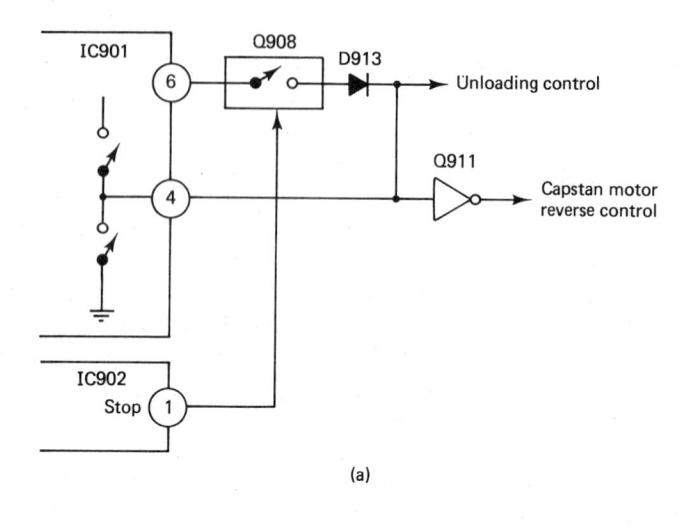

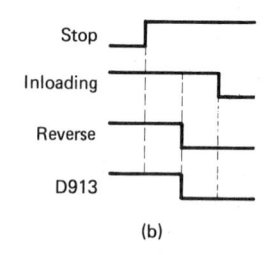

**FIGURE 4-49.** Tape protection control circuit and associated waveforms.

about 50 ms, a reverse rotation instruction is applied to the capstan motor. Next, a reverse rotation instruction is given to the supply reel, and the loading motor is reverse rotated to remove slack. This moves the guide base to the stop position, and feeds tape into the cassette.

The supply reel rotates at slow speed before the loading motor starts rotation and, after that, at fast speed in the reverse direction. The capstan motor is first braked and stopped by the reverse instruction, and then the capstan motor begins to rotate in the reverse direction. However, in the forward scanning mode, the capstan motor rotates at 10 times normal speed (10 times EP), so reversal is delayed due to the flywheel effect of the motor when the reverse instruction is first given. Under these conditions, it is possible that when the supply reel starts to rotate in the reverse direction, the capstan motor could still be rotating in the forward direction. In such a case, the tape could be pulled in both directions simultaneously, resulting in tape breakage.

To protect the tape from such breakage, unloading is performed to release the compression of the capstan and pressure roller (pinch roller), simultaneously when a reverse instruction is given. This is accomplished by the circuit of Fig. 4-49. When the stop instruction at pin 1 of IC902 is applied to Q908, the unloading control signal at pin 6 of IC901 is applied to pin 4 of IC901 and to the capstan motor reverse control circuit simultaneously. Therefore, unloading and capstan reverse control are performed simultaneously.

# 5

# Mechanical Operation of Typical VCRs

This chapter describes operation for the mechanical sections of typical VCRs. Both Beta and VHS systems are covered, using the VCRs described in Chapters 3 and 4 as examples. Since operation of the mechanical sections (tape transport, safety devices, loading/unloading, etc.) are closely related to the system control functions, it is essential that you review the corresponding system control sections of Chapters 3 and 4 when studying mechanical operation as described in this chapter. Chapter 6 describes adjustment procedures for the mechanical section, as well as electrical adjustments.

By studying the mechanical operation found here, you should have no difficulty in understanding the mechanical operations of similar VCRs. This understanding is essential for logical troubleshooting and service, no matter what type of mechanical equipment is involved. For example, if you know that a particular solenoid is actuated to pull a certain rod in a given mode of operation, and you observe that the solenoid does not actuate in the given mode, you have pinpointed a failure. The origin of the trouble may be electronic (no actuating signal is present) or mechanical (the solenoid is jammed), but you have a starting point for troubleshooting. The descriptions given here should also help you interpret the mechanical sections of VCR service literature.

The writers of VCR service literature use several techniques to show mechanical operation. We use the two most popular methods in this chapter so as to familiarize you with what to expect from a VCR manual. One technique involves showing the physical location of all mechanical parts by means of photos or drawings, and then describing the operation of each part, for each mode of operation in step-by-step detail. We use this first technique to show Beta mechanical operation. However, to simplify the sometimes elaborate drawings found in many well-prepared VCR service manuals, we show only those parts being described. The other technique involves the use of timing charts, where the sequence of operation and movement of related parts are given by a combination of charts and rotation drawings. We use this second technique to show VHS mechanical operation. Most of the mechanical operation descriptions found in VCR service manuals are variations of these two techniques.

## 5-1 TYPICAL BETA MECHANICAL OPERATION

The following paragraphs describe mechanical operation of the Beta VCR covered in Chapter 3. Operation of the corresponding system control is discussed in Sec. 3-5.

### 5-1.1 Major Elements of the Mechanical Section

The mechanical section or mechanism of the VCR can be divided into two blocks: *drum block* and *chassis block*. The drum block is composed primarily of the tape travel system (drum with the rotating video heads, erase head, audio/control head, tape guide, capstan, pinch roller, loading ring, etc.). The tape travel or drum block system is adjusted at the time of manufacture and assembled into a complete functioning unit. The chassis block is composed of the supply and take-up reel base, brake, capstan motor, reel motor, solenoids, cassette holder, and related parts.

The drive system for the mechanical section is composed of three d-c motors: the direct-drive (DD) *head motor* for video head rotation (installed on the drum); the *capstan motor,* which drives the capstan for tape feed; and the *reel motor,* which drives the reel base for take-up and rewinding of the cassette tape. The capstan motor and reel motor use belts for drive transmission. Turning of the loading ring during loading and unloading is executed by the capstan motor via the capstan flywheel.

The mechanism also has the following solenoids: the *pinch roller solenoid* for pinch roller attraction; the *brake solenoid* for brake release, and the *roller solenoid* for roller attraction. Mechanism switching for the various operating modes is executed by combined operation of the three solenoids.

### 5-1.2 Tape Run System

Figure 5-1 shows mechanical parts of the tape run system. As shown, the tape run system consists essentially of the drum, the tape guides around the drum, and the tape guides installed on the loading ring. The tape guide (2) of the forward

## 224 Mechanical Operation of Typical VCRs

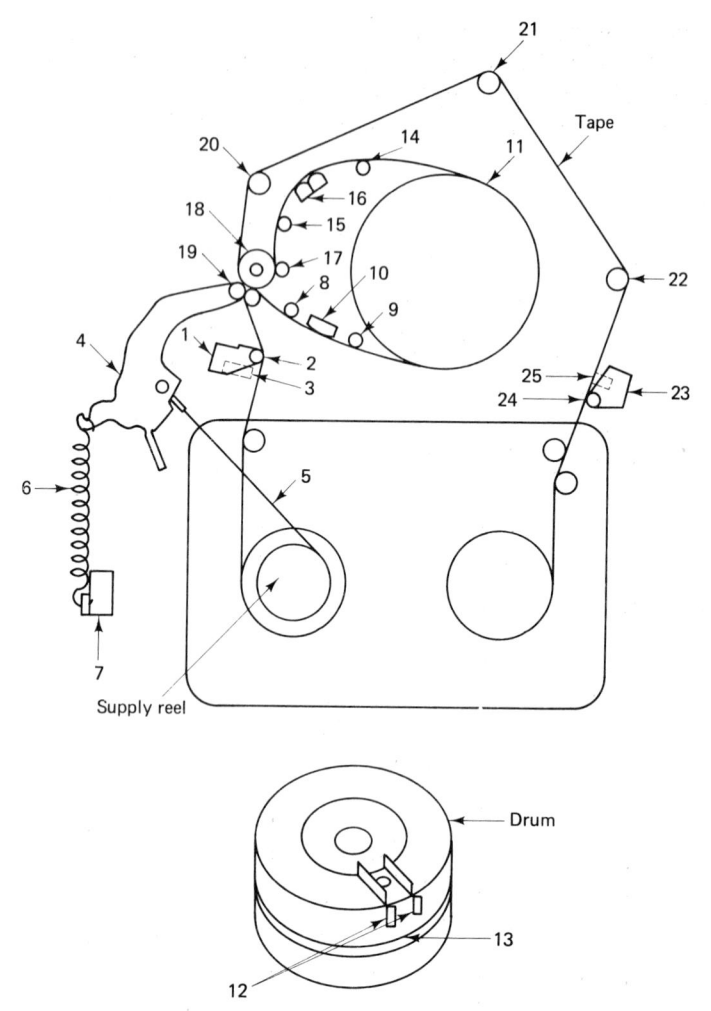

**FIGURE 5-1.** Tape run system.

sensor assembly (1) is arranged so that a constant distance is kept between the forward sensor assembly (3) and the tape.

The back-tension lever (4) guides the tape so that the tape runs in the direction toward the drum, and the tension of brake band (5) is adjusted so that a constant braking force acts on the supply reel base for constant back-tension on the tape. The back-tension is adjusted by the spring (6), and the installation fitting (7) of this spring (6) plays an important role in the stability of the tape run. The perpendicularity of the back-tension lever (4) has a large influence on the tape run from the cassette to the drum.

The erase head roller (8) absorbs lengthwise vibrations of the tape, caused between the back-tension lever (4) and the erase head (10), and thus removes jitter at the time of playback. The tape input guide (9) controls the upper edge of the tape, and guides the tape so that correct tape run is executed. In the record mode, the erase head (10) erases all signals from the passing tape.

The tape is wound around the drum (11) at an angle along the groove (13). The tape wraps the drum by 180° + alpha (where alpha is the amount of overlap). The two rotating video heads (arranged at 180° from each other) scan the tape and execute recording and playback of the video signals. The tape push lever (12), installed on the top of the drum, uses a spring to push the tape (at the approximate center of the drum) in a downward direction. As a result, the tape can run correctly along the groove (13) at the bottom of the drum. The tape outlet guide A (14) regulates the upper edge of the tape, while the tape outlet guide B(15) regulates the lower edge of the tape, so that the tape coming out from the drum (11) has a stable run.

The tape inlet guide (9) and the tape outlet guide A (14) regulate the angle (and amount of wrap) at which the tape is wound around the drum.

The audio/control head (16) executes recording, and playback, of the audio signal in the REC, PLAY, and AUD DUB modes. The control portion of head (16) executes recording and playback of the signal for control of video head rotation. For head (16) installation height and angle are adjusted at the factory. Azimuth adjustment and tracking position adjustment are performed after assembly of the mechanism.

The capstan (17) has a large influence on the tape speed and the stability of tape run. For this reason, the capstan (17) has a precision finish to eliminate out-of-roundness, rectangularity, and surface roughness. The pinch roller (18) is pushed at a right angle against the capstan (17), to provide the tape with a uniform pressure and to ensure stable tape run.

Guide (19), loading guide A(20), and loading guide B(21) are roller-type guides to reduce the friction load. The loading guide C(22) compensates for the plane angle between the cassette plane and the loading ring plane, and holds the tape so as to guide the tape from the loading guide B(21) to the cassette. The tape guide (24) of the rewind sensor assembly (23) is arranged so that a constant distance is kept between the rewind sensor (25) and the tape.

### 5-1.3 Operation in the Loading Mode

Figure 5-2 shows operation of the mechanical parts in the loading mode. When the cassette is inserted into the cassette holder (1) and is pushed down, the synchronization gears (2), (3) engage with the racks on the cassette holder (1) and are turned in the arrow direction. The damper gear (4), engaged with the synchronization gears (3), also turns in the arrow direction. The damper unit (5) is installed on the same shaft as the synchronization gears (2). When the synchronization gears (2) rotate, the damper unit shifts in the arrow direction, is disengaged by the stopper (6), the rotation of the damper gear (4) becomes unbraked, and the cassette holder (1) can be pushed in quickly. The cassette holder

226   Mechanical Operation of Typical VCRs

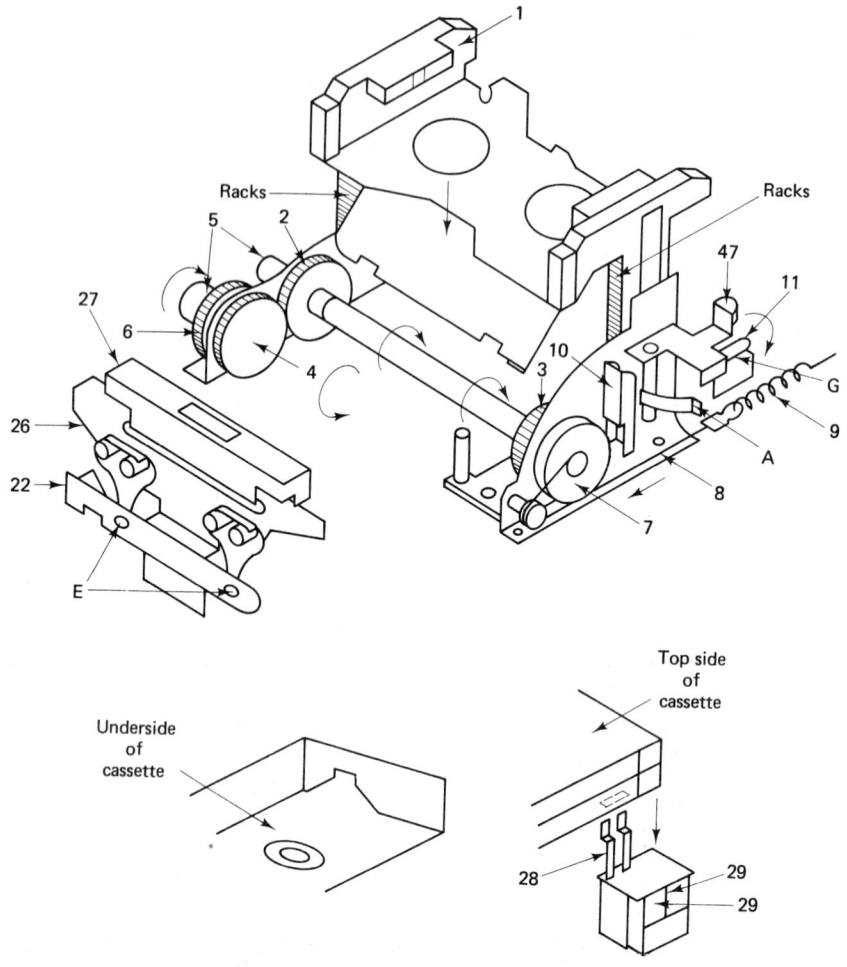

**FIGURE 5-2.**   Loading mode operation

lock cam (7) is installed on the end of the synchronization gears (2), (3), and rotates in coincidence with the rotation of the synchronization gears (2), (3). The ejection spring (9) is pulled via the attached wire (8). The cassette holder (1) is locked by catching of the cassette holder lock lever (10) at the cam part.

When the cassette holder lock lever (10) is shifted, the cassette holder pulley lock lever (11), held at part A, is shifted in the arrow direction, and this shifts the eject link (12) held by this and part B. The cam (13) is installed on the eject link (12), and the eject switches (14), (15) come ON by movement of the eject link (12). The eject link (12) is connected to the loading lever (16) at part C and moves the lever in the arrow direction. The loading lever (16) cancels the control of the loading gear assembly (18), pulled by spring (17). The loading gear assembly (18)

Typical Beta Mechanical Operation  227

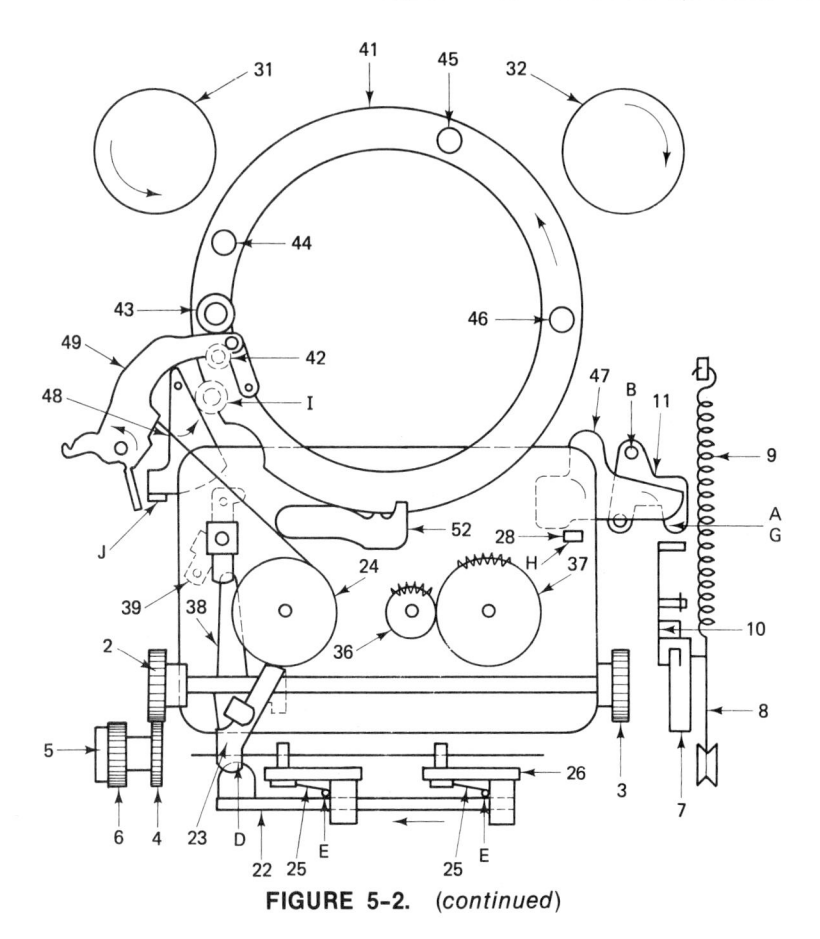

**FIGURE 5-2.** (continued)

is turned in the arrow direction, and the loading roller (19) is pushed against the flywheel (20).

The eject link (12) also moves the eject operation link (22) by the eject lever (21) in the arrow direction. At part D, the eject operation link (22) shifts the eject brake lever (23) in the arrow direction, and in order to release the brake from the supply reel base (24), the eject operation lever (25) is turned at part E to return the eject operation slide (26) and the EJECT lever (27) in the arrow direction.

When the cassette holder (1) is locked, the cassette pushes the cassette switch actuator (28) and the cassette switch (29) becomes ON. Also, the eject switches (14), (15) become ON, and head motor (30), capstan motor (31), reel motor (32), and roller solenoid are engaged by these switches. When, at this time, the cassette holder (1) is locked without insertion of a cassette, the cassette switch (29) does not become ON, so that only eject operation becomes possible.

The roller solenoid (33) attracts in the arrow direction and shifts the for-

228  Mechanical Operation of Typical VCRs

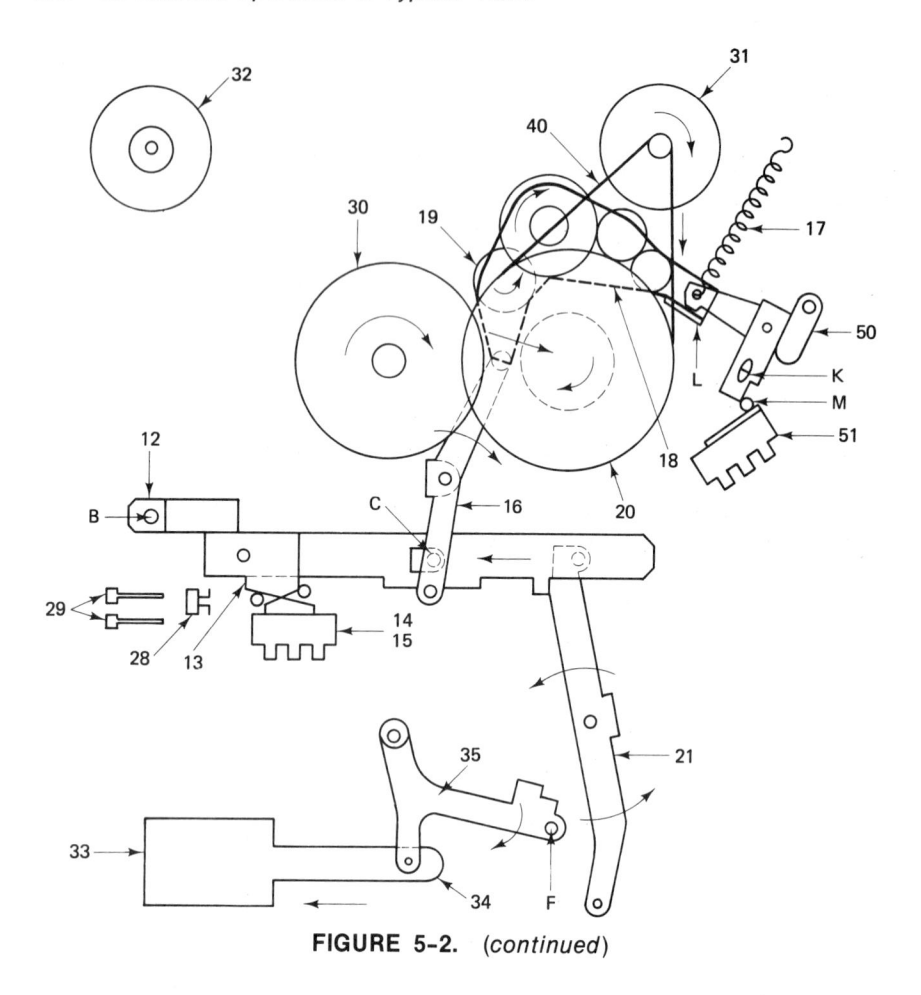

**FIGURE 5-2.** (continued)

ward roller lever (35) via the link (34). The forward gear (36) is removed from the take-up reel base (37), and the drive of the reel motor (32) is not transmitted to the take-up reel base (37).

The forward roller lever (35) turns the supply reel base brake lever (38) at part F in the arrow direction and releases the brake of the supply reel base (24). However, the soft brake lever (39) is applied at the supply reel base (24) for slight braking to prevent tape slack at the time of loading.

The rotation force of the capstan motor (31) is transmitted via the belt (40) to the flywheel (20) and further from the loading roller (19) to the loading gear assembly (18). The loading ring (41) (engaged with the loading gear) is turned in the arrow direction, and guide (42), pin roller (43), and loading guides (44), (45), and (46) pull the tape from the cassette to execute loading.

The cam part of the loading ring (41) turns the unloading end-sensing lever (47), the cassette holder lock lever (10) is pushed at part G, the cassette holder (1) is pushed down farther from the prelock position, and the cassette is locked. The unloading end-sensing lever is controlled by the cassette switch actuator (28) at part H, and when the cassette holder (1) is locked without insertion of a cassette, this actuator (28) does not move to release the control, so that the unloading end-sensing lever (47) locks the loading ring (41) at the cam part to prevent unnecessary turning.

The cam I of the loading ring (41) shifts the loading end-sensing lever (48), which controls the spring-pulled back-tension lever (49) at part J. The shifting by rotation of the loading ring (41) loosens the control release of the back-tension lever (49) for shifting in the arrow direction to pull the tape from the cassette.

When the loading end-sensing lever (48) engages in the recessed part of cam I of the loading ring (41), the back-tension lever (49) is released from the control by part J. The loading end-sensing lever (48) turns the loading gear cancellation lever (50), connected at part K, is turned in the arrow direction. The loading gear cancellation lever (50) pushes the loading gear assembly (18) at part L to cancel the pressure of the loading roller (19) and flywheel (20), and to stop rotation of the loading ring (41). At the same time, the loading end switch (51) is actuated to ON by part M, and the energizing of motors and solenoids is stopped. The loading ring (41) is stopped by stopper (52). Supply reel base brake operation and forward gear pushing operation are executed by attraction of the roller solenoid (33). The STOP mode is reached with completion of loading.

### 5-1.4  Operation in the PLAY Mode

Figure 5-3 shows operation of the mechanical parts in the PLAY mode. In the PLAY mode, the head motor (1), capstan motor (2), and reel motor (3) are running, and the pinch roller solenoid (4) is attracted.

When the pinch roller solenoid (4) is attracted, the pinch roller pressure lever (8) and the pressure operation lever (9) are shifted in the arrow direction via the links (5), (6), and (7). The pinch lever (10) is pushed by part A of the pinch roller pressure lever (8), and the pinch roller (11) is pushed against the capstan (12).

The pressure spring (13) is applied to the pressure operation lever (9), and pressure is applied by the force of this spring. At part B, the pinch roller pressure lever (8) turns the brake release lever (14), the coupled soft brake lever (15) is turned in the arrow direction, and soft brake release is executed. At part C, the soft brake lever (15) shifts the supply reel brake lever (16) in the arrow direction and executes brake release for the supply reel base (17).

The rotation of the capstan motor (2) is transmitted by the belt (18) to the flywheel, and the tape is transported with constant speed by the capstan (12) and the pinch roller (11). The capstan motor (2) is controlled so that the tape runs with a speed of 20 cm/sec in BII mode and a speed of 1.33 cm/sec in BIII mode.

The rotation of the reel motor (3) is transmitted to the forward pulley (23) via belt (20), intermediate pulley (21), and belt (22). The forward gear (24), in-

**230** Mechanical Operation of Typical VCRs

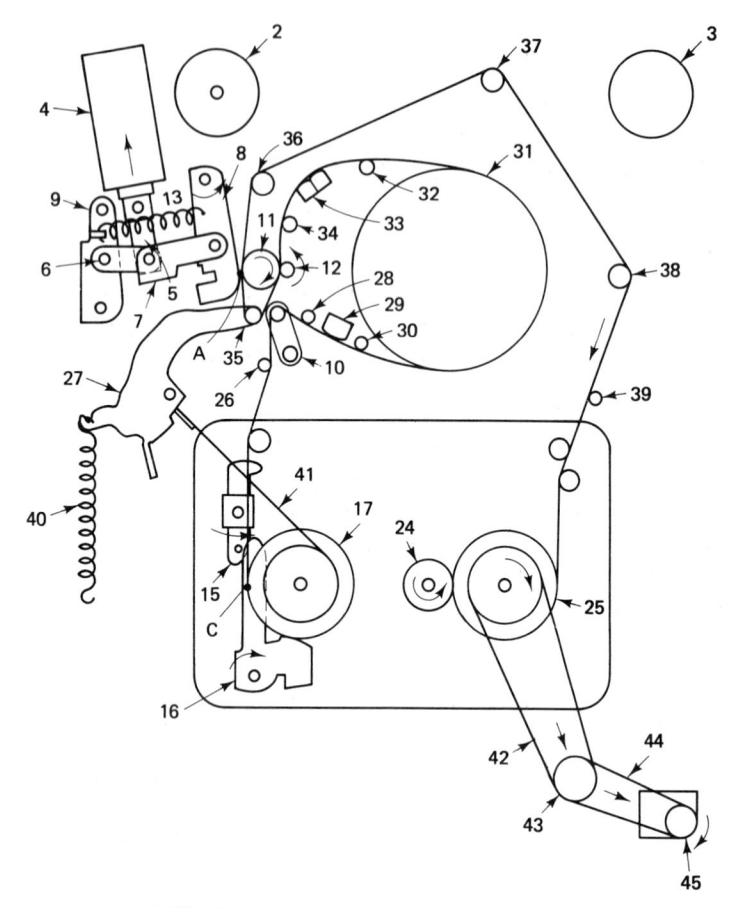

**FIGURE 5-3.** PLAY mode operation

stalled as one body with forward pully (23), turns the take-up reel base (25) in the arrow direction, and the tape paid out by capstan (12) and pinch roller (11) is taken up. The intermediate pulley (21) has a built-in slip mechanism, which controls the rotation force of the reel motor (3) so that the tape is taken up with a tension force that does not damage the tape.

The tape is pulled from the cassette and passes the forward sensor guide (26), back-tension lever (27), erase head roller (28), erase head (29), inlet guide (30), drum (31), outlet guide (32), audio control head (33), and outlet guide (34). The tape is then transported by pinch roller (11) and capstan (12), passing guide (35), loading guides (36), (37), (38), and rewind sensor guide (39) to be taken up into the cassette.

The back-tension lever (27) pulls the brake band (41) with the spring (40) to brake the supply reel base (17) so that a constant back-tension acts on the tape.

*Typical Beta Mechanical Operation* 231

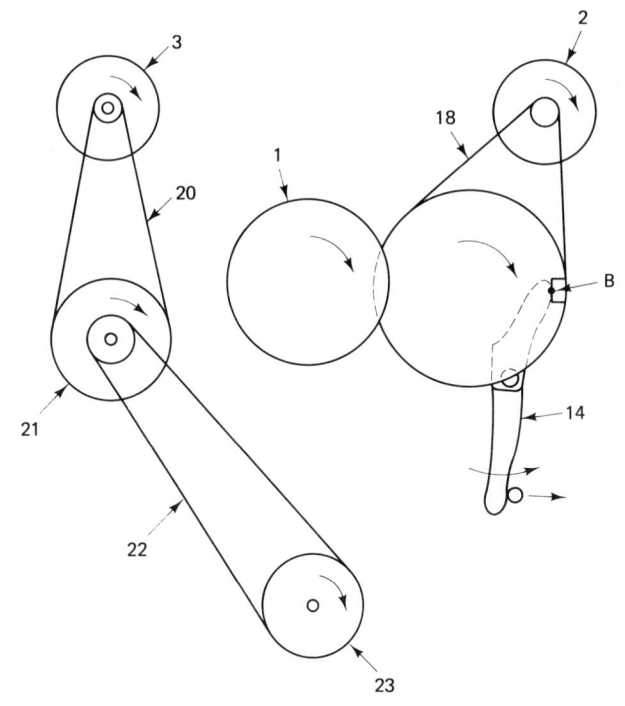

**FIGURE 5-3.** *(continued)*

The rotation of the take-up reel base (25) is transmitted via belt (42), pulley (43), and belt (44) to the tape counter (45) for tape position display.

### 5-1.5 Operation in the STILL Mode

Figure 5-4 shows operation of the mechanical parts in the STILL mode. The STILL mode is reached from the PLAY mode when the capstan motor (1) is stopped, and the reel motor (2) is stopped a little later to remove tape slack. When the capstan motor is stopped, the tape run by capstan (3) and pinch roller (4) are stopped. When the reel motor (2) is stopped, tape take-up by the take-up reel base (5) is stopped, so that the tape is stopped. The video heads of the drum (6) play back the video signal from the tape, and a STILL (or freeze frame) picture is obtained on the TV screen.

### 5-1.6 Operation in the REC (Record) Mode

Figure 5-5 shows operation of the mechanical parts in the REC mode. Operation of the mechanism in the REC modes is the same as in the PLAY mode (Sec. 5-1.4), but the signals on the passing tape are erased by the erase head (1), and the

*232 Mechanical Operation of Typical VCRs*

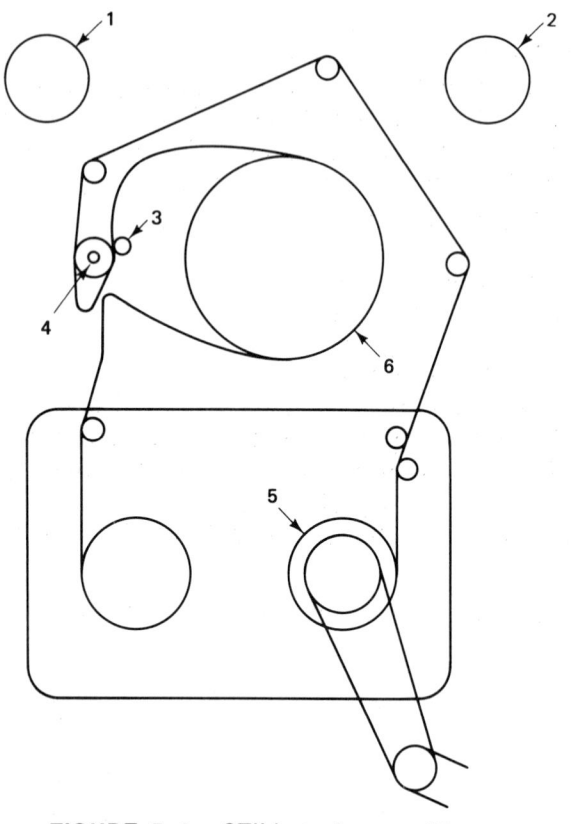

**FIGURE 5-4.** STILL mode operation.

video signal is recorded on tape by the video heads, while the audio signal and control signal are recorded by the audio/control head (3).

When a cassette is used, where the tab (4) has been removed to prevent undesired erasing, the actuator (6) of the erase prevention switch (5) is not pushed, so that the switch (5) does not become ON, and REC operation cannot be executed.

### 5-1.7  Operation in the PAUSE Mode

Figure 5-6 shows operation of the mechanical parts in the PAUSE mode. The PAUSE mode is reached from the REC mode (Sec. 5-1.6) by stopping the pinch roller solenoid (1), stopping pressure of the pinch roller (2) so that tape feed is stopped, and bringing the tape to a stop by stopping the reel motor (3) so that tape take up by the take-up reel base (4) is stopped.

By stopping the pinch roller solenoid (1), the pinch roller pressure lever (5)

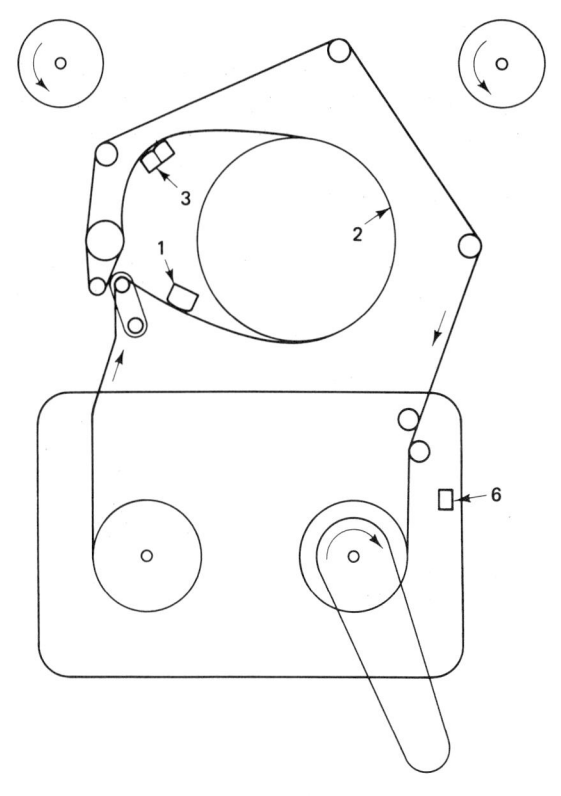

**FIGURE 5-5.** REC (record) mode operation

is shifted in the arrow direction; brake release lever (6), soft brake lever (7), and supply reel base brake lever (8) are turned in the arrow direction; the brake is applied to the supply reel base (9); and tape slack is prevented.

## 5-1.8 Operation in the AUD DUB (Audio Dubbing) Mode

Figure 5-7 shows operation of the mechanical parts in the AUD DUB mode. Operation of the mechanism in the AUD DUB mode is the same as in the PLAY mode (Sec. 5-1.4), but the audio signal of the passing tape is erased by the audio erase head (1), while the new or dubbed audio signal is recorded by the audio/control head (2). The video heads on the drum (3) are in the normal playback condition for the video signal. The recorded video signal is displayed on the TV screen.

When a cassette is used, where the tab (4) has been removed to prevent undesired erasing, the erase prevention switch (5) is not pushed, so that the switch (5) does not become ON, and AUD DUB operation cannot be executed.

234  Mechanical Operation of Typical VCRs

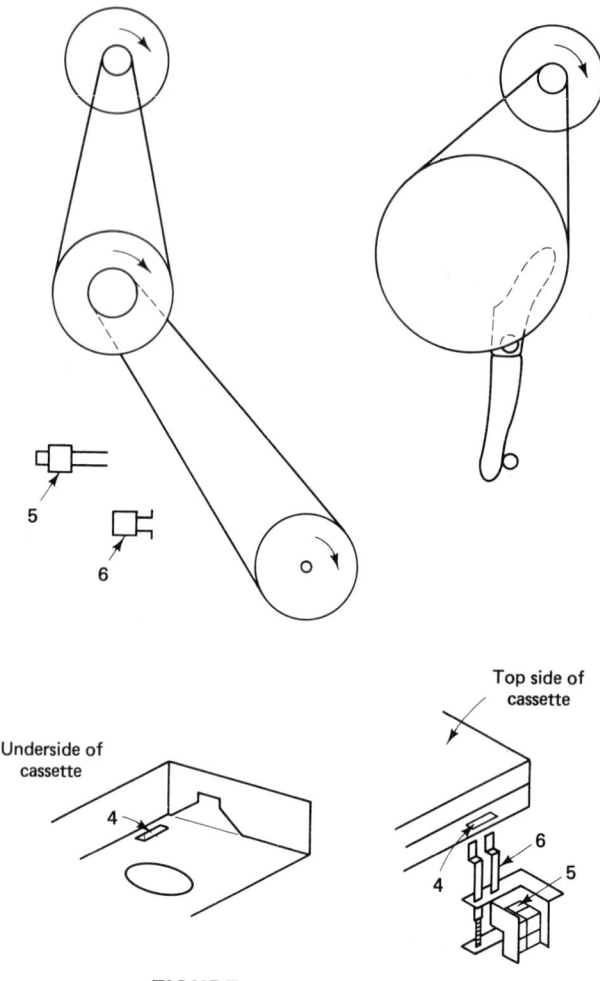

**FIGURE 5-5.** *(continued)*

### 5-1.9  Operation in the F FWD (Fast Forward) Mode

Figure 5-8 shows operation of the mechanical parts in the F FWD mode. In the F FWD mode, the head motor (1) and the reel motor (2) run, and the brake solenoid (3) is attracted. When the brake solenoid (3) is attracted, the brake is released via link (4) and lever (5), and link (6) is shifted in the arrow direction. At part A, the brake cancellation link (6) shifts the back-tension lever (7) in the arrow direction, and the brake band (8) is released from the supply reel base (9).

The back-tension lever (7) is controlled by part B of the back-tension lever stopper (10), so that the lever does not move out of the specified range, to prevent

*Typical Beta Mechanical Operation* 235

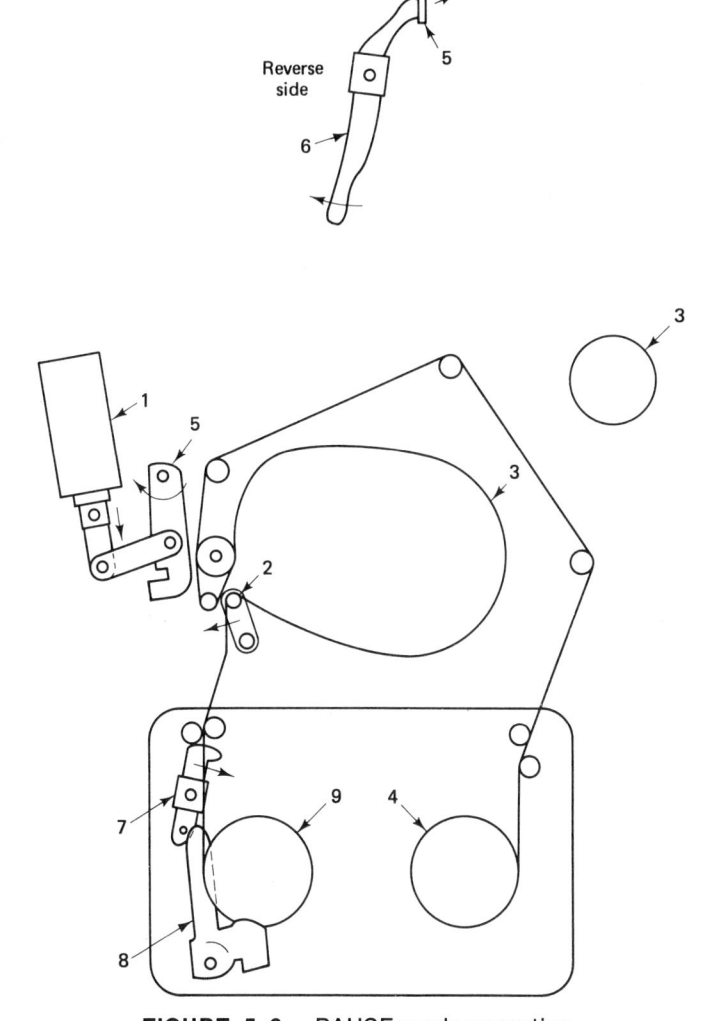

**FIGURE 5-6.** PAUSE mode operation.

large movement of the back-tension lever (7) from load changes during tape run, and to obtain a stable tape run.

The brake release link (6) pushed the soft brake lever (11) at part C, the supply reel base brake lever (12), coupled with the soft brake lever (11), is shifted in the arrow direction, and the brake is released. The brake release link (6) shifts the clutch operation lever (13) at part D in the arrow direction. The clutch operation lever (13) shifts the clutch operation levers (14) and (15) in the arrow direction. The clutch operation lever (15) pushes down the clutch (18), installed in the intermediate pulley (17) of the intermediate pulley assembly (16).

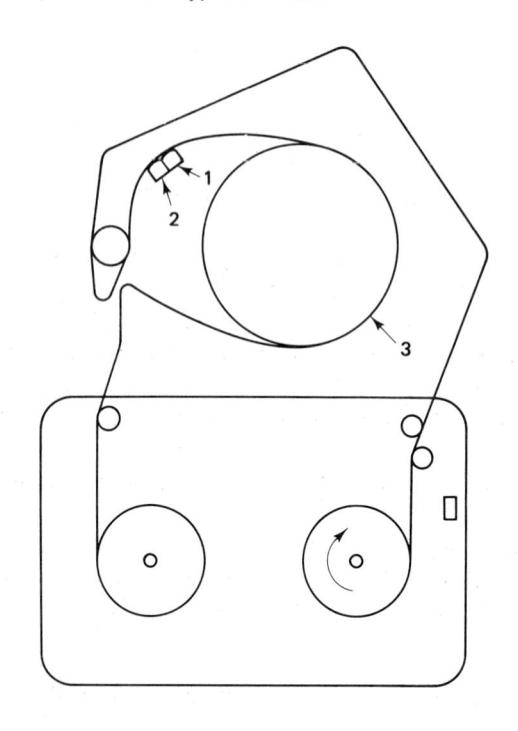

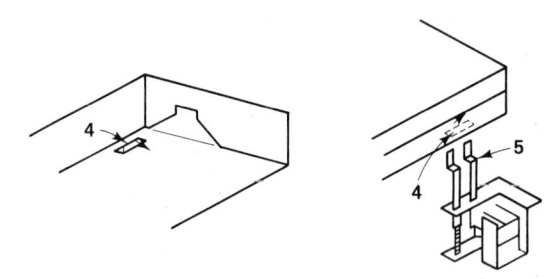

**FIGURE 5-7.** AUD DUB (audio dubbing) mode operation.

Clutch (18) and intermediate pulley (19) are connected, and the slip mechanism of the intermediate pulley (19) is kept from operating. Without operation of the slip mechanism, the intermediate pulleys (17) and (19) turn as one body, so that the intermediate pulley assembly can transmit the large rotation force required for F FWD mode. The rotation of the reel motor (2) is transmitted to the forward pulley (22) via belt (20), intermediate pulley assembly (16), and belt (21). The take-up reel base (24) is turned by the forward gear (23) in the arrow direction, and the tape is taken up. The rotation of the take-up reel

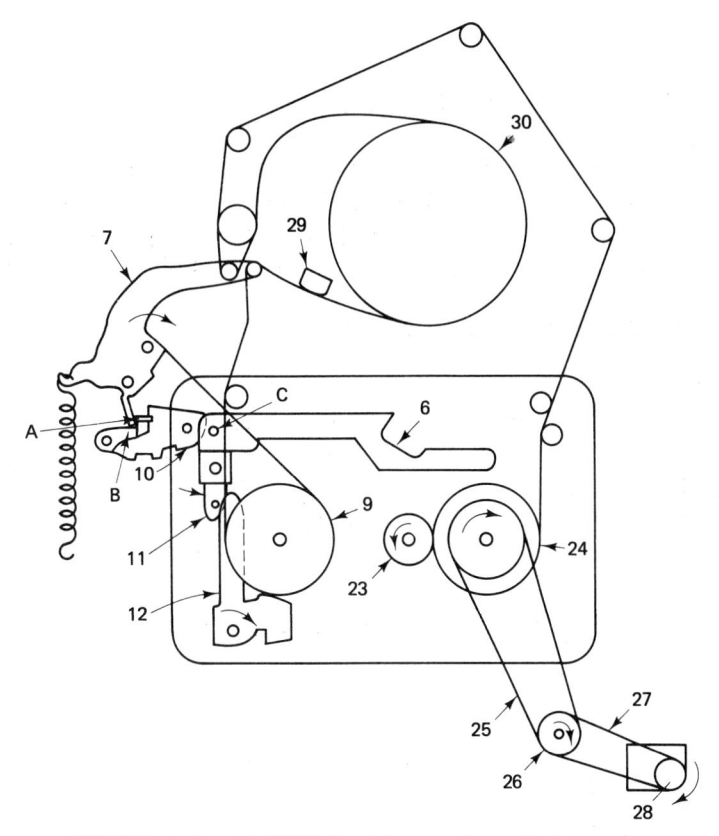

**FIGURE 5-8.** F FWD (fast forward) mode operation.

base (24) turns the tape counter (28) via belt (25), pulley (26), and belt (27) in the arrow direction to display the tape position.

In comparison to the PLAY mode, the tape path in F FWD has a smaller tape winding angle for back-tension lever (7) and erase head (29) to reduce the friction load. Also, the head motor (1) is running and the friction load at the drum (30) during tape run is reduced.

### 5-1.10  Operation in the F-SEARCH (Forward Search) Mode

Figure 5-9 shows operation of the mechanical parts in the F-SEARCH mode, which is essentially the same as the F FWD mode (Sec. 5-1.9) as far as the head motor (1), reel motor (2), and brake solenoid (3) are concerned. However, in F-SEARCH the video heads on drum (4) execute playback of the video signal, and the speed of reel motor (2) becomes about one-half normal speed, as does the tape take-up speed.

238  *Mechanical Operation of Typical VCRs*

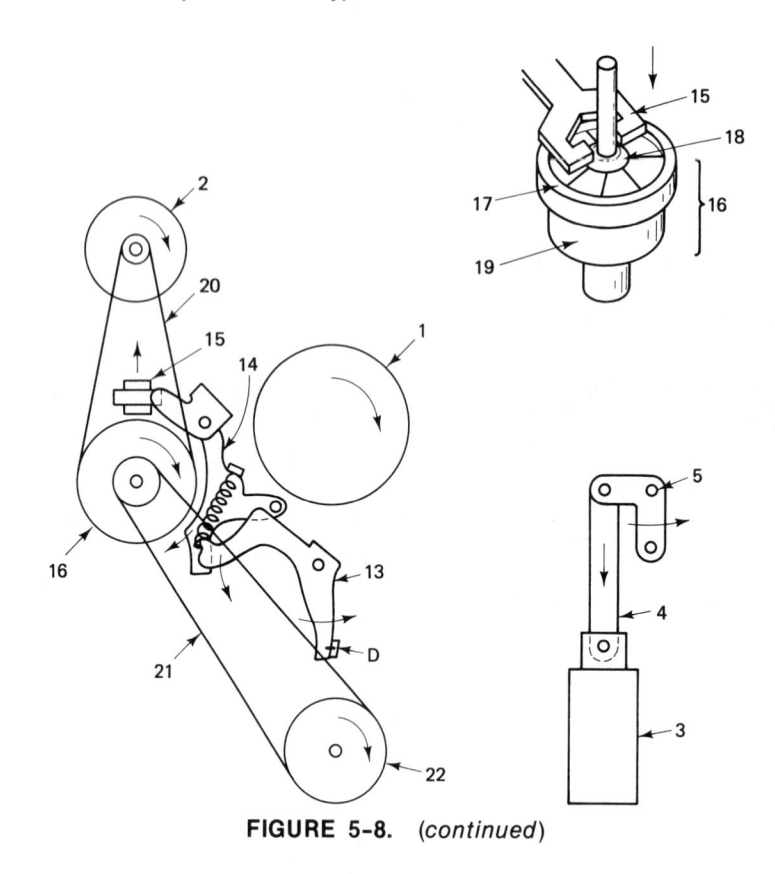

**FIGURE 5-8.** *(continued)*

### 5-1.11  Operation in the REW (Rewind) Mode

Figure 5-10 shows operation of the mechanical parts in the REW mode. Head motor (1), reel motor (2), roller solenoid (3), and brake solenoid (4) operate in the REW mode. When the roller (3) is attracted, the FWD roller lever (6) is shifted in the arrow direction at part A via the link (5), the forward gear (7) is released from the take-up reel base (8), and the take-up reel base (8) is freed. The rewind roller stopper is shifted in the arrow direction at part B via the link (5), and the control of the rewind roller lever (10) at part C is released.

When the brake solenoid (4) is attracted, link (11), lever (12), and brake release link (13) are shifted in the arrow direction, in the same way as for the attraction operation in F FWD mode (Sec. 5-1.9). At part D, the brake release link (13) pushes the back-tension lever (14) and, at part E, the soft brake lever (15) and the supply reel base brake lever (16) in the arrow direction, so that the brake of the supply reel base (17) is released. At part F, the brake release link (13) shifts

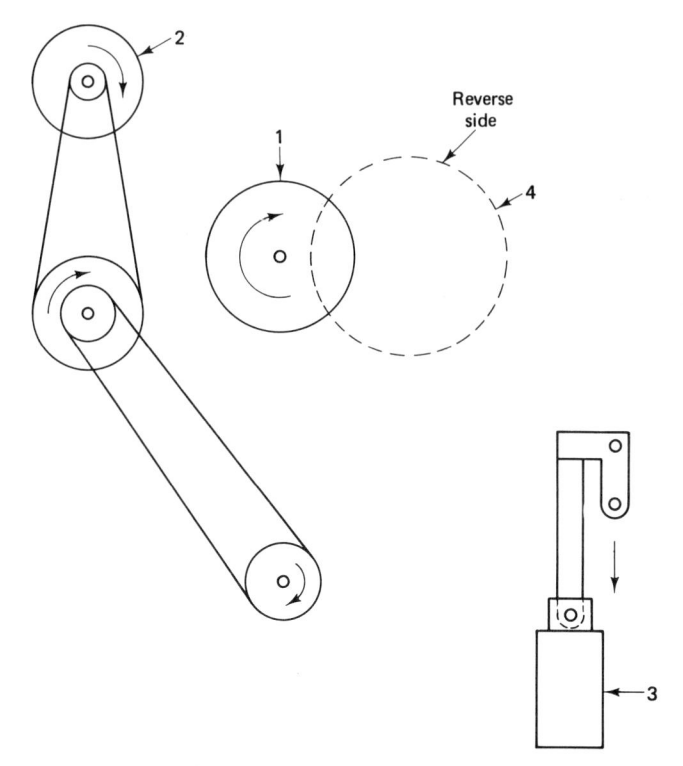

**FIGURE 5-9.** F-SEARCH (forward search) mode operation.

the clutch operation levers (18), (19), and (20) so that the slip mechanism of the intermediate pulley assembly (21) is locked for transmission of a large rotation force, in the same way as in the F FWD mode.

At part G, the brake release link (13) cancels the control of the rewind roller lever (10). The control by part C of the rewind roller stopper (9) and part G of the brake release link (12) is canceled for the rewind roller lever (10), which is shifted in the arrow direction by spring (22). The rewind roller (23) is pushed against the supply reel base (17) and the roller part installed on the forward gear (7).

The rotation of the reel motor (2) is transmitted to the supply reel base (17) via belt (24), intermediate pulley assembly (21), belt (25), forward pulley (26), forward gear (7), and rewind roller (23), and the supply reel base (17) is turned in the arrow direction for rewinding of the tape.

For rewinding operation, the take-up reel base is turned in the arrow direction via the tape, and this rotation turns the tape counter (30) in the arrow direction via belt (27), pulley (28), and belt (29) for display of the tape position. For tape run, the friction load is reduced by reduction of the tape winding angle for drum (31), back-tension lever (14), and erase head (32). In the same way as in F

240  Mechanical Operation of Typical VCRs

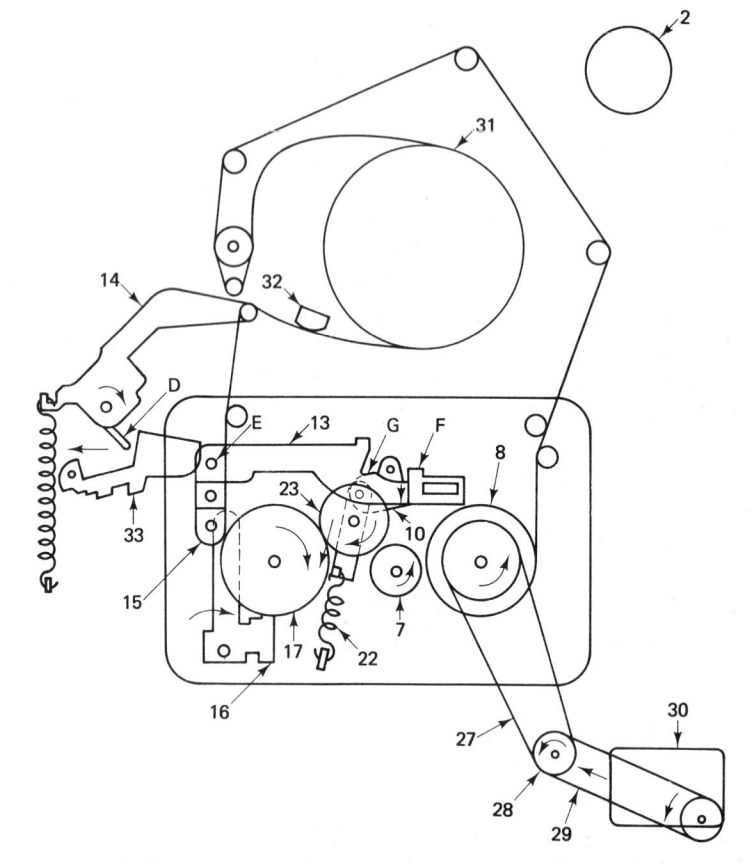

**FIGURE 5-10.** REW (rewind) mode operation.

FWD mode, the position of the back-tension lever (14) is controlled by the back-tension lever stopper (33) to prevent large movement of the back-tension lever (14) by load changes during tape run.

### 5-1.12  Operation in the R-SEARCH (Reverse Search) Mode

Figure 5-11 shows operation of the mechanical parts in the R-SEARCH mode, which is essentially the same as the REW mode (Sec. 5-1.11), as far as the head motor (1), reel motor (2), roller solenoid (3), and brake solenoid (4) are concerned. However, in R-SEARCH, the video heads on drum (5) execute playback of the video signal, and the speed of reel motor is reduced to about one-half speed, so that the tape rewinding speed is also about one-half of normal.

Typical Beta Mechanical Operation   241

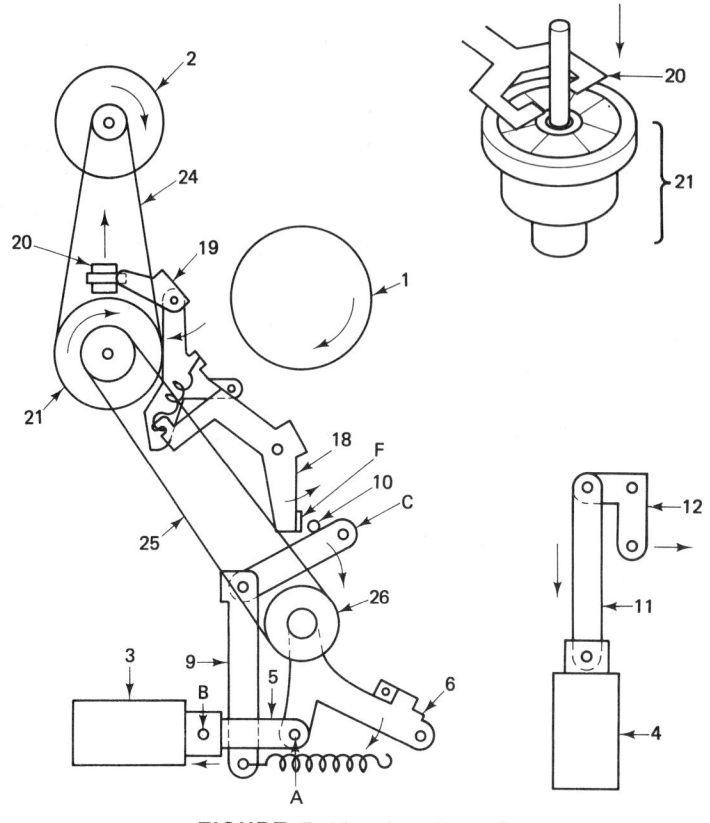

**FIGURE 5-10.** (continued)

## 5-1.13. Operation in the AUTO REW (Automatic Rewind) Mode

When the forward sensor detects the tape end trailer tape in the PLAY, REC, or F FWD modes, REW mode is entered via the STOP mode, and the tape is rewound automatically.

## 5-1.14 Operation in the Tape Counter Memory Mode

When the tape counter memory switch is set to ON in the REW mode, REW mode is stopped when the tape has been rewound to the position 9999 of the tape counter, STOP mode is reached, and the tape is stopped.

242  Mechanical Operation of Typical VCRs

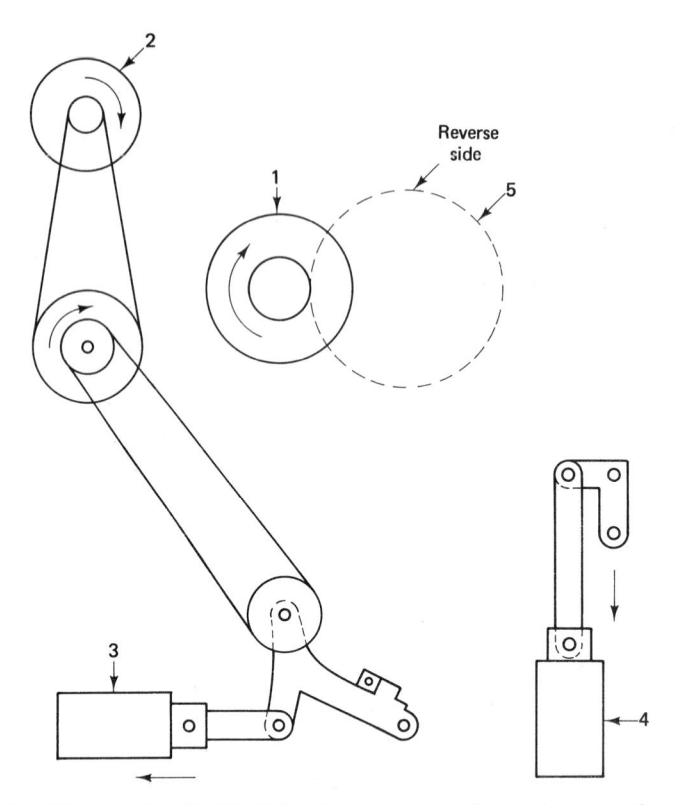

**FIGURE 5-11.** R-SEARCH (reverse search) mode operation.

### 5-1.15 Operation in the Timer Record Mode

The timer record mode is set in the program timer memory, and when the timer reaches the set time, all parts are energized automatically, the REC mode is executed, the STOP mode is reached after the set timer REC time, power is stopped, and the stand-by condition is reached.

### 5-1.16 Operation in the EJECT Mode

The EJECT mode, which can be executed in any mode by pushing the eject button, is essentially the reverse of the loading mode described in Sec. 5-1.3 and illustrated in Fig. 5-2.

## 5-2 TYPICAL VHS MECHANICAL OPERATION

The following paragraphs describe mechanical operation for one of the VHS VCRs covered in Chapter 4. Operation of the corresponding system control is discussed in Sec. 4-6.

Typical VHS Mechanical Operation 243

Figure 5-12 is the *timing chart* and related *rotation drawing* for the rewind mode of operation. The physical relationships of the parts called out in the blocks of the timing charts (identified by a circled index number) are shown by means of photographs in the VCR service manual. To use these illustrations,

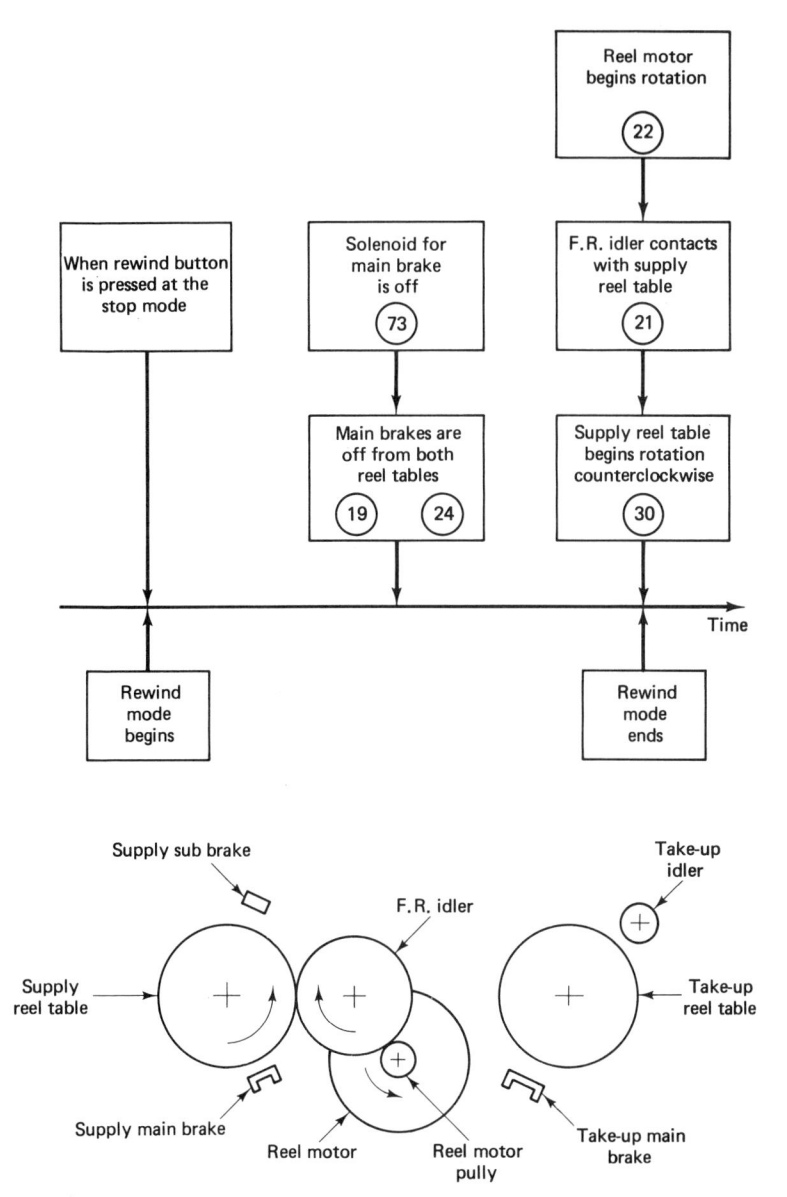

**FIGURE 5-12.** Timing chart and rotation drawing for VHS REWIND mode.

find the timing and rotation illustration for the particular mode of operation. Read the timing blocks in sequence from left to right. Use the photos to locate the parts.

As an example, assume that you are interested in the rewind mode. As shown in Fig. 5-12, after the REWIND button has been pressed (after reaching the STOP mode), the first two functions are: solenoid (73) for the main brake is OFF, and the main brakes (19) and (24) are OFF. In the next sequence, reel motor (22) begins rotation, the F.R (forward-reverse) idler (21) contacts with the supply reel table, and supply reel table (30) begins rotation counterclockwise. The rewind mode ends when the tape has been fully wound on the supply reel.

# 6

# Typical Adjustment, Cleaning, Lubrication, and Maintenance Procedures

This chapter describes typical adjustment, cleaning, lubrication, and maintenance procedures for VCRs. Both Beta and VHS systems are described. The systems covered are those described in Chapters 3 and 4, using the test equipment and tools described in Chapter 2. Keep in mind that these specific procedures apply directly to the VCRs of Chapters 3 and 4. When servicing other VCRs, *you must follow* manufacturer's service instructions exactly. Each type of VCR has its own electrical and mechanical adjustment points and procedures, which may or may not be different from procedures for other VCRs.

In the absence of manufacturer's instructions, and to show you what typical adjustments involve, we describe complete electrical and mechanical adjustment procedures for both Beta and VHS, as recommended by the manufacturers. Using these examples, you should be able to relate the procedures to a similar set of adjustment points on most VCRs. Where it is not obvious, we also describe the purpose of the procedure.

The waveforms measured at various test points during adjustment are also included here. By studying these waveforms you should be able to identify typical signals found in most VCRs, even though the signals may appear at different points for your particular unit.

## 6-1 BETA ELECTRICAL ADJUSTMENTS

The following paragraphs describe complete adjustment procedures for the electrical section of the Beta VCR described in Chapter 3. The electrical locations for the majority of adjustment controls and measurement points are given in the block diagrams of Chapter 3. The following paragraphs make reference to the illustrations in Chapter 3 that show the adjustment and measurement points.

### 6-1.1 Power Supply Circuit Adjustments

Two of the voltage outputs produced by the power supply are adjustable. One output is the 12 V supplied to the various circuits, while the other output is 12 V supplied to the servo and reel motors. The procedures are as follows:

1. Using a digital voltmeter (Sec. 2-5), measure the voltage between TP1350 and chassis ground. TP1350 monitors the voltages to the various circuits. Adjust VR1350 for a reading of 12, +0.1, −0 volts. Set the VCR to the STOP mode.
2. Using a digital voltmeter, measure the voltage between TP1351 and chassis ground. TP1351 monitors the voltages to the servo and reel motors. Adjust VR1351 for a reading of 12, +0.1, -0 volts.

### 6-1.2 Tuner Circuit Adjustments

Figure 6–1 shows the basic test connections for adjustment of the tuner circuits. Figures 3–52 through 3–55 show electrical locations for the adjustment and measurement points. Note that there are tuner circuits located on the tuner PC board that require adjustment, although there is only one adjustment control on the VHF tuner. As shown in Fig. 6–1, the output of a sweep/marker generator (Sec. 2-2) is applied through one of two input probes (probe A and probe B) to

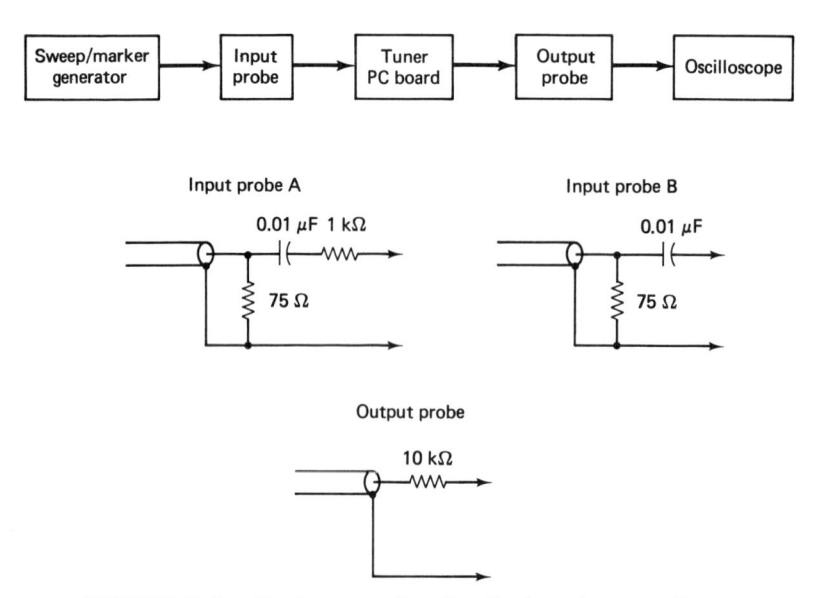

**FIGURE 6-1.** Basic tuner circuit adjustment connections.

measurement points on the tuner PC board, while the output of the tuner circuits is applied to an oscilloscope (Sec. 2-4) through an output probe. The procedures are as follows:

**VIF (Video IF) Adjustment**

1. Connect an external d-c power source between ground and TP906. The power source, which functions here as a variable AGC voltage, must be adjustable from +2 to +6 V, with the positive applied to TP906. Set the VCR to STOP mode.
2. Set the output level of the sweep/marker generator at the video IF frequency, at an amplitude of -30 dB, and connect the generator output to the TP terminal of the VHF tuner through input probe A.
3. Connect the output signal at TP909 on the PC board to the oscilloscope through the output probe and compare the waveform to that shown in Fig. 6-2.
4. Turn the T902 core to the right so that the waveform of the VIF does not change shape.
5. Adjust the AGC voltage so that the waveform level is equal to 1.0 V peak to peak (p-p).
6. Adjust the T901 upper core so that the absorption point agrees with S in Fig. 6-2 (the 41.25-MHz marker).
7. Adjust the T901 lower core so that C in Fig. 6-2 (the 42.17-MHz marker) becomes maximum.
8. Adjust the converter transformer of the VHF tuner so that P of Fig. 6-2 (the 45.75-MHz marker) becomes 35 ± 5%.
9. Adjust the T901 lower core so that C in Fig. 6-2 (the 42.17-MHz marker) becomes 50 ± 5%.
10. Repeat steps 6 and 9 as necessary.

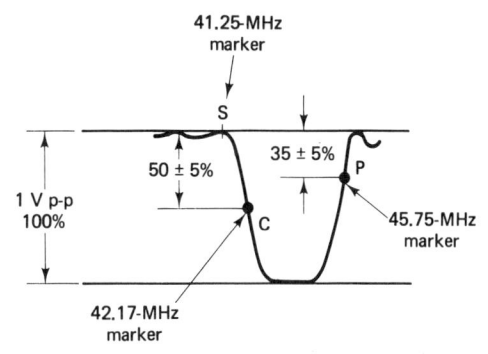

**FIGURE 6-2.** VIF adjustment waveform.

## Detection Transformer Adjustment

1. Connect an external d-c power source between ground and TP906. The power source, which functions here as a variable AGC voltage, must be adjustable from +2 to +6 V, with the positive applied to TP906. Set the VCR to STOP mode.
2. Set the output level of the sweep/marker generator at the video IF frequency, at an amplitude of −30 dB, and connect the generator output to the TP terminal of the VHF tuner through input probe A.
3. Connect the output signal at TP909 on the PC board to the oscilloscope through the output probe and compare the waveform to that shown in Fig. 6-3.
4. Adjust the AGC voltage so that the waveform level is equal to 1.0 V (p-p).
5. Adjust the T902 core so that P of Fig. 6-3 (the 45.75-MHz marker) becomes 45%.

## AFT (Automatic Fine Tuning) Adjustment

1. Connect an external d-c power source of +3.5 V between ground and TP906 (with positive at TP906) as a fixed AGC voltage.
2. Set the output level of the sweep/marker generator at the video IF frequency at an amplitude of −20 dB and connect the generator output to the TP terminal of the VHF tuner through input probe A. Set the VCR to STOP mode.
3. Connect the output signal at TP908 on the PC board to the oscilloscope through the output probe and compare the waveform to that shown in Fig. 6-4.
4. Adjust T903 so that P of Fig. 6-4 (the 45.75-MHz marker) is positioned on the baseline.

## FM Detection Characteristic Adjustment

1. Connect an external d-c power source of +12 V between ground and TP906 (with positive at TP906) as a fixed AGC voltage. Set the VCR to the STOP mode.

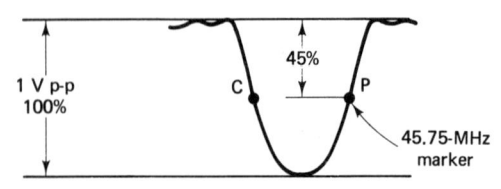

**FIGURE 6-3.** Detection transformer adjustment waveform.

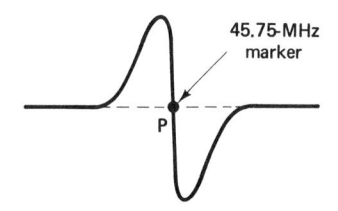

Beta Electrical Adjustments    249

**FIGURE 6-4.** AFT adjustment waveform.

2. Set the output level of the sweep/marker generator at the sound IF frequency (SIF) at an amplitude of −30 dB and connect the generator output to TP907 through input probe B.
3. Connect the output signal at TP911 on the PC board to the oscilloscope through the output probe and compare the waveform to that shown in Fig. 6-5.
4. Adjust T904 so that S of Fig. 6-5 (the 4.5-MHz marker) is positioned on the baseline.

### Video Output Adjustment

1. Connect an NTSC color bar generator (Sec. 2-3) to the ANTENNA terminal. The generator output should be on VHF channel 2, at 60 dB or higher, with 87.5% modulation. Set the VCR to the STOP mode.
2. Set the VHF channel selector switch to channel 2 and tune the VCR for best reception by observing the picture on the TV screen.
3. Monitor the waveform of the video output signal at TP909 on the PC board and adjust VR902 so that the output level is equal to 1.0 ± 0.2 V (p-p).

### RF AGC Adjustment

1. Connect an NTSC color bar generator (Sec. 2-3) to the ANTENNA terminal. The generator output should be on VHF channel 2, at 75 dB$\mu$ ± 1 dB$\mu$, with 87.5% modulation. Set the VCR to the STOP mode.
2. Set the VHF channel selector switch to channel 2 and tune the VCR for best reception by observing the picture on the TV screen.
3. Measure the voltage at the AGC terminal of the tuner and adjust VR901 so that the voltage is 4.5 ± 0.1 V.

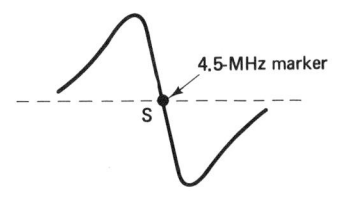

**FIGURE 6-5.** FM detection characteristic adjustment waveform.

**Audio Output Adjustment**

1. Connect a TV signal generator to the ANTENNA terminal. The generator output should be on VHF channel 2, at 60 dB$\mu$ or higher, with 60% modulation at an audio frequency of 400 Hz. Set the VCR to the STOP mode.
2. Set the VHF channel selector switch to channel 2 and tune the VCR for best reception by observing the picture on the TV screen.
3. Monitor the waveform of the audio output signal at TP911 on the PC board and adjust VR903 so that the output level is equal to 0.75 ± 0.25 V (p-p) (-10 ± 3 dBm).

### 6-1.3 System Control Circuit Adjustments

Figures 3-38 through 3-44 show electrical locations for the majority of the system control adjustment and measurement points. The procedures are as follows:

**Dew-Sensor Adjustment**

1. Disconnect plug P1201 connecting the dew sensor and the W6 PC board. Connect an 82-k$\Omega$ resistor between pin 1 and pin 2 of S1201. Set the VCR to STOP mode.
2. Turn VR1201 fully to the right and connect an oscilloscope to TP1201.
3. Monitor the square wave of oscillation at TP1201 and slowly turn VR1201 to the left. Set VR1201 at the point where the oscillation is just stopped.
4. Remove the 82-k$\Omega$ resistor, and reconnect plug 1201.

**FWD (Forward)-Sensor Adjustment**

1. Set the VCR to the STOP mode with the tape not run to either end.
2. Monitor the waveform at TP1202.
3. Adjust VR1201 until the amplitude of the waveform at TP1201 is 2.5 ± 0.1 V (p-p).

**REW (Rewind)-Sensor Adjustment**

1. Set the VCR to the STOP mode with the tape not run to either end.
2. Monitor the waveform at TP1203.

3. Adjust VR1203 until the amplitude of the waveform at TP1203 is 2.5 ± 0.1 V (p-p).

**Timing-Phase Adjustment**

1. Load a blank cassette and set the VCR in RECORD to record a television broadcast. Set the tape speed selector to BII.
2. Connect an oscilloscope to TP1205. Use the oscilloscope in the external triggering mode and connect the external triggering probe to TP1204.
3. Observe the d-c up level at TP1205 while pushing the PAUSE button several times at intervals of 1 s and adjust VR1205 so that the phase difference between the leading edge of the d-c level and triggering point is equal to 25 ms, as shown in Fig. 6-6. The oscilloscope triggering slope should be "+."
4. Observe the d-c down level at TP1205 while pushing the PAUSE button several times at intervals of 1 s and adjust VR1204 so that the phase difference between the trailing edge of the d-c level and triggering point is equal to 8 ms, as shown in Fig. 6-7. The oscilloscope triggering slope should be "-."
5. Connect the oscilloscope to TP407 (Fig. 3-15). Use the oscilloscope in the external triggering mode and connect the external triggering probe to TP401.
6. Push the PAUSE button several times. Place the VCR in the PLAY mode and measure the period of the control pulse at TP407.
7. If the period is ± A ms, readjust VR1204 so that the delay time is 8 ± A ms.
8. Repeat the procedure several times and confirm that the tolerance of the control pulse period remains within ± 5 ms.

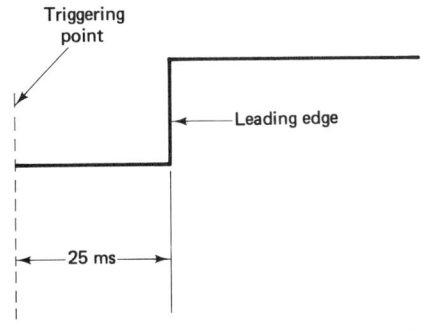

**FIGURE 6-6.** Timing phase adjustment waveform (leading edge).

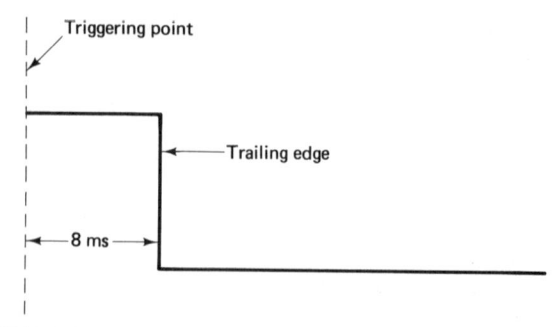

**FIGURE 6-7.** Timing phase adjustment waveform (trailing edge).

### Reel Motor Voltage Adjustment

1. Load a cassette and set the VCR to the PLAY mode.
2. Measure the voltage across the reel motor terminals with a multimeter and adjust VR1206 so that the voltage is equal to 3.0 ± 0.1 V dc.
3. Set the VCR to the F FWD mode.
4. Measure the voltage across the reel motor terminals with a multimeter and adjust VR1207 so that the voltage is equal to 8.0, +0.3, -0 V dc.

### Speed of Reel Motor Rotation Adjustment

1. Load a cassette and set the VCR to the F-SEARCH mode.
2. Connect a frequency counter (Sec. 2-6) to TP406 (Fig. 3-16).
3. Adjust VR1208 so that the frequency is 600 ± 40 Hz.

## 6-1.4 Servo Circuit Adjustments

Figure 3-15 shows electrical locations for the majority of the servo system adjustment and measurement points. The procedures are as follows:

### Forward Head Servo Lock

1. Load a cassette and set the VCR to the RECORD mode to record a television broadcast.
2. Measure the voltage at TP402 with the oscilloscope.
3. Adjust VR409 so that the voltage is equal to 3.0 ± 0.1 V.

### Head Switching Position

1. Play back an alignment tape and observe the composite synchronizing signal at TP413. Use the oscilloscope in the external triggering mode.

Apply the RF switching pulse (TP401) to the external triggering terminal.

2. Turn the tracking control so that maximum reproduced composite synchronizing signal is obtained.
3. Adjust VR402 so that the phase difference between the trailing edge of the RF switching pulse and the front edge of the vertical synchronizing signal is 7H (±0.5H), with the triggering slope "-," as shown in Fig. 6-8.
4. Adjust VR401 so that the phase difference between the leading edge of the RF switching pulse and the front edge of the vertical synchronizing signal is 7H (±0.5H), with the triggering slope " + ," as shown in Fig. 6-9.

**Forward Lock Position**

1. Load a cassette and set the VCR into the RECORD mode to record a television broadcast.
2. Observe the waveform of the composite synchronizing signal at TP413. Use the oscilloscope in the external triggering mode and supply the RF switching pulse (TP401) to the external triggering terminal, with the triggering slope "-."
3. Adjust VR403 so that the phase difference between the trailing edge of the RF switching pulse and the front edge of the vertical synchronizing signal is 7H (±0.5H), as shown in Figs. 6-8 and 6-9.

**F-SEARCH Lock**

1. Load a cassette and set the VCR into the F-SEARCH mode.
2. Measure the voltage at TP402 with the oscilloscope.
3. Adjust VR404 so that the voltage is equal to 4.0 ± 0.1 V.

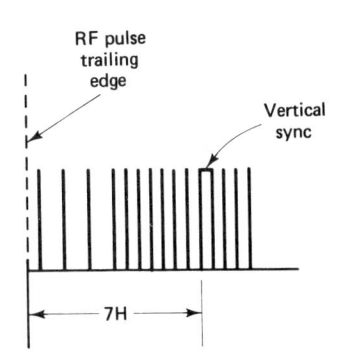

**FIGURE 6-8.** Head switching position adjustment waveform (trailing edge).

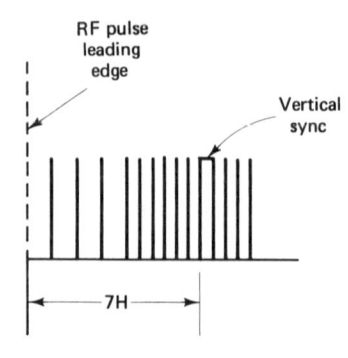

254  Typical Adjustment and Maintenance Procedures

**FIGURE 6-9.** Head switching position adjustment waveform (leading edge).

### R-SEARCH Lock

1. Load a cassette and set the VCR into the R-SEARCH mode.
2. Measure the voltage at TP402 with the oscilloscope.
3. Adjust VR404 so that the voltage is equal to 4.0 ± 0.1 V.

### Forward Capstan Servo Lock (1)

1. Load a cassette and set the VCR to the RECORD mode to record a television broadcast. Set the tape speed selector to BII.
2. Measure the voltage at TP408 with the oscilloscope.
3. Adjust VR410 so that the voltage is equal to 6.0 ± 0.1 V.

### Forward Capstan Servo Lock (2)

1. Load a cassette and set the VCR to the RECORD mode to record a television broadcast. Set the tape speed selector to BIII.
2. Measure the voltage at TP408 with the oscilloscope.
3. Adjust VR411 so that the voltage is equal to 6.0 ± 0.1 V.

### Tracking (1)

1. Load a cassette and set the VCR in the PLAY mode.
2. Turn the tracking control to the center position. A click stop is provided at the tracking control center.
3. Observe the square wave at TP412. Use the oscilloscope in the external triggering mode. Connect the external trigger probe to TP404. Set the triggering slope to "+."
4. Adjust VR405 so that the phase difference is 2 milliseconds between the trailing edge of the pulse (TP404) and the trailing edge of the pulse (TP412), as shown in Fig. 6-10.

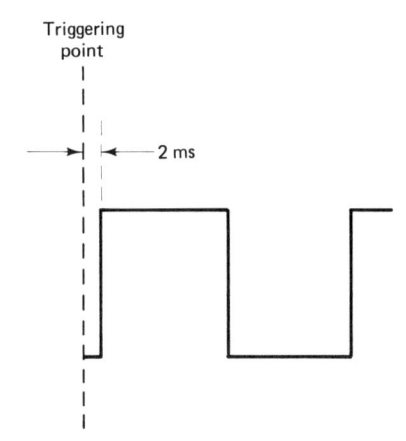

**FIGURE 6-10.** Tracking (1) adjustment waveform.

### Tracking (2)

1. Load a cassette and set the VCR to the RECORD mode to record a television broadcast. Set the tape speed selector to BII. Play back the recorded cassette.
2. Turn the tracking control to the center position, using the click stop.
3. Observe the playback control pulse at TP406 with an oscilloscope. Use the oscilloscope in the external triggering mode. Connect the external trigger probe to TP404. Set the triggering slope to "-."
4. Adjust VR407 so that the trailing edge of the playback control pulse coincides with the triggering point, as shown in Fig. 6-11.

### Tracking (3)

1. Load a cassette and set the VCR to the RECORD mode to record a television broadcast. Set the tape speed selector to BIII. Play back the recorded cassette.

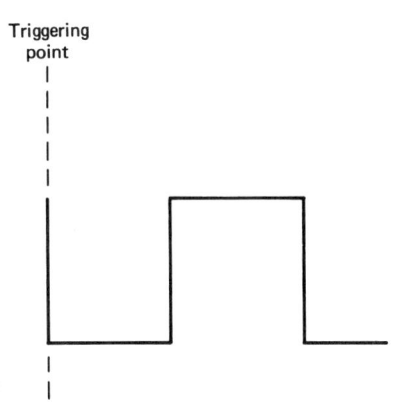

**FIGURE 6-11.** Tracking (2) adjustment waveform.

256  Typical Adjustment and Maintenance Procedures

2. Turn the tracking control to the center position, using the click stop.
3. Observe the playback control pulse at TP406 with an oscilloscope. Use the oscilloscope in the external triggering mode. Connect the external trigger probe to TP404. Set the triggering slope to "−."
4. Adjust VR406 so that the trailing edge of the playback control pulse coincides with the triggering point, as shown in Fig. 6-11.

### 6-1.5  Video Circuit Adjustments

Figure 3-2 shows electrical locations for the majority of the video system adjustment and measurement points. The procedures are as follows:

**3.58-MHz Oscillation Frequency**

1. Connect a frequency counter to TP9. Set the VCR to the PLAY mode.
2. Adjust T8 (with an alignment tool or special screwdriver) so that the oscillation frequency is 3.579545 MHz ± 5 Hz.

**$44\frac{1}{4}$-fH VCO Oscillation Frequency**

1. Load a blank cassette and set the VCR in the PLAY mode. Apply an external 8.5 ± 0.1 V dc to pin 11 of Q4.
2. Connect the frequency counter to TP8.
3. Adjust VR23 so that the oscillation frequency is 688.373 ± 1.5 kHz.

**Comb Filter**

1. Connect a color bar generator to the VIDEO IN terminal of the VCR. Set the VCR to STOP.
2. Connect the oscilloscope to TP1 and observe the color signal superimposed on the luminance signal, as shown in Fig. 6-12.
3. Adjust T3 and VR11 so that the color signal level is minimum.

**Color Video Level**

1. Connect a color bar generator to the VIDEO IN terminal of the VCR. Set the VCR to STOP.
2. Connect the oscilloscope to TP10 and observe the video signal level, as shown in Fig. 6-13.
3. Connect the oscilloscope to TP11 and observe the video signal level, as shown in Fig. 6-13.
4. Adjust VR12 so that the video signal levels of TP10 and TP11 are equal (Fig. 6-13).

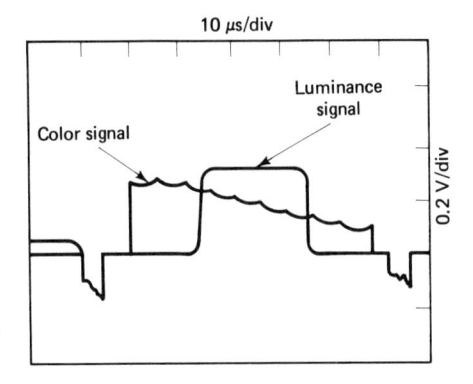

**FIGURE 6-12.** Comb filter adjustment waveforms.

### AGC (Automatic Gain Control)

1. Connect a color bar generator to the VIDEO IN terminal of the VCR. Set the VCR to STOP. Use the standard NTSC signal 1.0 V p-p (including a 0.29 V p-p sync signal) for the color bar signal. The VIT signal from a broadcast station can be used for this adjustment. The VIT signal is located on line 16, 17, or 18 of off-the-air TV signals. Be sure that the pulse representing peak white is used in this case. The peak white pulse is the most positive peak of the VIT signal.
2. Connect the oscilloscope to TP11 and observe the video signal level, as shown in Fig. 6-14.
3. Adjust VR14 so that the video signal level is maximum.
4. Adjust VR1 so that the video signal level is 1.1 + 0.05, -0 V p-p.
5. Readjust VR14 so that the video signal level is 1.05 ± 0.05 V p-p.

### ACC (Automatic Color Control)

1. Connect a color bar generator to the VIDEO IN terminal of the VCR. Set the VCR to STOP.

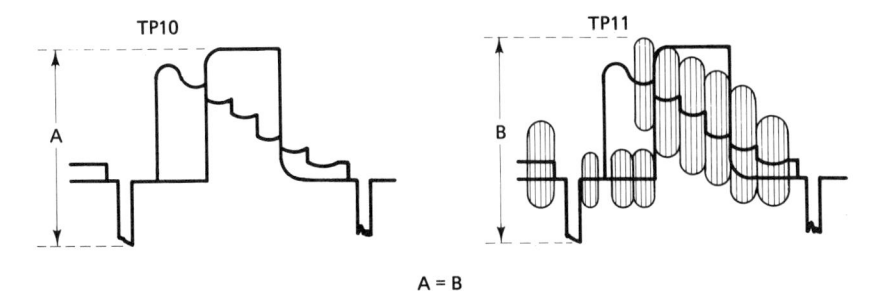

**FIGURE 6-13.** Color video level adjustment waveforms.

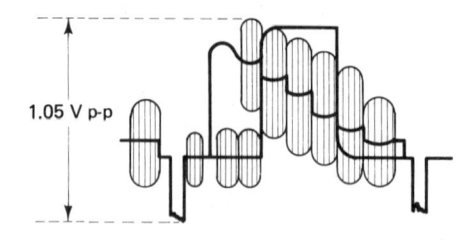

## 258 Typical Adjustment and Maintenance Procedures

**FIGURE 6-14.** AGC adjustment waveforms.

Connect the oscilloscope to TP14 and observe the waveform of the color signal, as shown in Fig. 6-15.

Adjust VR21 so that the color signal level is equal to 1.0 ± 0.01 V p-p.

### 4.27-MHz Bandpass Filter

1. Connect a color bar generator to the VIDEO IN terminal of the VCR. Set the VCR to STOP.
2. Connect the oscilloscope to TP7 and observe the 4.27-MHz sine-wave signal.
3. Adjust T9 and T10 so that the 4.27-MHz sine-wave signal is maximum.

### 4.27-MHz Carrier Leakage

1. Connect a color bar generator to the VIDEO IN terminal of the VCR. Set the VCR to STOP.
2. Connect the oscilloscope to pin 24 of Q3 and observe the waveform of the color signal, as shown in Fig. 6-16.
3. Adjust VR22 so that the 4.27-MHz carrier leakage on the baseline is minimum.

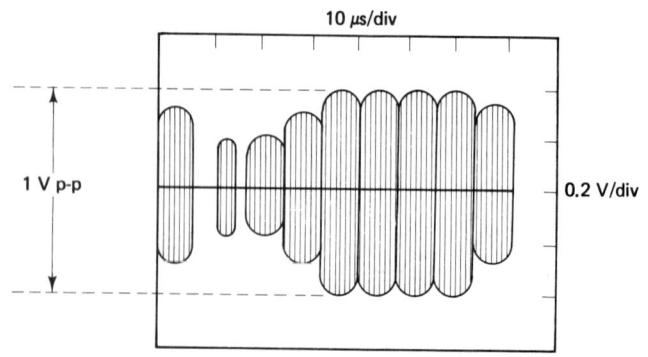

**FIGURE 6-15.** ACC adjustment waveforms.

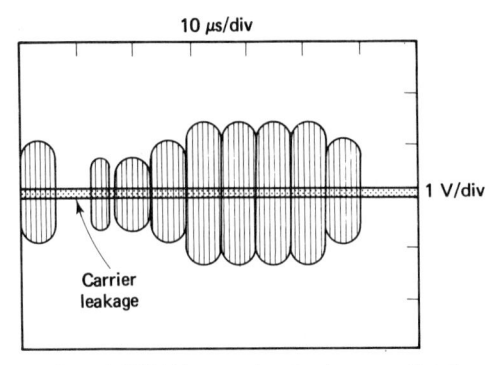

**FIGURE 6-16.** 4.27-MHz carrier leakage adjustment waveforms.

### E-E Video Level

1. Connect a color bar generator to the VIDEO IN terminal of the VCR. Set the VCR to STOP. Terminate the VIDEO OUT terminal with a 75-Ω resistor.
2. Connect the oscilloscope to TP12 and observe the video signal as shown in Fig. 6-17.
3. Adjust VR20 so that the video signal level is equal to 1.0 ± 0.05 V p-p.

### Playback Level of Luminance Signal

1. Play back a standard tape or an alignment tape that contains a color bar signal.
2. Connect the oscilloscope to pin 21 of Q1 and measure the luminance signal level.
3. Adjust VR2 so that the luminance signal level is equal to 0.8 ± 0.05 V p-p.

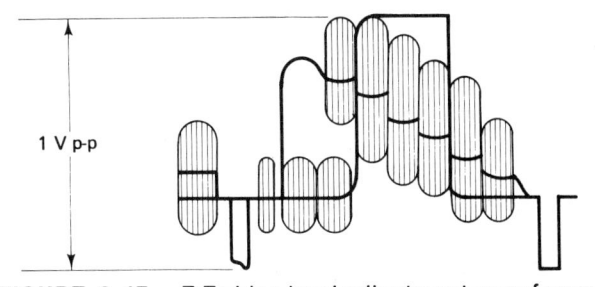

**FIGURE 6-17.** E-E video level adjustment waveforms.

## FM Frequency Deviation

1. Connect a color bar generator to the VIDEO IN terminal of the VCR. Set the VCR to RECORD and record the color bar signal.
2. Connect the oscilloscope to TP3 and measure the luminance signal level as shown in Fig. 6-18.
3. Adjust VR17 so that the luminance signal level is equal to 0.7 ± 0.1 V p-p.

## Carrier Frequency

1. Remove all signal inputs to the VCR. Connect a plug to the VIDEO IN terminal but do not apply a video signal. This plug is necessary because an automatic input selector turns to the tuner input side if the plug is omitted. Apply an external 6.0 ± 0.1 V dc to pin 3 of Q1.
2. Load a cassette and set the VCR to RECORD.
3. Adjust VR15 so that the carrier frequency measured at TP5 (with a frequency counter) is 3.6 ± 0.04 MHz.

## White Clip

1. Connect a color bar generator to the VIDEO IN terminal of the VCR. Set the VCR to RECORD and record the color bar signal.
2. Measure the voltage across C104 with a multitester or digital meter and adjust VR18 so that the voltage is equal to 0.4 V.

## Black Clip

1. Connect a color bar generator to the VIDEO IN terminal of the VCR. Set the VCR to RECORD and record the color bar signal.
2. Connect the oscilloscope to TP3 and observe the waveform, as shown in Fig. 6-19.

**FIGURE 6-18.** FM frequency deviation adjustment waveforms.

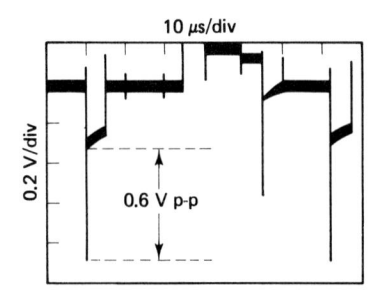

**FIGURE 6-19.** Black clip adjustment waveforms.

3. Adjust VR19 so that the level of the "shoot" extending downward from the front edge of the synchronizing signal level is equal to 0.6 V p-p.

**$\frac{1}{2}$fH Carrier Level Setting**

1. Connect a color bar generator to the VIDEO IN terminal of the VCR. Set the VCR to RECORD and record the color bar signal.
2. Connect the oscilloscope to TP2 and observe the carrier setting waveform.
3. Adjust VR16 so that the carrier setting level is equal to 6 V p-p.

**Head Resonance**

1. Set VR3 and VR4 fully to the right and set VR7, VR8, and VR9 to the center position.
2. Play back an alignment tape using the RF sweep portion of the tape. Connect the oscilloscope to TP4 and observe the waveform (envelope) of the playback FM signal, as shown in Fig. 6-20.
3. Adjust VR8 (channel 1) and VR7 (channel B) so that both of their resonant frequencies are equal to 5.1 ± 0.1 MHz.

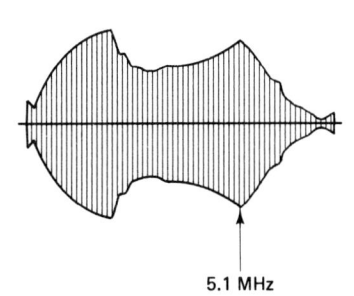

**FIGURE 6-20.** Head resonance adjustment waveform.

### Frequency Characteristics of Playback

1. Repeat the procedure described for head resonance. If the resonant level at 5.1 MHz is greater than the levels at 2 to 3.5 MHz, adjust VR3 (channel A) and VR4 (channel B) so that each channel has a flat frequency response within the range from 2 to 5.1 MHz, as shown in Fig. 6-21.
2. Adjust VR9 so that each channel has a flat frequency response within a range 2 to 3.5 MHz.
3. Repeat the procedures described for head resonance, and for this paragraph, as required to obtain the waveform shown in Fig. 6-21.

**Playback Balance.** After adjusting for proper head resonance and for proper frequency characteristics of playback, adjust VR9 so that the signal level at 3.58 MHz is equal for both channel A and channel B, as shown in Fig. 6-21.

### Dropout Compensator Sensitivity

1. Turn VR10 fully to the left, and play back a recorded tape that contains considerable drop out. Monitor the results on a TV screen.
2. Turn VR10 to the right until the dropout on the monitor picture is eliminated. Leave VR10 in that position.

### Frequency Characteristics of FM Recording Current

1. Connect a color bar generator to the VIDEO IN terminal of the VCR. Turn the generator color off, and set the generator controls to produce a black-and-white staircase signal (Sec. 2-3.3, Fig. 2-12). Load a cassette and set the VCR to RECORD.
2. Connect the oscilloscope to TP5 and observe the waveform of the FM recording current, as shown in Fig. 6-22.
3. Adjust VR5 so that the amplitude of the overlapped waves are equal.

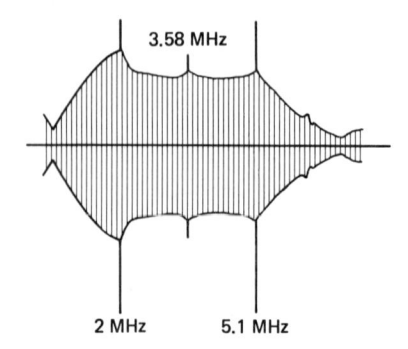

**FIGURE 6-21.** Playback frequency characteristics adjustment waveform.

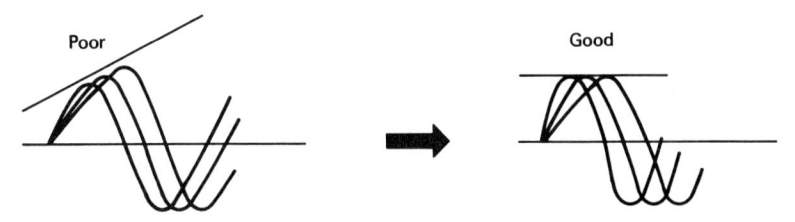

**FIGURE 6-22.** FM recording current frequency characteristics adjustment waveforms.

**FM Recording Current Level.** After adjusting for proper frequency characteristics of the FM recording current, adjust VR6 so that the FM recording current level is 160 mV p-p.

**Color Recording Current Level**

1. Connect a color bar generator to the VIDEO IN terminal of the VCR. (Set the generator to produce color bars.) Load a cassette and set the VCR to RECORD. Set the tape speed selector to BIII.
2. Connect the oscilloscope to TP5 and observe the waveform of the color recording current, as shown in Fig. 6-23.
3. Adjust VR25 so that the color burst level is equal to the maximum color recording current level (as shown for BIII in Fig. 6-23).
4. Adjust VR13 so that the color recording current level is 50 mV p-p.
5. Set the tape speed selector to BII and adjust VR24 so that the color recording current level of BII mode is equal to the color recording current level of BIII (Fig. 6-23).

## 6-1.6 Audio Circuit Adjustments

Figure 3-46 shows electrical locations for the majority of the audio system adjustment and measurement points. The procedures are as follows:

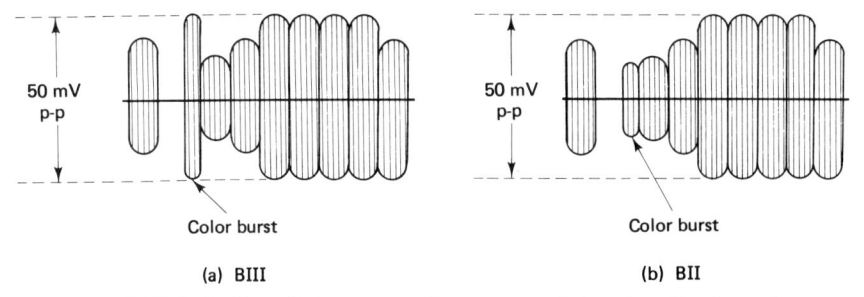

**FIGURE 6-23.** Color recording current level adjustment waveforms.

**EE Output Level**

1. Connect an audio frequency generator (Sec. 2-2) to the AUDIO IN terminal of the VCR. Set the generator output at 1 KHz, -10 dB (0.25 V rms). Set the VCR to STOP.
2. Connect an oscilloscope to TP701 and observe the output signal.
3. Adjust VR704 so that the output signal level is -5 dB (1.24 V p-p).

**Oscillation Frequency (1)**

1. Load a cassette and set the VCR to RECORD.
2. Connect a frequency counter (Sec. 2-6) to TP704 and adjust T701 so that the oscillation frequency is 70 ± 1 kHz.

**Oscillation Frequency (2)**

1. Load a cassette and set the VCR to AUD DUB (audio dubbing).
2. Connect a frequency counter to TP704 and adjust the L705 core so that the oscillation frequency is 70 ± kHz.

**Bias Leakage (1)**

1. Connect an audio generator to the AUDIO IN terminal of the VCR. Set the generator output at 1 kHz, -10 dB (0.25 V rms).
2. Load a cassette and set the VCR to RECORD.
3. Connect an oscilloscope to TP701 and observe the waveform of the output signals, as shown in Fig. 6-24.
4. Adjust the core of L702 so that the 70-kHz bias leakage on the output signal is minimum.

**Bias Leakage (2).** Repeat the procedure for bias leakage (1), but monitor the waveform at TP703 and adjust L703 for minimum bias leakage.

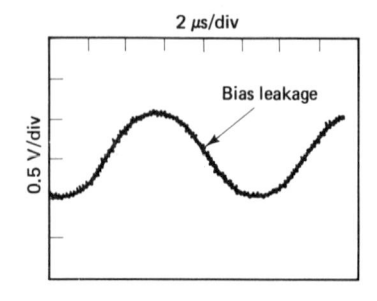

**FIGURE 6-24.** Bias leakage adjustment waveform.

## Bias Current

1. Load a cassette and set the VCR to RECORD, but do not apply an input signal. Connect a terminated audio plug to the AUDIO IN terminal.
2. Connect the oscilloscope to TP2502 (hot) and TP2501 (ground), and observe the bias waveform as shown in Fig. 6-25.
3. Adjust VR705 so that the level of the bias waveform is 51 mV p-p.

## Audio Recording Current

1. Connect an audio generator to the AUDIO IN terminal of the VCR. Set the generator output at 1 kHz, -10 dB (0.25 V rms).
2. Load a cassette and set the VCR to RECORD.
3. Connect the oscilloscope to TP703 and observe the waveform of the recording current.
4. Adjust VR706 so that the level of the recording current is 1.7 V p-p.

## Audio Playback Output Level

1. Connect an audio generator to the AUDIO IN terminal of the VCR. Set the generator output at 333 Hz, -10 dB (0.25 V rms).
2. Load a cassette and set the VCR to RECORD. Set the tape speed selector to BII. Play back the recorded cassette.
3. During playback, connect an oscilloscope to TP701 and observe the output signal.
4. Adjust VR701 so that the output signal is equal to -5 dB (1.23 V p-p).

## Playback Equalizer (1)

1. Connect an audio generator to the AUDIO IN terminal of the VCR. Set the generator output at 333 Hz, -10 dB (0.25 V rms).
2. Set the tape speed selector BII. Load a cassette and record a 333-Hz signal for 10 s. Change the audio generator frequency to 1 kHz and record the signal for 10 s.
3. Connect an oscilloscope to TP701 and play back the cassette.

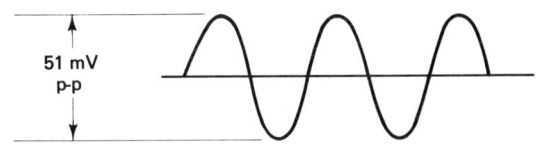

**FIGURE 6-25.** Bias current adjustment waveforms.

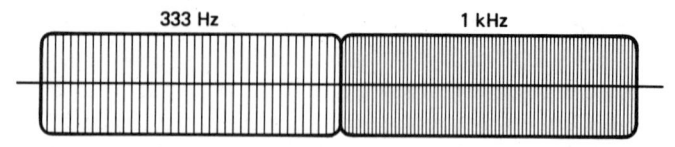

*266 Typical Adjustment and Maintenance Procedures*

**FIGURE 6-26.** Playback equalizer adjustment waveforms.

4. Observe the output signal and adjust VR702 so that the 1-kHz signal level is equal to the 333-Hz signal level, as shown in Fig. 6-26.

## 6-2 BETA MECHANICAL ADJUSTMENTS

The following paragraphs describe complete adjustment procedures for the mechanical section of the Beta VCR described in Chapters 3 and 5. The tools required for these mechanical adjustments are discussed in Sec. 2-8.

### 6-2.1 Reel Height Adjustment

If the reel height is different from the required standard value, the tape may be wound unevenly inside the cassette. When this tape is used again, the uneven winding may cause fluctuation in the tape tension, resulting in abnormal tape running. Therefore, reel height must be precisely adjusted.

1. Turn the power switch to OFF.
2. Remove the cassette holder.
3. Mount the cassette standard plate gauge VJ-0008 and slide the gauge in the direction indicated by the arrow, as shown in Fig. 6-27.
4. Adjust the height of the reel with spacers (0.2*T*), so that the gauge will be stopped by the B surface but not by the A surface. If the gauge is stopped by the A surface, remove one spacer and try again.
5. After the adjustment is completed, install the cassette holder assembly.

### 6-2.2 Back-Tension Lever Positioning

The following steps apply to back-tension lever positioning in the unloading state.

1. Remove the cassette cover.
2. Place the back-tension lever location gauge VJ-0084 in the position shown in Fig. 6-28.
3. If there is a clearance larger than 0.5 mm between the gauge and the back-tension lever guide, adjustment is necessary. First, loosen the cap

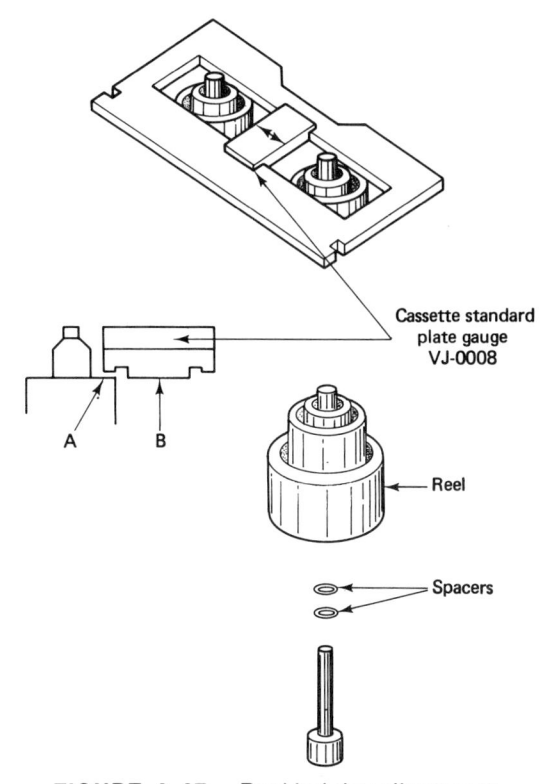

**FIGURE 6-27.** Reel height adjustment.

bolt (a) with the hexagonal wrench VJ-0024 and insert the eccentric screwdriver VJ-0018 into a hole noted with an arrow. Then move the back-tension operation lever in the direction indicated by the arrow for the clearance adjustment. (The clearance will decrease if the lever is moved towards A.)

4. After the adjustment, be sure to fasten the cap bolt (a) tightly.

The following steps apply to back-tension lever positioning in the loading state.

1. Remove the cassette holder.
2. Turn the power switch to ON.
3. Start the loading operation while holding down the cassette switch actuator with your fingertip.
4. Place the back-tension lever location gauge VJ-0084 in the position shown in Fig. 6-29.

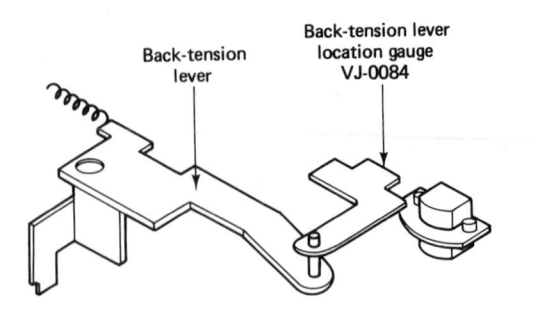

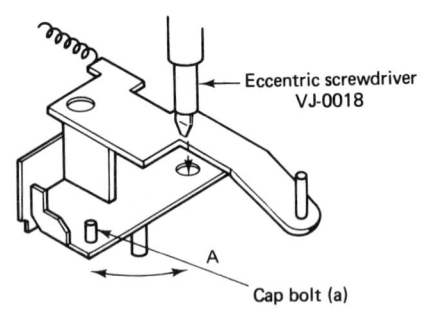

**268** Typical Adjustment and Maintenance Procedures

**FIGURE 6-28.** Back-tension lever positioning

5. If some clearance is noted between the gauge and the back-tension lever guide, loosen the screw (a) and adjust the clearance by sliding the brake band bracket in the direction indicated by the arrow. (The clearance will decrease if the bracket is shifted in the A direction.)
6. After the adjustment, be sure to fasten the screw (a) tightly.

### 6-2.3 Take-Up Reel Gear Clearance

The transmission of the driving force from the fast-forward roller (Fig. 6-30) to the take-up reel is made by means of the gear attached to the take-up reel. The amount of gear backlash (determined by the amount of gear clearance) is an important factor for the transmission of normal rotating force. Use the following procedure if it becomes necessary to adjust backlash. Note that increased clearance produces increased backlash.

1. Remove the cassette holder.
2. Loosen screw (a) (Fig. 6-30) that holds the fast-forward roller stopper.

Beta Mechanical Adjustments 269

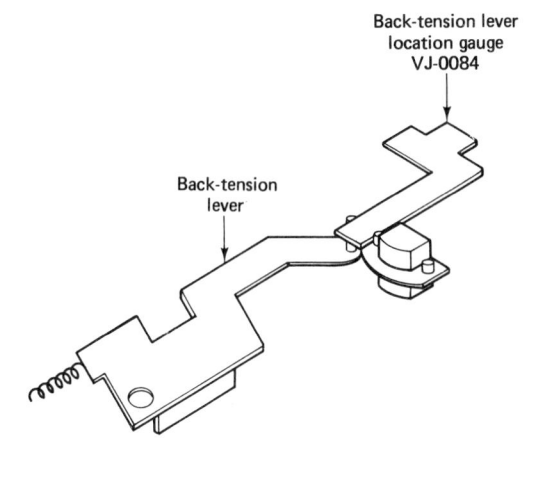

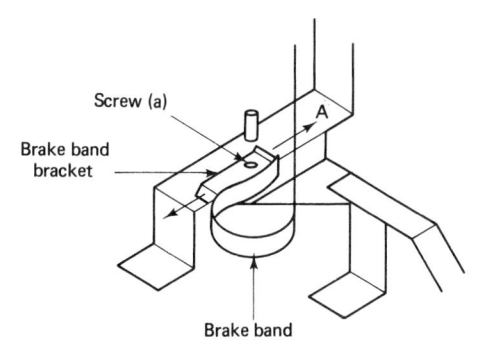

**FIGURE 6-29.** Back-tension lever positioning (in the loading state).

3. Insert the gear spacer VJ-0084 between the bottom surface of the take-up reel gear and the fast-forward roller. (Note that tool VJ-0084 is also used for back-tension lever positioning, as described in Sec. 6-2.2.)
4. Insert the eccentric screwdriver VJ-0018 into the hole marked with an arrow and move the fast-forward roller stopper to adjust the clearance.
5. After the adjustment, be sure to fasten screw (a) tightly.

### 6-2.4 Take-Up Reel Torque

In the PLAY mode, the tape is fed by the supply reel and taken up by the take-up reel. However, if the take-up torque of the take-up reel is not sufficient, the tape will not be taken up completely, resulting in slack tape. On the other hand, if the take-up torque is excessive, abnormal tension is applied to the tape and can cause the tape to stretch. *Both tape slack and tape stretching must be avoided on this or any VCR.*

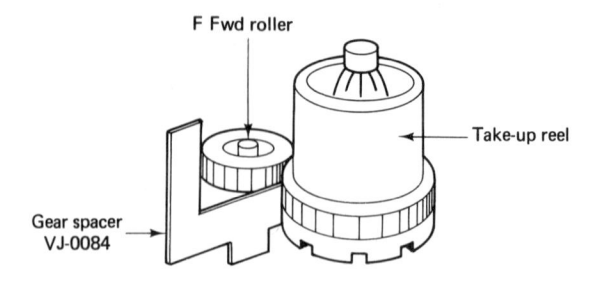

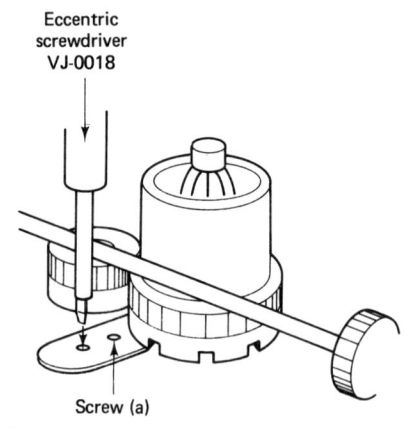

270  *Typical Adjustment and Maintenance Procedures*

**FIGURE 6-30.** Take-up reel gear clearance.

1. Turn the power switch OFF.
2. Remove the cassette holder assembly.
3. Turn the power switch ON and press the cassette switch actuator to start loading.
4. Place the tension gauge for reel VJ-0038 on the take-up reel, as shown in Fig. 6-31, and put the string on the spring tension gauge VJ-0012.
5. Push the PLAY button while pressing the cassette switch actuator. Make sure that the measured tension is within the range 20 to 55 g.
6. If the measured tension is not within the correct range, replace the clutch pulley assembly as described in Sec. 6-4.8.
7. Reinstall the cassette holder assembly.

## 6-2.5  F FWD and REW Torques

In the F FWD and REW modes, the tape fed by the supply reel or the take-up reel is taken up by the other reel. As in the case of the PLAY mode, excessive or insufficient torque can cause tape stretching or breakage in the fast-forward and re-

*Beta Mechanical Adjustments* **271**

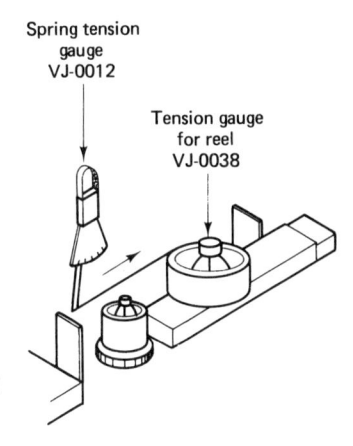

**FIGURE 6-31.** Take-up reel torque.

wind modes. Torque must be measured in both modes, and the clutch pulley assembly replaced if the torque is out of tolerance.

To measure torque in the fast-forward mode, repeat the procedures of Sec. 6-2.4, except use spring tension gauge VJ-0014 to measure for a torque of 280 g, or greater, as shown in Fig. 6-32.

To measure torque in the rewind mode, repeat the procedures of Sec. 6-2.4, except use spring tension gauge VJ-0014 to measure for a torque of 280 g, or greater, as shown in Fig. 6-33.

## 6-2.6 Capstan Pulley Height

If the heights of the capstan pulley and the capstan flywheel are different, wow may increase and the capstan belt can run off of the pulley. It is therefore essential that the capstan pulley height must be adjusted precisely.

1. Put the VCR on a flat surface with the tuner side down. Open up the circuit board to gain access to the belt.
2. Take off the capstan belt shown in Fig. 6-34.

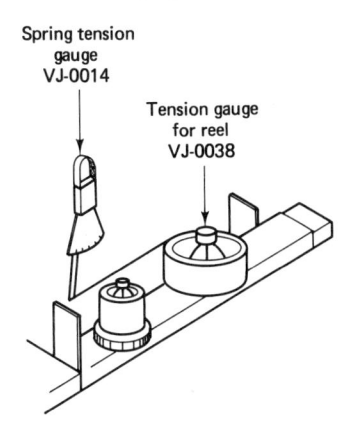

**FIGURE 6-32.** Fast forward torque.

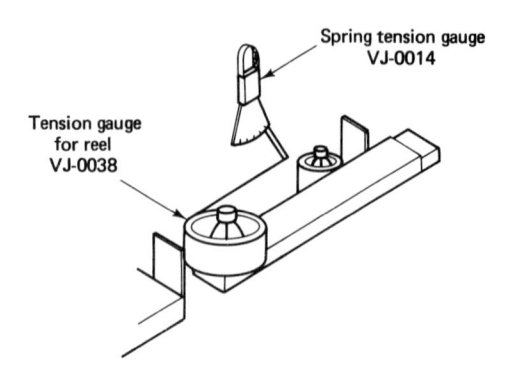

272  Typical Adjustment and Maintenance Procedures

**FIGURE 6-33.** Rewind torque.

3. Loosen the two screws (a) that hold the capstan pulley, using the hexagonal wrench VJ-0022.

4. Adjust the capstan pulley position so that the distance between the motor bracket and the edge of the capstan pulley becomes 16.9 ± 0.2 mm, as shown in Fig. 6-34. After adjustment, fasten the two screws (a).

5. Install the capstan belt and make sure that the capstan belt moves properly by rotating the capstan flywheel several times clockwise with your hand or the reel motor pulley several times. *Be very careful not to twist the belt.*

### 6-2.7  Reel Motor Pulley Height

If the heights of the reel motor pulley and the clutch pulley are not correctly adjusted, the reel belt may vibrate. This may sometimes cause abnormal sound and possible defacement or other damage to the belt. It is therefore essential that the capstan pulley height must be adjusted precisely.

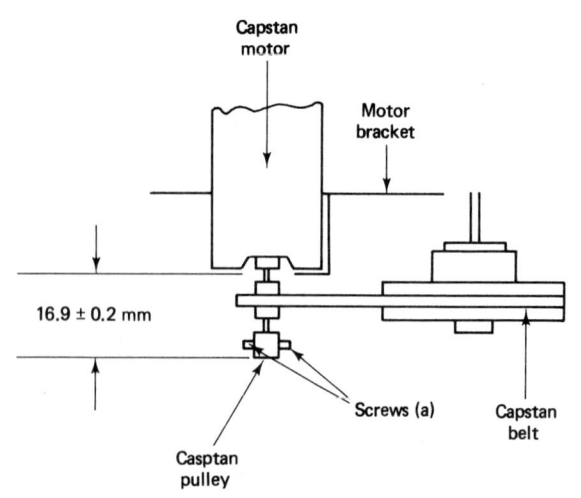

**FIGURE 6-34.** Capstan pulley height.

1. Put the VCR on a flat surface with the tuner side down. Open up the circuit board to gain access to the belt.
2. Take off the reel belt shown in Fig. 6-35.
3. Loosen the two screws (a) that hold the reel motor pulley, using the hexagonal wrench VJ-0022.
4. Adjust the reel motor pulley position so that the distance between the chassis and the edge of the reel motor pulley becomes 12.4 ± 0.2 mm, as shown in Fig. 6-35.
5. Install the reel belt on the pulley and make sure that the belt moves properly when rotating the pulley. Be careful not to twist the belt.

### 6-2.8 Tape Path

The tape path for most VCRs is critical to proper operation. For this reason, the position and height of the tape guides and heads have been precisely adjusted at the factory. Since these components greatly affect normal tape running, *never touch the components unless necessary*. If it becomes necessary to replace any of the tape path components shown in Fig. 6-36, check operation of the VCR using an alignment tape and a known good monitor TV. If the playback is good, quit while you are ahead! If you have playback problems, then (and only then) make the following adjustments.

1. Turn the power switch ON.
2. Connect a good monitor TV to the VCR.
3. Connect an oscilloscope to TP4 (Fig. 3-2) on the W1 circuit board. Play back the alignment tape VJ-0037 and observe the waveform (envelope) on the oscilloscope. Adjust the tracking control so that the waveform (envelope) has a maximum amplitude on the oscilloscope. Figures 6-37 through 6-39 show typical waveforms.

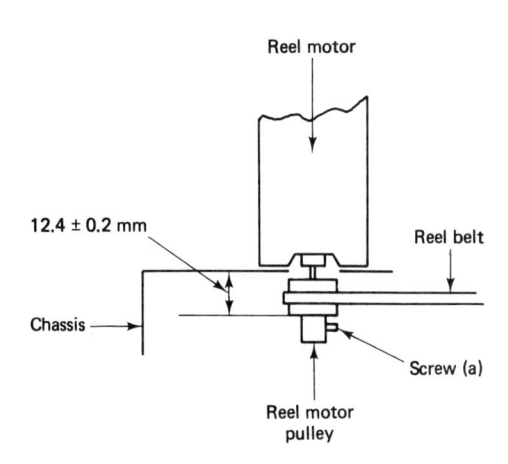

**FIGURE 6-35.** Reel motor pulley height.

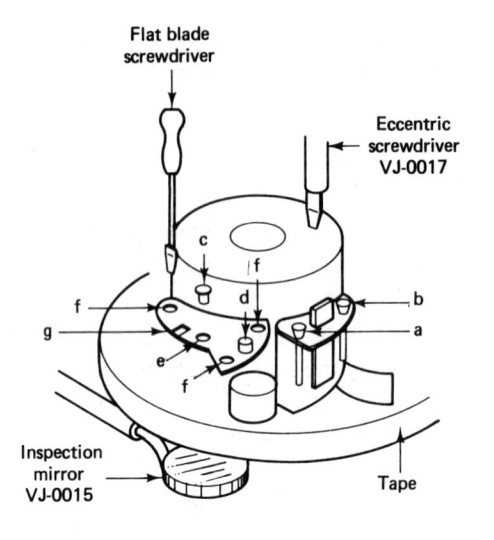

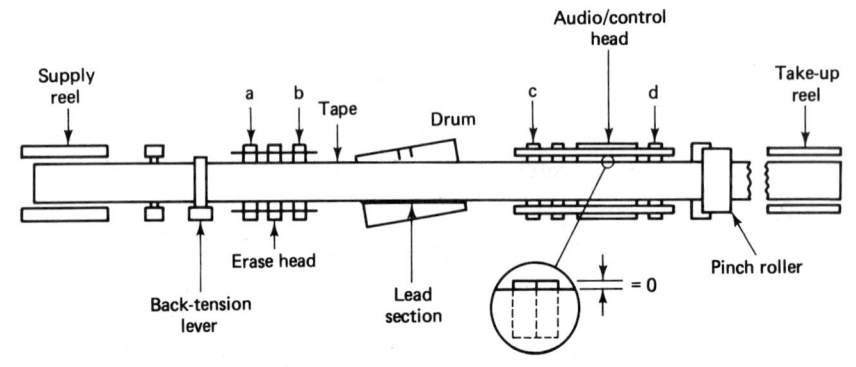

**FIGURE 6-36.** Tape path components.

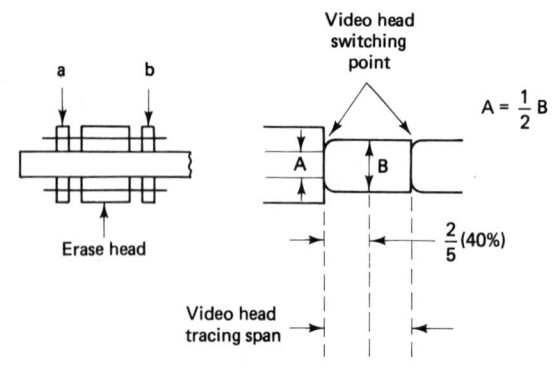

**FIGURE 6-37.** Erase head adjustment waveform.

$$A = \frac{1}{2} B$$

$$\frac{2}{5} (40\%)$$

274

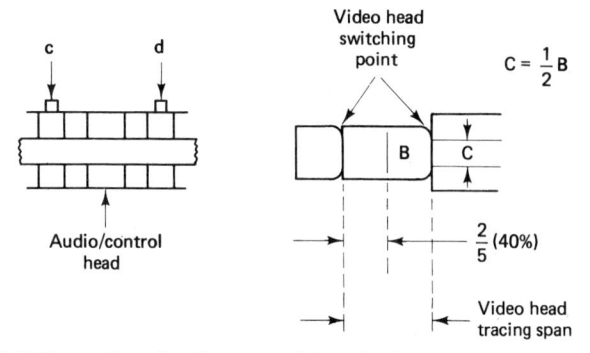

**FIGURE 6-38.** Audio/control head adjustment waveform.

4. Observe the running state of the tape around the back tension lever. When slack is noted in the tape at either the top edge or the bottom edge, *slightly* bend the back-tension lever (Fig. 6-36) with the bending jig VJ-0009 to eliminate the slack.

5. Adjust screw (b) using the eccentric screwdriver VJ-0017 so that the top edge of the tape does not hit against the guide at the side below screw (b).

6. Observe the waveform on the oscilloscope, and adjust screw (a) to make the waveform amplitude at A equal to one-half that at B, as shown in Fig. 6-37. Note that A is at the video head switching point, and B is two-fifths or 40% of the video head tracing span.

7. Using the inspection mirror VJ-0015 as shown in Fig. 6-36, check that slack does not develop along (a), (b), and the lead section during the following adjustment. Adjust screw (b) for the optimum amplitude at A, as discussed in step 6.

8. Adjust screw (c) so that the tape top edge does not hit against the guide below. Then adjust screw (d) to make the waveform amplitude at C equal to one-half that at B, as shown in Fig. 6-38.

9. During the adjustment of step 8, check with the inspection mirror that the tape bottom edge is steadily in contact with the flange shoulder below screw (d).

10. While checking the tape with the inspection mirror, adjust screw (c)

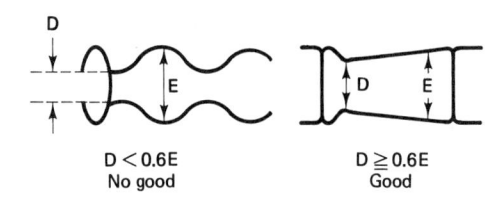

**FIGURE 6-39.** Acceptable ratios for video head waveform.

$D < 0.6E$
No good

$D \geq 0.6E$
Good

for the optimum waveform at C so that there is no slack in the tape along (c), (d), and the lead section.

11. Ideally, the center portion of the video head tracing span waveform should be flat, after all the adjustments are complete. Figure 6-39 shows acceptable ratios for the narrowest and widest portion of the envelope.

12. Disconnect the oscilloscope from TP4 and reconnect the oscilloscope to the AUDIO OUT terminal.

13. Play back the alignment tape VJ-0037 to observe an audio signal output waveform and adjust screw (e) for the maximum amplitude.

14. Change the oscilloscope connection from the AUDIO OUT terminal to TP4 and place the tracking control at the center position (click stop).

15. Loosen the three screws (f) (Fig. 6-36) slightly (about a quarter turn and insert the flat blade screwdriver into hole (g). Turn the flat blade screwdriver to the left and right, and find the best position where the waveform (envelope) shows the maximum amplitude on the oscilloscope. Then tighten the three screws (f).

16. Turn the tracking control to the right or left, and make sure that the waveform (envelope) on the oscilloscope changes symmetrically.

17. Check operation of the VCR by recording and playing back a program. If the playback is satisfactory, after making all the adjustments described in this section, you have made the adjustments correctly (or you are very lucky!).

## 6-3 CLEANERS, LUBRICATION OILS, AND MAINTENANCE TIMETABLES

Figure 6-40 is the recommended maintenance timetable for the VCR described in Chapters 3 and 5 and is typical for the great majority of VCRs. However, *never lubricate or clean any part not recommended by the manufacturer.* Most VCRs use sealed bearings that do not require lubrication. A drop or two of oil in the wrong places can cause damage! Clean off any excess or spilled oil. In the absence of a specific recommendation, use a light machine oil such as sewing machine oil. Note that only the supply reel, take-up reel, fast-forward roller, clutch pulley, and rewind idler require lubrication, as shown in Fig. 6-40.

Although there are spray cans of head cleaner, most manufacturers recommend alcohol and cleaning sticks or wands for all cleaning. Methyl alcohol does the best cleaning job but can be a health hazard. Isotropyl alcohol is usually satisfactory for most cleaning.

Cleaning and Lubrication Procedures 277

| Names of components \ Intervals (hours) | 500 | 1000 | 1500 | 2000 | 2500 | 3000 | 3500 | 4000 | 5000 |
|---|---|---|---|---|---|---|---|---|---|
| Video heads | △ | △ | △ | △ | △ | △ | △ | △ | △ |
| Audio/control head | △ | △ | △ | △ | △ | △ | △ | △ | △ |
| Pinch roller | △ | △ | △ | △ | △ | △ | △ | △ | △ |
| Erase head | △ | △ | △ | △ | △ | △ | △ | △ | △ |
| Supply reel | | | | △ ▲ | | | | △ ▲ | |
| Take-up reel | | | | △ ▲ | | | | △ ▲ | |
| F Fwd roller | | ▲ | | △ ▲ | | △ ▲ | | △ ▲ | △ ▲ |
| Clutch pulley | | | | ▲ | | ▲ | | ▲ | ▲ |
| REW idler | | ▲ | | △ ▲ | | △ ▲ | | △ ▲ | △ ▲ |
| F FWD roller | | △ | | △ | | △ | | △ | △ |
| Capstan assembly | | △ | | △ | | △ | | △ | △ |
| Loading gear | | △ ▲ | | △ ▲ | | △ ▲ | | △ ▲ | △ ▲ |

△ Cleaning

▲ Lubrication

**FIGURE 6-40.** Recommended maintenance timetable.

## 6-4 CLEANING AND LUBRICATION PROCEDURES

The following paragraphs describe the procedures to partially disassemble and reassemble the mechanical section of the VCR to perform cleaning and lubrication.

### 6-4.1 Video Head Cleaning

1. Turn the power switch to OFF, or pull out the power cord.
2. Rotate the video head disk by hand to a position convenient for cleaning the video head, as shown in Fig. 6-41. Moisten a cleaner stick with alcohol, lightly press the buckskin portion of the stick against the video head drum, and move the head disk by turning the motor back and forth.
3. Clean both video heads (on opposite sides of the drum) in the same way.

**CAUTION:** Do not move the cleaner stick vertically while in contact with the heads. Always clean the heads in the same direction as the tape path. Cleaning across the tape path can damage the heads.

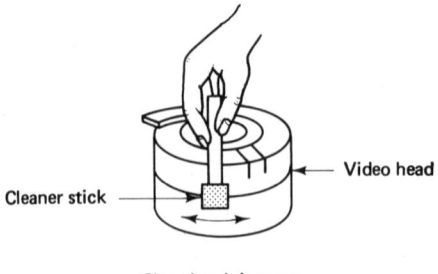

*278 Typical Adjustment and Maintenance Procedures*

Cleaner stick — Video head

Clean heads in same direction as tape path

**FIGURE 6-41.** Video head cleaning.

### 6-4.2  Audio/Control and Erase Head Cleaning

Moisten the cleaner stick with alcohol, press the stick against each head surface, and clean the heads by moving the stick horizontally, as shown in Fig. 6-42.

### 6-4.3  Pinch Roller Cleaning

1. Wipe off the surface of the pinch roller (Fig. 6-43) with a piece of soft cloth moistened with alcohol.
2. In case scratches or other damage are found on the pinch roller, remove screw (a) and replace the pinch roller. Scratches on the pinch roller can damage tape.

### 6-4.4  Tape Path Cleaning

Clean the drum surface and each surface of tape guides with a soft cloth moistened with alcohol. When cleaning the drum surface, be careful not to touch the video head with the cleaning cloth. Rotate the video head disk by hand to move the head away from the spot to be cleaned.

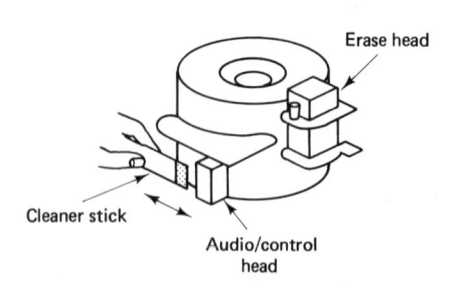

Erase head

Cleaner stick

Audio/control head

Clean heads in same direction as tape path

**FIGURE 6-42.** Audio/control and erase head cleaning.

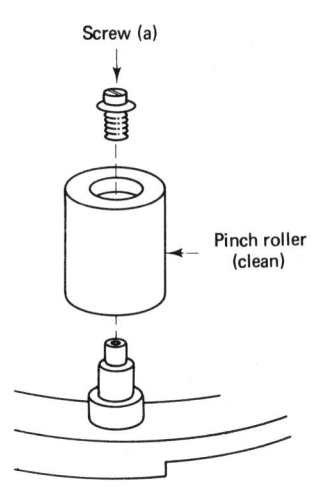

**FIGURE 6-43.** Pinch roller cleaning.

### 6-4.5 Supply Reel Cleaning and Lubrication

1. Remove the cassette holder assembly.
2. Remove screw (a), which fixes the brake band, and move the brake band from the supply reel (Fig. 6-44). Note that the brake band is fastened at the other end.
3. Remove screw (b), which fixes the supply reel, and detach the supply reel. Take care not to bend the brake band. In some cases there may be spacers for height adjustment under the bottom of the supply reel. Be careful not to lose the spacers. Make a note of how many (if any) spacers are used.
4. Using a soft cloth moistened with alcohol, clean off the old oil adhering to the reel shaft and apply a few drops of new oil to the reel shaft.
5. Replace the supply reel. When installing the supply reel, be careful not to bend the brake band.
6. Adjust the height of the reel if a new supply reel is used. Refer to Sec. 6-2.1.
7. Perform the back-tension lever positioning adjustment described in Sec. 6-2.2

### 6-4.6 Take-Up Reel Cleaning and Lubrication

1. Remove the cassette holder assembly.
2. Remove the counter belt.
3. Remove the take-up reel by unfastening screw (a), as shown in Fig. 6-45. In some cases, there may be spacers for height adjustment under

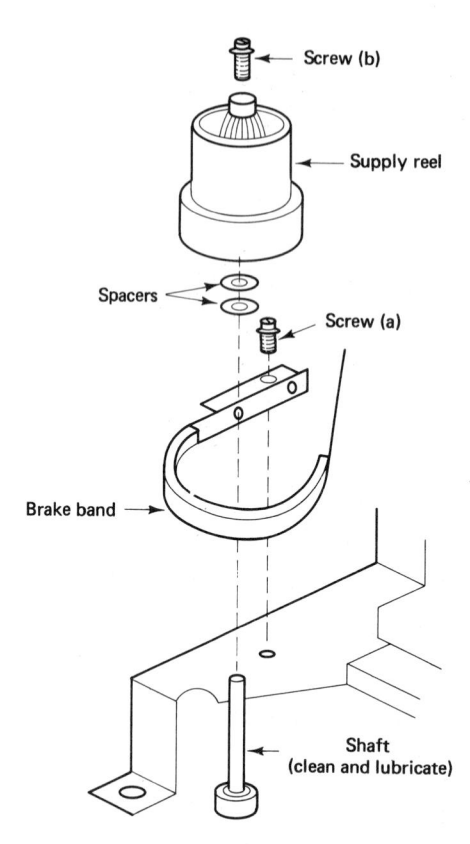

**280** Typical Adjustment and Maintenance Procedures

**FIGURE 6-44.** Supply reel cleaning and lubrication.

the bottom of the supply reel. Be careful not to lose the spacers. Make a note of how many (if any) spacers are used.

4. Using a soft cloth moistened with alcohol, clean off the old oil adhering to the reel shaft, and apply a few drops of new oil to the reel shaft.
5. Replace the take-up reel.
6. Adjust the height of the reel if a new take-up reel is used. Refer to Sec. 6-2.1.
7. Reinstall the counter belt, being careful not to twist the belt.
8. Reinstall the cassette holder assembly and the cassette holder.

### 6-4.7 Fast-Forward Roller Cleaning and Lubrication

1. Remove the cassette holder assembly.
2. Place the VCR in an upright position with the tuner side down, and open up the W1 circuit board.

*Cleaning and Lubrication Procedures* 281

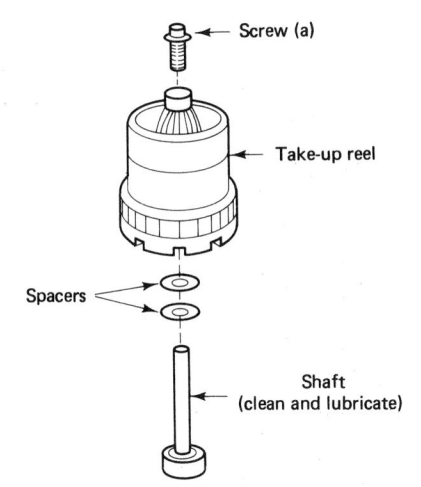

**FIGURE 6-45.** Take-up reel cleaning and lubrication.

3. Disengage the reel belt (b) from the fast-forward pulley, as shown in Fig. 6-46.
4. Using the hexagonal wrench VJ-0022 to remove screws (a), and remove the fast-forward pulley. Be careful not to lose the polyslider (or a plastic washer) used in the pulley assembly.

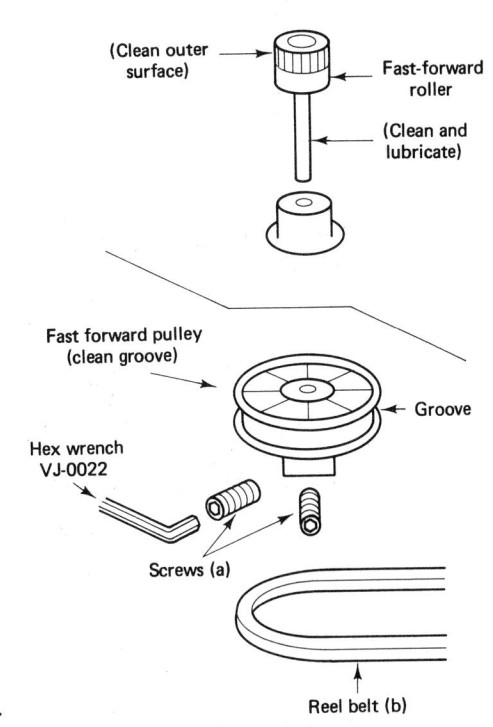

**FIGURE 6-46.** Fast-forward roller cleaning and lubrication.

5. Clean the groove of the pulley with a piece of soft cloth moistened in alcohol.
6. Pull off the fast-forward roller (by pulling upward). Be careful not to lose the polyslider used under the roller.
7. Clean the outer surface of the roller with a soft cloth moistened in alcohol.
8. Using a piece of soft cloth moistened in alcohol, clean off the old oil adhering to the roller shaft and apply a few drops of new oil to the roller shaft. Be careful not to damage the geared portion of the roller (which is made of plastic).
9. Reinstall the related components in the reverse order of disassembly, being careful not to twist the reel belt.

## 6-4.8 Clutch Pulley Cleaning and Lubrication

1. Remove the E-ring and polyslider from the clutch pulley, as shown in Fig. 6-47.
2. Place the VCR in an upright position with the tuner side down, and open up the W1 circuit board.
3. Disconnect the reel belt A and reel belt B from the clutch pulley.
4. Pull off the clutch pulley (downward), being careful not to lose the polyslider.
5. Clean each groove of the pulley with a soft cloth moistened with alcohol.
6. Using a soft cloth moistened with alcohol, clean off the old oil adhering to the pulley shaft, and apply a few drops of new oil to the shaft.
7. With the C-shaped portion of the clutch actuating lever forced against the chassis, insert the clutch pulley shaft into the bracket, and end-engage both belts A and B with the pulley, being careful not to twist the belts.
8. Fit the polyslider onto the pulley shaft and put on the E-ring. Apply a drop of new oil to the area where the E-ring is seated.

## 6-4.9 Rewind Idler Cleaning and Lubrication

1. Remove the cassette holder assembly.
2. Unfasten screw (a) to remove the loading ring stopper, as shown in Fig. 6-48.
3. Remove the E-ring before removal of the rewind idler.
4. Clean the outer surface of the idler with a soft cloth moistened with

*Cleaning and Lubrication Procedures* **283**

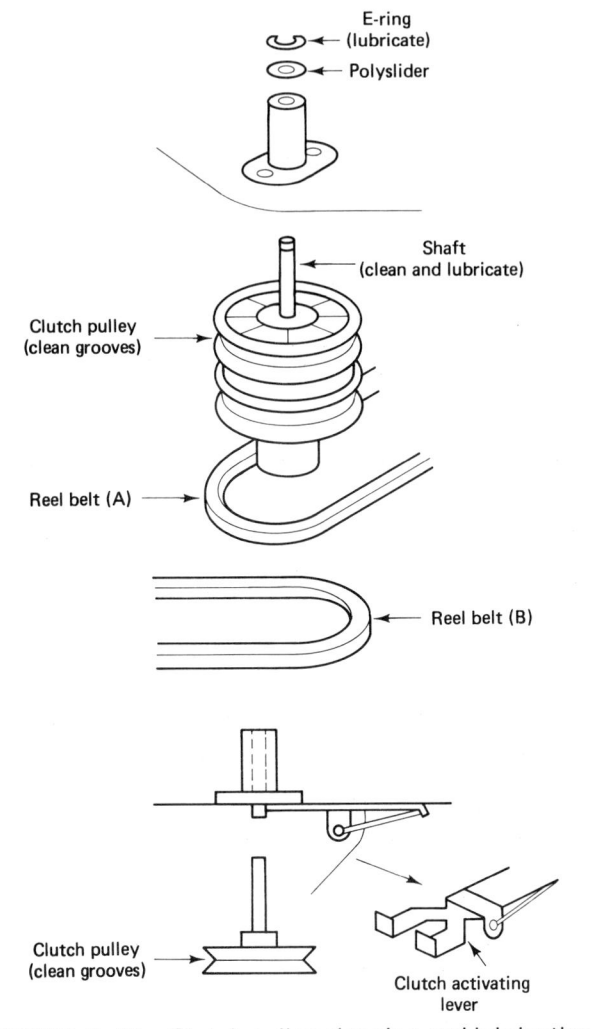

**FIGURE 6-47.** Clutch pulley cleaning and lubrication.

alcohol, and wipe off the old oil adhering to the idler shaft. Apply two or three drops of lubricating oil to the shaft.

5. Replace the rewind idler or install a new one.
6. Reinstall the components in the reverse order of disassembly.

## *6-4.10 Capstan Assembly Cleaning*

1. Place the VCR in an upright position with the tuner side down, and open up the W1 circuit board.
2. Remove the capstan belt.

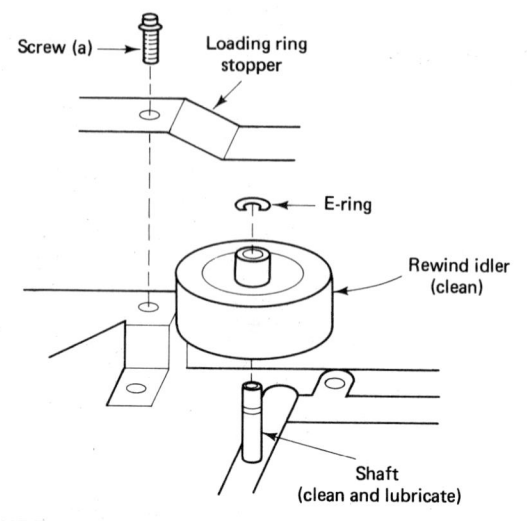

**FIGURE 6-48.** Rewind idler cleaning and lubrication.

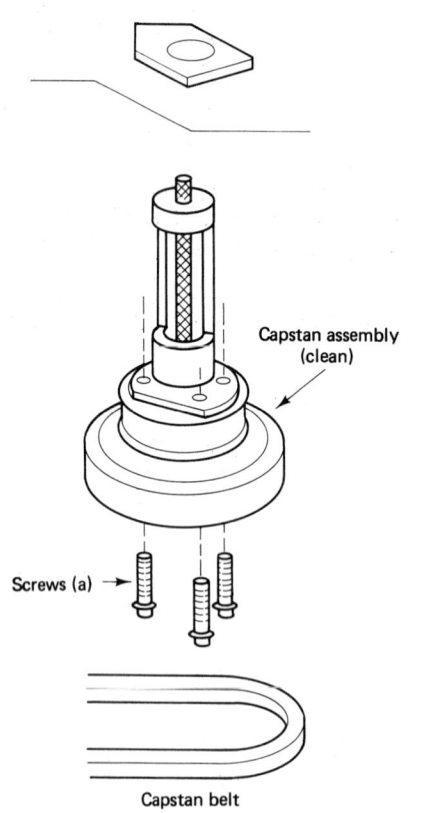

**FIGURE 6-49.** Capstan assembly cleaning.

3. Remove the three screws (a) with a screwdriver inserted through the corresponding hole found on the capstan flywheel, as shown in Fig. 6-49.
4. Rotate the bearing flange clockwise (seen from the VCR bottom) by about 5° to pull off the capstan assembly.
5. Clean both the spindle and the periphery of the capstan flywheel with a soft cloth soaked with alcohol.
6. Reinstall the components in the reverse order of disassembly.

### 6-4.11 Loading Gear Cleaning and Lubrication

1. Place the VCR in an upright position with the tuner side down, and open up the W1 circuit board.
2. Remove the capstan motor assembly.
3. Remove the two screws (a) to remove the loading gear assembly, as shown in Fig. 6-50.
4. Clean the geared portion of the loading gear assembly with a soft cloth moistened with alcohol.
5. Reinstall the components in the reverse order of disassembly.
6. On reinstallation, make sure that the teeth of the loading gear assembly are correctly engaged with the teeth of the loading ring.

### 6-4.12 Loading Ring Support Roller Cleaning and Lubrication

1. Remove the cassette holder assembly.
2. Remove one screw (a) to detach the loading ring stopper, as shown in Fig. 6-51.
3. Remove one screw (b) to detach the roller location bracket.

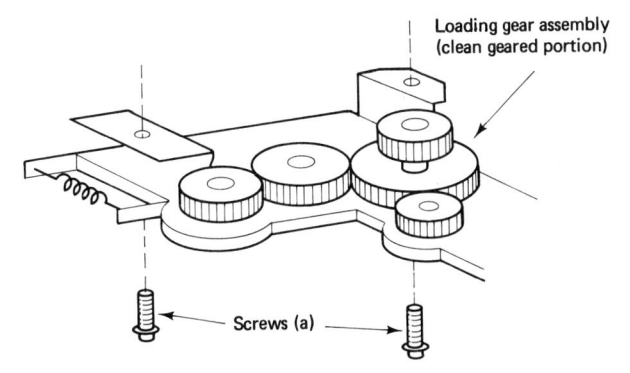

**FIGURE 6-50.** Loading gear cleaning and lubrication.

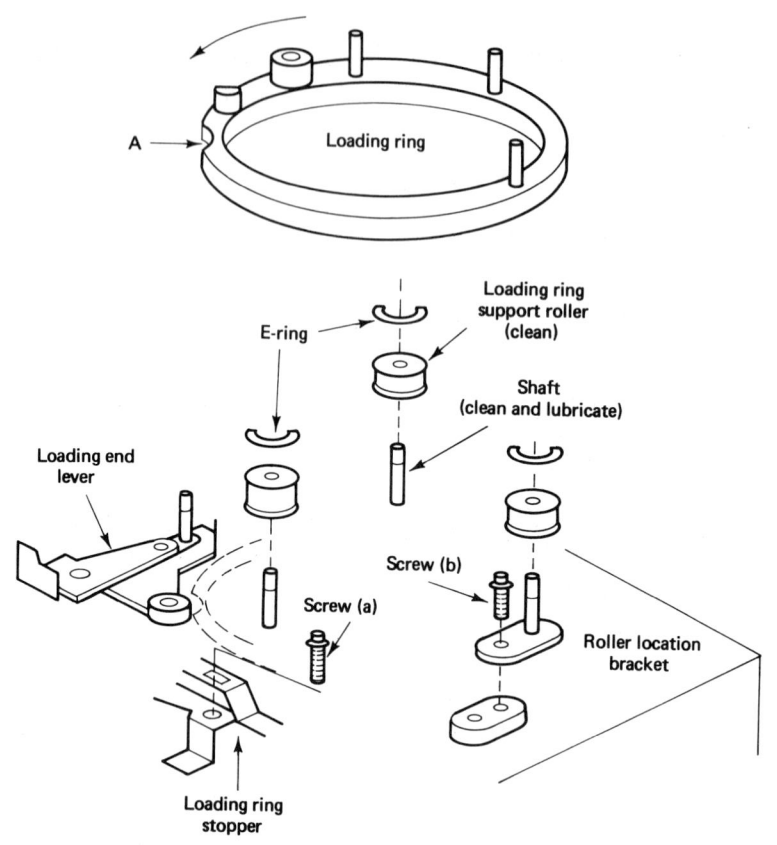

286  *Typical Adjustment and Maintenance Procedures*

**FIGURE 6-51.** Loading ring support roller cleaning and lubrication.

4. Rotate the loading ring counterclockwise by hand until the concave portion (marked with an A) of the loading ring faces the pinch roller pressing lever, as shown by the arrows in Fig. 6-51.
5. Push the loading end lever in the arrow direction and raise the loading ring. Carefully place the loading ring on a smooth, flat surface.
6. Disassemble the three loading ring support rollers by removing each corresponding E-ring.
7. Clean the inner surface of each loading ring support roller, and each corresponding shaft, with a soft cloth moistened with alcohol. Then lubricate each shaft with one or two drops of oil.
8. Reinstall the components in the reverse order of disassembly.
9. On reinstallation of the roller location bracket, make the concave portion of the bracket face the concave portion of the drum base before tightening the screw (b).

## 6-5 VHS ELECTRICAL ADJUSTMENTS

The following paragraphs describe complete adjustment procedures for the electrical sections of the VHS VCR described in Chapter 4. The test equipment required for these electrical adjustments are discussed in Chapter 2.

### 6-5.1 Servo Circuit Adjustments

The following paragraphs describe adjustment of the servo system discussed in Sec. 4-5.

**9.5V Power Supply Adjustment.** The purpose of this adjustment is to set the voltage of the servo system power supply.

1. Connect a d-c voltmeter (digital or multitester) to PG509-2 on the servo system PC board.
2. Turn the power ON and adjust R659 for a reading of 9.5 ± 0.1 V.

**Capstan Speed Adjustment.** The purpose of this adjustment is to make the tape speed approximately 33.4 mm/sec in the SP mode, 16.7 mm/sec in the LP mode, and 11.1 mm/sec in the EP mode.

1. Connect the color bar generator and oscilloscope to the servo system PC board as shown in Fig. 6-52.
2. Insert a blank tape and record a color bar signal at each of the adjustment tape speeds SP, LP, and EP. You can also make this adjustment by tuning in a local TV program, but the color bar generator provides a more stable test signal.
3. Connect the short circuit shown in Fig. 6-52a and adjust the capstan speed controls (Fig. 4-31) for each of the three tape speeds so that the sample pulse is locked on the trapezoidal waveform, or the sample pulse flows as slowly as possible on the trapezoid waveform, as shown in Fig. 6-52b. Note that Fig. 6-52 shows the three capstan speed controls for each of the three tape speeds.
4. Remove the short circuit and check that the sample pulse falls approximately on the center of the trapezoid leading edge, as shown in Fig. 6-52c.

**Cylinder Speed Adjustment.** The purpose of this adjustment is to make the cylinder speed 1800 rpm.

1. Connect the color bar generator and oscilloscope to the servo system PC board as shown in Fig. 6-52, except connect the oscilloscope to PG509-7 instead of PG509-5.

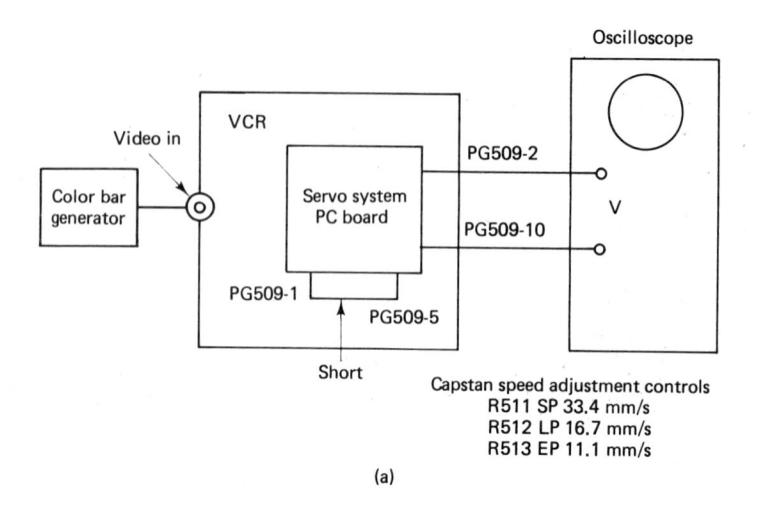

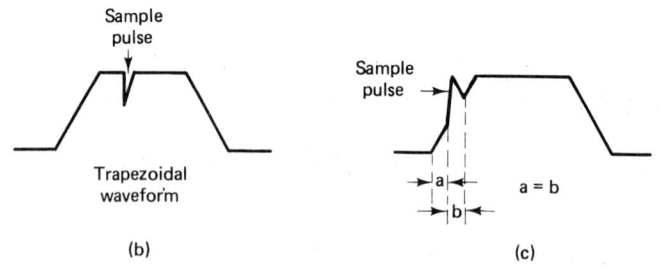

288  Typical Adjustment and Maintenance Procedures

**FIGURE 6-52.** Capstan speed adjustment connections and waveforms.

2. Insert a blank tape and record a color bar signal in the SP mode.
3. Connect the short circuit shown in Fig. 6-52a, and adjust the cylinder speed control R507 so that the sample pulse is locked on the trapezoidal waveform, or the sample pulse flows as slowly as possible on the trapezoidal waveform, as shown in Fig. 6-52b.
4. Remove the short circuit and check that the sample pulse fall approximately on the center of the trapezoid trailing edge, as shown in Fig. 6-53.

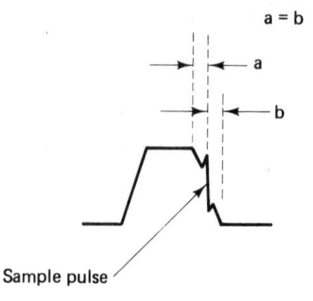

**FIGURE 6-53.** Cylinder speed adjustment waveform.

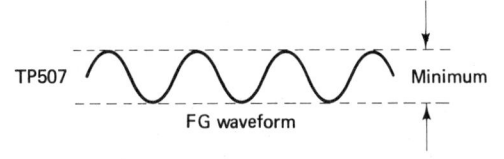

**FIGURE 6-54.** Cylinder FG level adjustment waveform.

### Cylinder FG Level Adjustment

1. Connect the oscilloscope to PG509-6 on the servo system PC board.
2. Insert a blank tape and load the VCR.
3. Adjust the cylinder FG level control R506 (Fig. 4-31) so that the ripple level is minimum, as shown in Fig. 6-54.

**Head Switching Point Adjustment.** This adjustment positions the video head switching point so that head switching occurs 6.5 horizontal lines (6.5H) before the start of the vertical sync pulse, or three lines before the start of the equalizing pulse, as shown in Figs. 6-55 and 6-56.

1. Connect an oscilloscope to the VIDEO OUTPUT of the VCR, and apply a synchronization signal to the oscilloscope from PG509-8 on the servo system PC board.
2. Insert an alignment tape and play back a color bar signal in the SP mode.

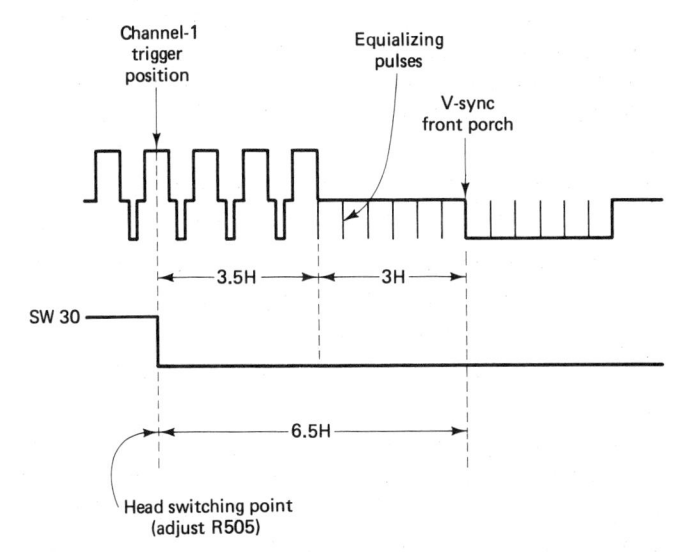

**FIGURE 6-55.** Channel 1 head switching point adjustment waveforms.

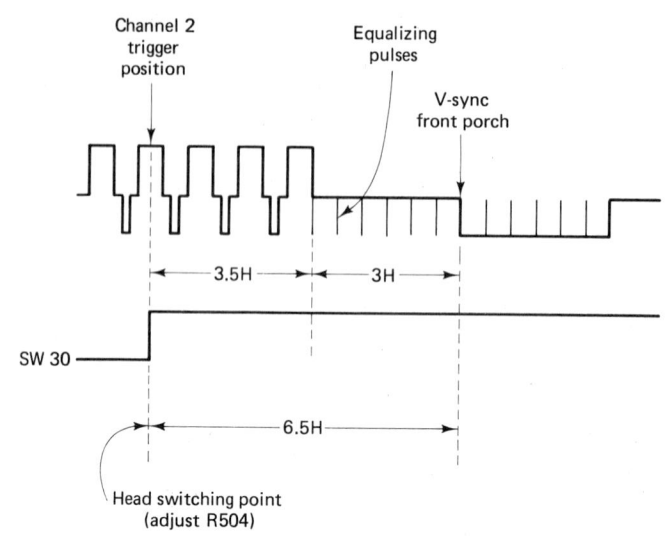

290  Typical Adjustment and Maintenance Procedures

**FIGURE 6-56.** Channel 2 head switching point adjustment waveforms.

3. Set the synchronizing range of the oscilloscope to external and set the sync slope to "−" for channel 1 adjustment and to "+" for the channel 2 adjustment.

4. Adjust the PG shifter control R505 (for channel 1) and R504 (for channel 2) shown in Fig. 4-21 so that the start of the trigger position is 6.5H before the start of the vertical sync pulses, as shown in Figs. 6-55 and 6-56.

It is difficult to count 6.5H. An alternative procedure is to adjust R504 and R505 so that the equalizing pulse is approximately 220 $\mu$s (3.5H) after the trigger position, as shown in Figs. 6-55 and 6-56.

If head switching occurs too soon, a narrow noise band may appear at the bottom of the picture. If the head switching pulse is late, noise can be introduced during vertical sync, possibly resulting in vertical sync problems.

**Record Shifter Adjustment**

1. Connect an oscilloscope to the VIDEO OUTPUT of the VCR and apply a synchronization signal to the oscilloscope from PG509-8 on the servo system PC board.

2. Connect a color bar generator to the VIDEO INPUT on the front panel or tune in a local TV program.

3. Insert a blank tape and record a color bar signal, or the TV program, in the SP mode.

4. Check the head switching point as shown in Figs. 6-55 and 6-56. If necessary, adjust R502 as described for "head switching point adjustment."

**Tracking Preset Adjustment (SP PRESET)**

1. Connect the oscilloscope to TP202 on the WYC PC board (the luminance and chroma circuits).
2. Connect the color bar generator to the VIDEO INPUT on the front panel or tune in a local TV program.
3. Insert a blank tape and record a color bar signal, or the TV program, for 1 to 2 minutes in the SP mode.
4. Set the TRACKING control RV502 to the center click or detent position. Adjust the tracking preset control R503 so that the FM wave level at TP202 is maximum, as shown in Fig. 6-57.

**SLOW Speed Adjustment**

1. Connect the color bar generator and oscilloscope to the servo system PC board as shown in Fig. 6-52 (same as for capstan speed adjustment).
2. Insert a blank tape and record a color bar signal in the EP mode for 1 to 2 minutes. You can also make this adjustment by tuning in a local TV program, but the color bar generator provides a more stable test signal.
3. Connect the short circuit shown in Fig. 6-52a. Play back, in the SLOW mode, the 1- to 2-minute program just recorded. Adjust the SLOW speed control R513 so that the pair of sampling pulses are locked on the two periodic trapezoidal waveforms, or the pair of sampling pulses flow as slowly as possible, as shown in Fig. 6-58. Remove the short circuit.

**CUE Speed Confirmation**

1. Connect a frequency counter to pin 8 of IC503 on the servo system PC board.
2. Connect a color bar generator to the VIDEO INPUT on the front panel, or tune in a local TV program.

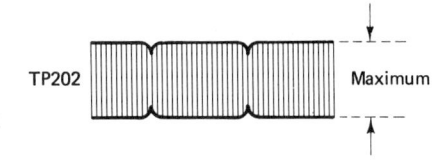

**FIGURE 6-57.** Tracking preset adjustment waveform.

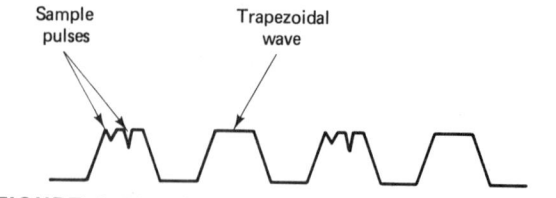

## 292  Typical Adjustment and Maintenance Procedures

**FIGURE 6-58.** Slow-speed adjustment waveforms.

3. Insert a blank tape and record the color bar signal, or the TV program, in the EP mode.
4. Play back, in the CUE mode, the program just recorded. Check that the frequency at pin 8 of IC503 is 2500 ± 150 Hz.

### CUE Offset Confirmation

1. Connect a frequency counter to pin 8 of IC502 on the servo system PC board.
2. Connect a color bar generator to the VIDEO INPUT on the front panel, or tune in a local TV program.
3. Insert a blank tape and record the color bar signal, or the TV program, in the EP mode.
4. Play back, in the CUE mode, the program just recorded. Check that the frequency at pin 8 of IC502 is 122 ± 1 Hz.

### REVIEW Offset Confirmation

1. Connect a frequency counter to pin 8 of IC502 on the servo system PC board.
2. Connect a color bar generator to the VIDEO INPUT on the front panel, or tune in a local TV program.
3. Insert a blank tape and record the color bar signal, or the TV program, in the EP mode.
4. Play back, in the REVIEW mode, the program just recorded. Check that the frequency at pin 8 of IC502 is 118 ± 1 Hz.

### Frame Advance Speed Adjustment

1. Insert a blank tape and record a color bar signal, or a TV program, in the EP mode.
2. Play back, in the PAUSE mode, the program just recorded.
3. Connect a jumper between PG509-10 and PG509-11.

4. Adjust the frame-advance speed control R510 so that the noise band appearing on the monitor TV flows 18 to 20 times in 10 seconds when the frame advance button is kept pressed in the PAUSE mode.

**QUICK Preset Confirmation**

1. Connect a color bar generator to the VIDEO INPUT on the front panel, or tune in a local TV program.
2. Insert a blank tape and record the color bar signal, or the TV program, in the EP mode.
3. Play back, in the QUICK mode, the program just recorded.
4. Adjust the quick preset control so that the noise band on the monitor TV is as shown in Fig. 6-59 (near the top of the screen).

## 6-5.2 Luminance and Color Circuit Adjustments

The following paragraphs describe adjustment of the color and luminance circuits discussed in Secs. 4-1 through 4-4.

**Head Resonance and Head Q Adjustment**

1. To adjust channel 1, set the VCR to play, without a cassette installed.
2. Connect an oscilloscope to TP-2C (FM output).
3. Connect a video sweep generator (Sec. 2-2) to TP-2A through the resistance pad shown in Fig. 6-60a. Adjust the sweep generator output to a level that does not distort the oscilloscope waveform.
4. Apply 1-MHz and 4.08-MHz markers to the sweep signal.
5. Adjust VC2A1 channel 1 resonance (Fig. 1-42) for maximum (or resonance) at 4.08 MHz.

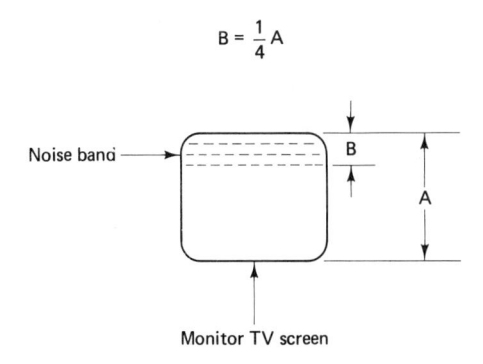

**FIGURE 6-59.** Quick preset confirmation adjustment.

**294** *Typical Adjustment and Maintenance Procedures*

Decoupling resistance pad

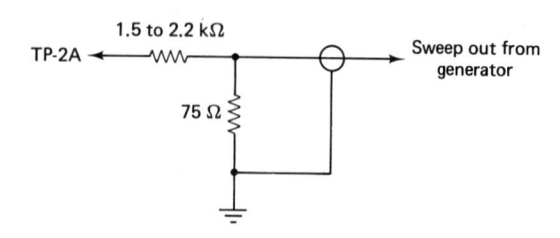

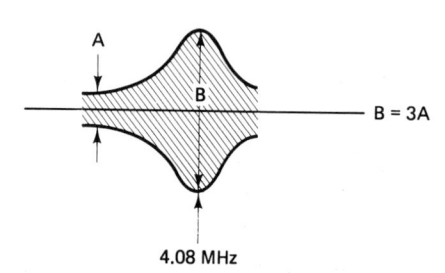

TP-2A ← —[1.5 to 2.2 kΩ]—•—→ Sweep out from generator

75 Ω

B = 3A

4.08 MHz

**FIGURE 6-60.** Head resonance and head Q adjustment connections and waveform.

6. Adjust VR2A1 channel 1 head Q so that the ratio of B to A in Fig. 6-60b is 3.

7. For channel 2, change the video sweep signal to TP-2B (channel 2 video head input).

8. Adjust VC2A0 channel 2 resonance for maximum (or resonance) at 4.08 MHz.

9. Adjust VR2A0 channel 2 head Q so that the ratio of B to A in Fig. 6-60b is 3.

    The following steps require a dual-trace oscilloscope with an ADD function. Dual-trace oscilloscopes are discussed in Sec. 2-4. If you are not familiar with the ADD function (or other features) of a dual-trace oscilloscope, your attention is invited to the author's best-selling *Handbook of Oscilloscopes, Revised and Enlarged* (Englewood Cliffs, N.J.: Prentice-Hall, Inc., 1982).

10. Connect oscilloscope channel A to TP-2C (FM output) and channel B to TP-2H. Invert the polarity of channel 2 for the ADD function.

11. Supply a video sweep signal to TP-2A (channel 1 video head input) and TP-2B (channel 2 video head input) at the same time, through the decoupling pad shown in Fig. 6-60a.

12. Overlap the waveforms of the channel 1 and channel 2 signals on the

oscilloscope, and adjust for equal level using VR2A2 (FM balance). Set the oscilloscope vertical deflection for 0.2 V/division, with line synchronization.

13. If either Q of channel 1 or channel 2 head cannot be 3 or more, decrease the highest Q to even both characteristics.

### FM Channel Balance Adjustment

1. Play the color bar pattern of an alignment tape.
2. Connect an oscilloscope to TP-2C (FM output). Use the 30-Hz drum pulse to synchronize the oscilloscope.
3. Adjust VR2A2 (FM balance) so that channel 1 and channel 2 FM outputs are at the same level as illustrated in Fig. 6-61.

### FM (Reproduced) Level Adjustment

1. Connect the oscilloscope to TP-2L (FM input).
2. Adjust VR2F4 to 0.5 V p-p.

### FM Record Level Adjustment

1. Apply a color bar signal to the video input and set the VCR to record.
2. Connect an oscilloscope to TP-2M (record output).
3. Adjust VR2G3 (FM record level) (Fig. 1-41) so that the white part level of the color bar is 3.0 V p-p, as shown in Fig. 6-62.

### Color Record and Playback Level Adjustment

1. Apply a color bar signal to the video input and set the VCR to record.
2. Connect an oscilloscope to TP-2Q (VR2G4 terminal).
3. Adjust VR2G4 (color record level) so that the burst signal level is 80 mV p-p.

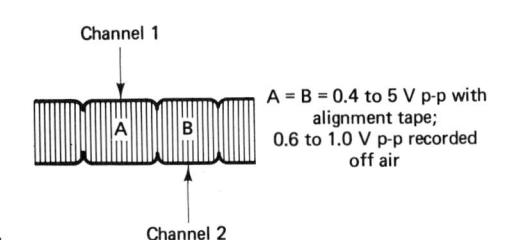

**FIGURE 6-61.** FM channel balance adjustment waveforms.

A = B = 0.4 to 5 V p-p with alignment tape; 0.6 to 1.0 V p-p recorded off air

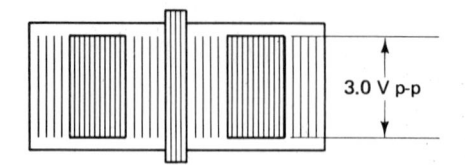

**FIGURE 6-62.** FM record level adjustment waveforms.

3.0 V p-p

4. In the play mode (recording and playback), connect the oscilloscope to TP-6A (playback color output) and check that the record level is the same as the playback level of the alignment tape.

**Self-recording FM Channel Balance Check**

1. Apply a color bar signal to the video input and set the VCR to record. Record and playback the color bar signal.
2. At the time of playback, connect the oscilloscope to TP-2C (FM output).
3. Check that the level difference between channel 1 and channel 2 is A/B ≤ 1.4 (the ratio of A to B is less than 1.4), as shown in Fig. 6-63.

**AGC Adjustment**

1. Apply a color bar signal to the video input and set the VCR to record.
2. Connect an oscilloscope to TP-2R and adjust VR201 for 2.0 V.

**Carrier Set and Deviation Adjustment.** This adjustment should be made *only* when there is a badly distorted waveform, or very bad signal-to-noise S/N ratio, or obvious misadjustment of the FM carrier and deviation, or when components of the FM modulation circuit are replaced. Before performing the following adjustments, several special connections are required. Connect a jumper between pins 1 and 3 of connector VM to keep the modulator and demodulator turned on at all times. Connect a signal generator (sine wave) to pin 2 of connector VS through a 1-kΩ resistor and a 0.01-μF capacitor.

1. Prior to this adjustment, turn the white clip adjust VR2F3 and dark clip adjustment VR2F9 fully clockwise.

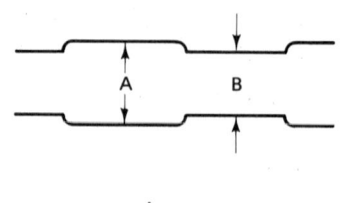

$\frac{A}{B} \leq 1.4$

**FIGURE 6-63.** Self-recording FM channel balance check.

2. Supply a staircase signal to the video input (Sec. 2-3.3, Fig. 2-12).
3. Connect an oscilloscope to pin 23 of IC2F3.
4. Set the frequency of the signal generator to 3.4 MHz ± 30 kHz.
5. Insert the cassette tape and place the VCR in SP (2-hour) record mode.
6. Adjust the FM carrier set adjustment VC2F2 for minimum carrier at the sync tips, as shown in Fig. 6-64.
7. Change the frequency of the signal generator from 3.4 MHz to 4.4 MHz ± 30 kHz.
8. Adjust the FM deviation adjust VR2G2 for minimum carrier at the white peak, as shown in Fig. 6-64.
9. Remove the jumper.
10. Connect an oscilloscope probe to TP-2G. Set the oscilloscope for 1 V/division and 20 $\mu$s/divison. Place the VCR in SP (2H) record mode and make a recording.
11. Play back the recorded tape.
12. Check that the video level is 2 ± 1 V p-p.

**White Clip and Dark Clip Adjustments**

1. Apply a color bar signal to the video input and set the VCR to record in EP.
2. Connect an oscilloscope to TP-2E.

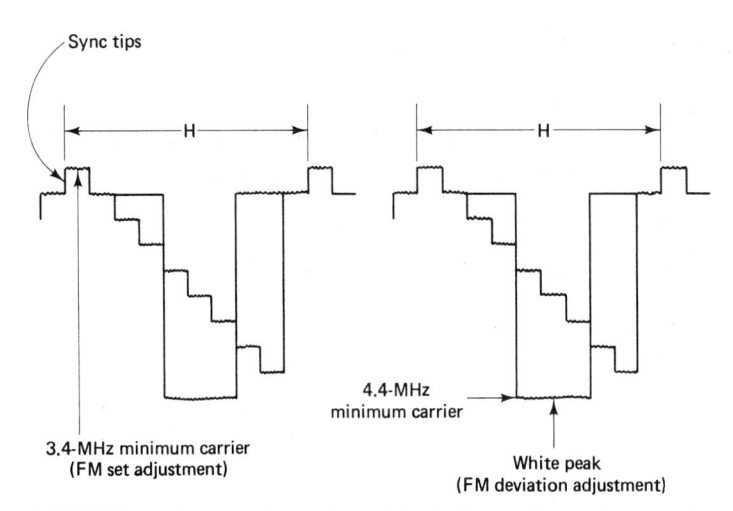

**FIGURE 6-64.** Carrier set and deviation adjustment waveforms.

3. Adjust VR2F3 (white clip) and VR2F9 (dark clip) for the waveform as specified in Fig. 6–65.

**E-E Output Level Adjustment**

1. Apply a color bar signal to the video input and set the VCR to the E-E mode.
2. Connect an oscilloscope to TP-2G (video output).
3. Adjust VR2G1 (EE level) for a 2 V p-p video signal.

**Limiter Balance 1**

1. Play the black-and-white (monoscope) portion of the alignment tape.
2. Observe the monitor TV and adjust VR2G7 (limit balance 1) for the least noise burst (fine noise that may appear along the fringe of a white picture).

**Limiter Balance 2**

1. Play the color bar portion of the alignment tape.
2. Connect an oscilloscope to TP-2K (demodulator output).
3. Adjust VR2G8 (limiter balance 2) and VR2G9 (carrier balance) so that the double waveform becomes single from sync tip to white peak, as shown in Fig. 6–66.

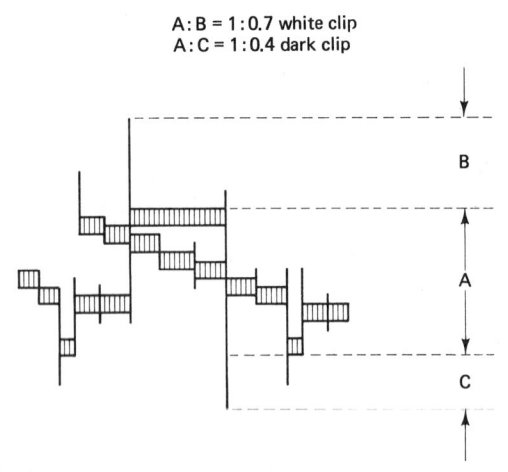

A : B = 1 : 0.7 white clip
A : C = 1 : 0.4 dark clip

**FIGURE 6-65.** White clip and dark clip adjustment waveforms.

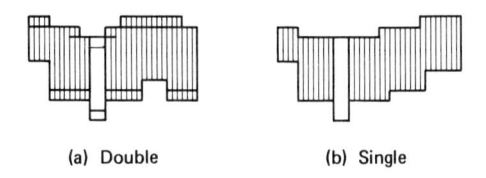

**FIGURE 6-66.** Limiter balance 2 adjustment waveforms.

(a) Double  (b) Single

### Reproduced Video Output Adjustment

1. Play the color bar portion of the alignment tape.
2. Connect an oscilloscope to TP-2G (video output).
3. Adjust VR2F6 (Y-level) so that the video signal is 2 V p-p, as shown in Fig. 6-67.

### Audio Interleave

1. Play the alignment tape in EP mode.
2. Connect an oscilloscope to TP-2S (vertical rate).
3. Adjust playback interleave VR2G5 (Fig. 4-7) so that the d-c level of the vertical synchronizing signals is in line.

### Duty Cycle Adjustment

1. Apply a color bar signal to the video input and set the VCR to the record mode.
2. Connect an oscilloscope to TP-6F. Use an external trigger from pin 3 of IC6F5.
3. Adjust VR6F3 so that the pulse width is 2 $\mu$s as shown in Fig. 6-68.

### AFC Adjustment

1. Apply a color bar signal to the video input and set the VCR to the E-E mode.
2. Connect the probes of a dual-trace oscilloscope to TP-6E and TP-6F.
3. Set the oscilloscope to the ADD mode.
4. Adjust VR6F4 for a waveform, as shown in Fig. 6-69.

**FIGURE 6-67.** Reproduced video ouput adjustment waveforms.

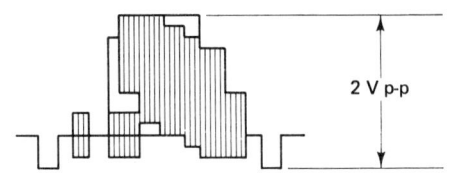

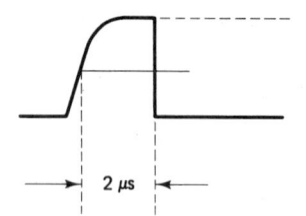

**FIGURE 6-68.** Duty cycle adjustment waveforms.

### Variable Crystal Oscillator (VXO) Adjustment

1. Reproduce the color bar of the alignment tape.
2. Ground TP-6D with a clip or short.
3. Connect a counter to TP-6G (subcarrier 3.58 MHz).
4. Adjust VXO VR6F2 (Fig. 1-43) so that the counter indicates 3.579545 MHz ± 10 Hz.

### Record Killer Operation Check

1. Apply a color bar signal to the video input and set the VCR to the record mode.
2. Check that the d-c voltage at TP-6H is more than 6 V.
3. Change the signal at the video input to a black-and-white signal. Leave the VCR in record mode.
4. Check that the d-c voltage at TP-6H is less than 0.3 V.

### Playback Killer Operation Check

1. Play back the gray scale of an alignment tape.
2. Check that the d-c voltage at TP-6H is less than 0.3. V.

### Crystal Oscillator Adjustment

1. Play back the color bar of an alignment tape.
2. Connect a counter to TP-6J (3.58-MHz crystal oscillator output).

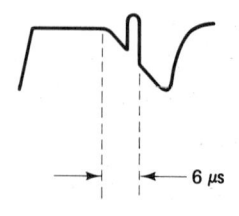

**FIGURE 6-69.** AFC adjustment waveforms.

3. Adjust 3.58-MHz oscillator VC6F0 (Fig. 1-43) so that the counter indicates 3.579545 MHz ± 10 Hz.

**Main Converter Balance Adjustment**

1. Play back the color bar of an alignment tape.
2. Connect an oscilloscope to TP-6C.
3. Adjust converter balance VR6F0 (Fig. 1-43) for minimum 4.21-MHz leakage component, as shown in Fig. 6-70.

**Reproduced Color Output Adjustment**

1. Play back the color bar of an alignment tape.
2. Connect an oscilloscope to TP-2G.
3. Adjust color output VR6F1 (Fig. 1-44) so that the burst amplitude is 0.5 V p-p as shown in Fig. 6-71.

## 6-5.3 Audio Circuit Adjustments

The following paragraphs describe adjustment of typical audio circuits.

**Audio/Control Head Azimuth Adjustment**

1. Connect an electronic voltmeter or VTVM to the audio output (rear panel).
2. Play back the 7-kHz audio signal portion of an alignment tape.

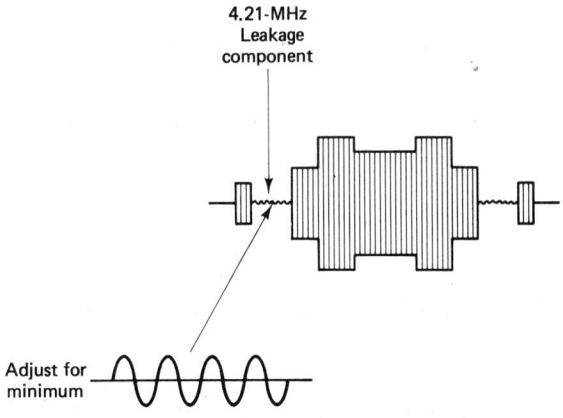

**FIGURE 6-70.** Main converter balance adjustment waveforms.

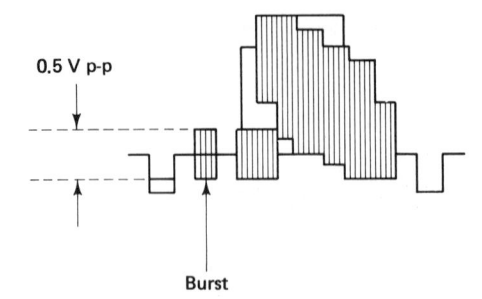

## 302 Typical Adjustment and Maintenance Procedures

**FIGURE 6-71.** Reproduced color output adjustment waveforms.

3. Adjust the azimuth adjustment screw shown in Fig. 6-72 for maximum audio output.

**Noise Canceler Adjustment**

1. Connect channel 1 of a dual-trace oscilloscope to TP411 and TP402 (ground). Connect channel 2 of the oscilloscope to the audio output (rear panel).
2. Insert a blank cassette and load the VCR.
3. Turn FG level control R482 fully clockwise, and adjust phase control R483 so that the phase of the channel 1 signal is 180° out of phase with the channel 2 signal, as shown in Fig. 6-73.
4. Adjust FG level control R482 for minimum amplitude, as shown in Fig. 6-73.

**Playback Gain Adjustment**

1. Connect an electronic voltmeter or VTVM to the audio output (rear panel).
2. Play back the 1-kHz, 0-dB audio signal portion of an alignment tape.

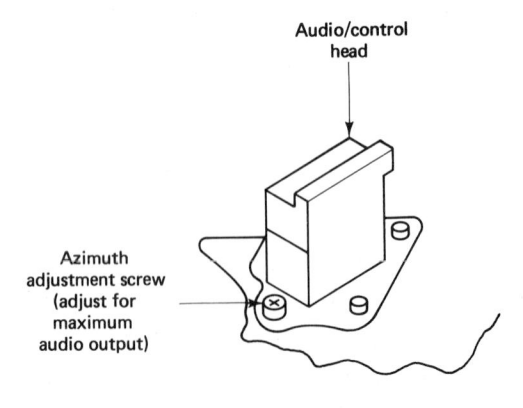

**FIGURE 6-72.** Audio/control head azimuth adjustment.

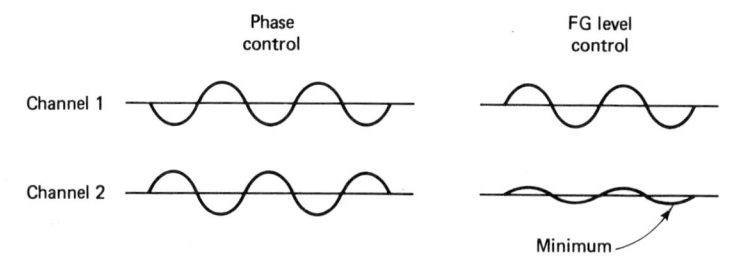

**FIGURE 6-73.** Noise canceler adjustment waveforms.

3. Adjust the playback gain control R481 so that the audio output becomes −4 dB, as indicated on the electronic voltmeter.

**Bias Trap Adjustment**

1. Connect an electronic voltmeter or VTVM to TP403 and TP404.
2. Connect a color bar generator to the video input.
3. Insert a blank cassette and record a color bar signal, or tune in a local TV program, in the SP mode.
4. Adjust the bias trap L403 for minimum output.

**Bias Level Adjustment**

1. Connect an electronic voltmeter or VTVM to the audio/control head as shown in Fig. 6-74.
2. Connect a color bar generator to the video input.
3. Insert a blank cassette and record a color bar signal, or tune in a local TV program, in the SP mode.
4. Adjust the bias level R484 for 4 mV on the electronic voltmeter in the SP mode.
5. Check for a bias level of 3 ± 0.2 mV in the LP and EP modes.

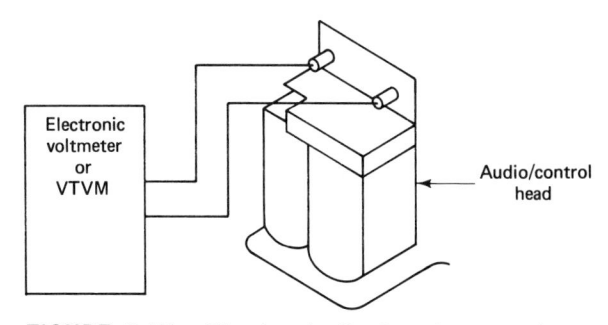

**FIGURE 6-74.** Bias level adjustment connections.

### 6-5.4 Tuner and IF Circuit Adjustments

The following paragraphs describe adjustment of typical tuner and IF circuits.

**Carrier Filter Adjustment (Minimum Output)**

1. Connect a pattern generator (Sec. 2-2.5) to the VHF antenna terminals.
2. Connect an oscilloscope to the test point TP of the IF circuits. This test point is between the tuner and the IF circuits and is similar to the "looker point" of a TV tuner. The test point makes it possible to monitor the output of the tuner and the input of the IF circuits.
3. Adjust the generator controls to produce a white pattern (crosshatch, etc.). This should result in a display on the oscilloscope as shown in Fig. 6-75.
4. Adjust the generator controls so that the input level at the IF terminal of the IF circuits is −41 ± 10 dBm.
5. Turn the core of carrier filter adjustment L805 gradually counterclockwise from the full clockwise position to a point where the spacing between the sync tip and white peak is minimum, as shown in Fig. 6-75.

**Carrier Filter Adjustment (1-V Output)**

1. Connect a pattern generator to the VHF antenna terminals.
2. Connect an oscilloscope to the video output terminal of the IF circuits. This test point makes it possible to monitor the output of the IF circuits and the input to the luminance and color recording circuits.
3. Adjust the generator controls to produce a white pattern (crosshatch, etc.). This should result in a display on the oscilloscope as shown in Fig. 6-76.

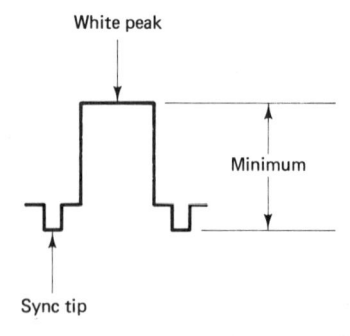

**FIGURE 6-75.** Carrier filter adjustment waveform (minimum output).

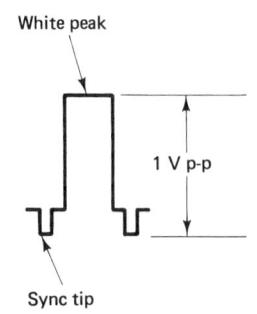

**FIGURE 6-76.** Carrier filter adjustment waveform (1 V output).

4. Adjust the generator controls so that the input level to the IF circuits is −41 ± 10 dBm (as measured at the input IF terminal).
5. Adjust R805 so that the spacing between the white peak and the sync tip is 1 V p-p, as shown in Fig. 6-76. Note that 1 V p-p output from the IF circuits to the luminance/color record circuits is typical for most VCRs, both VHS and Beta.

### 4.5-MHz Trap Adjustment

1. Connect a sweep generator (with markers) to the IF terminal of the IF circuits through the coupling circuit shown in Fig. 6-77.
2. Adjust the sweep generator controls to produce a sweep from about 40 to 50 MHz and to produce a marker at 41.25 MHz.
3. Connect an oscilloscope to the TP test point of the IF circuits.
4. Insert a 50-Ω resistor between pins 18 and 19 of IC801.
5. Connect pin 22 of IC801 to ground.
6. After obtaining the PIF (picture or video IF) waveform shown in Fig. 6-77, adjust L804 so that the 41.25 MHz portion is minimum.

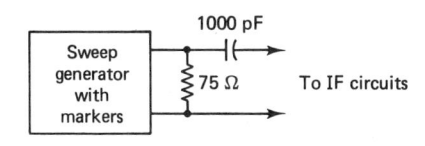

**FIGURE 6-77.** 4.5-MHz trap adjustment connections and waveform.

### AFC Adjustment

1. Connect a sweep generator (with markers) to the IF terminal of the IF circuits through the coupling circuit shown in Fig. 6–77.
2. Adjust the sweep generator controls to produce a sweep from about 40 to 50 MHz and to produce a marker at 45.75 MHz.
3. Connect the +8.3-V stabilized voltage power supply to pin 12 of IC801.
4. Connect pin 22 of IC801 to ground.
5. Connect an oscilloscope to the IF terminal of the IF circuits, and adjust the output of the sweep generator so that the waveform at the IF terminal is 1 V p-p (the output of the sweep generator is 1 V p-p).
6. Connect the oscilloscope to pin 16 of IC801.
7. After obtaining the waveform shown in Fig. 6–78, adjust L806 so that the output voltage at 45.75 MHz is 6.5 ± 0.1 V.

### SIF (Sound or Audio IF) Adjustment

1. Connect an AM signal generator to the contact point of R881 and L881 in the IF circuits through the coupling circuit shown in Fig. 6–79.
2. Connect an oscilloscope to the audio output terminal.
3. Set the AM generator controls to produce a signal at 4.5 MHz ± 5 kHz, modulated 30% by a 400-Hz tone, with an output level between −75 and −55 dBm.
4. Adjust L882 so that the signal amplitude at the audio output terminal is minimum.

### PIF (Picture or Video IF) Adjustment

1. Connect a sweep generator (with markers) to the TP test point of the tuner through the coupling circuit shown in Fig. 6–77.
2. Adjust the sweep generator controls to produce a sweep from about 40 to 50 MHz, and to produce markers at 41.25, 42.17, and 45.75 MHz.

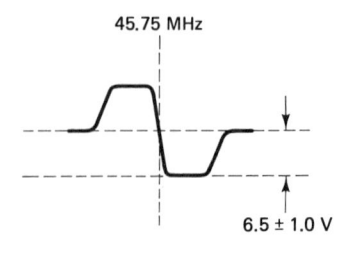

**FIGURE 6–78.** AFC adjustment waveform.

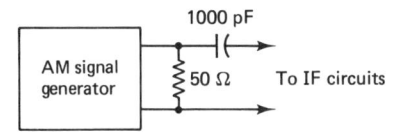

**FIGURE 6-79.** Sound or audio IF adjustment connections.

3. Connect an oscilloscope to the TP test point of the IF circuits.
4. Connect the +5.6-V stabilized voltage power supply to J751 (the AGC voltage input) of the tuner.
5. Connect pin 22 of IC801 to ground.
6. Connect a 50-Ω resistor between pins 18 and 19 of IC801.
7. Adjust L181 so that the PIF waveform level at each marker position shown in Fig. 6-80 is within the specification shown on the diagram.

**AGC Adjustment**

1. Connect a pattern generator to the VHF antenna terminals.
2. Connect an electronic voltmeter or VTVM to J751 (the AGC voltage input).
3. Adjust R810 so that the voltage at J751 is 9.2 ± 0.1 V when a nonmodulated signal with an input level of −49 dBm is applied at the VHF antenna terminals.

## 6-6 VHS MECHANICAL ADJUSTMENTS

The following paragraphs describe complete adjustment procedures for the mechanical section of the VHS VCR described in Chapters 4 and 5. The tools required for these mechanical adjustments are discussed in Chapter 2 (Sec. 2-8).

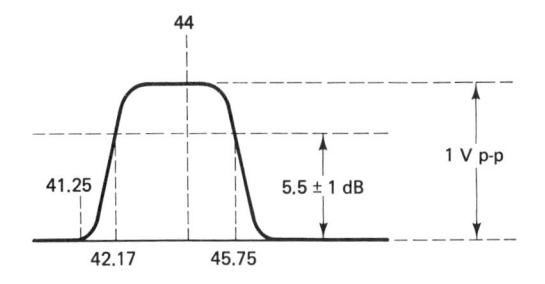

All frequencies in MHz

**FIGURE 6-80.** Picture or video IF adjustment waveform.

## 6-6.1 Reel Table Height

The height of the supply and take-up reel tables should be the same (within ±0.2 mm tolerance). If not, the tape may unwind in a disorderly fashion. The height of the reel tables is adjusted by changing the number of washers under each table. Check the height of the reel tables by installing the Height Reference Plate and the Reel Table Height Gauge as shown in Fig. 6–81, and adjust so that the reel tables are between A and B. Two sizes of washers are available for adjustment: 0.25 and 0.5 mm. These washers are used in combination to get the same reference height for both reel tables.

## 6-6.2 Main Brake Mounting Position

Adjust the two screws holding the main brake solenoid so that the chassis guide is 0.1 to 0.4 mm when the solenoid is ON (the plunger moves in the direction of the arrows), as shown in Fig. 6–82.

After this adjustment, check that the clearance between the brake conversion arm and the main brake arm is 1.5 ± 0.3 mm when the main brake is compressed against the supply and take-up reel tables, as shown in Fig. 6–83.

Then check that the clearance between the main brake and the supply and take-up reel tables is more than 0.3 mm when the latch solenoid is OFF (when the plunger moves in the opposite direction of the arrow), as shown in Fig. 6–84.

## 6-6.3 Tension Arm Position

Optimum back-tension is 35 to 45 g/cm. If back-tension is not within this range, check the pole position of the tension arm (Fig. 6–85). If the pole position is correct, but back-tension is not, adjust the back-tension. If the pole position is not correct, first adjust the pole position, then check the back-tension.

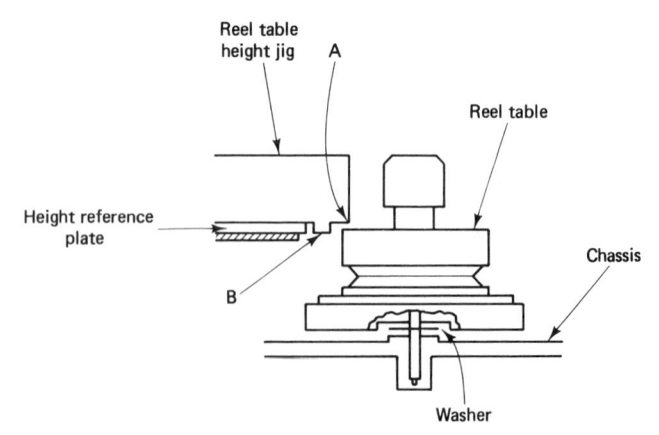

**FIGURE 6-81.** Reel table height.

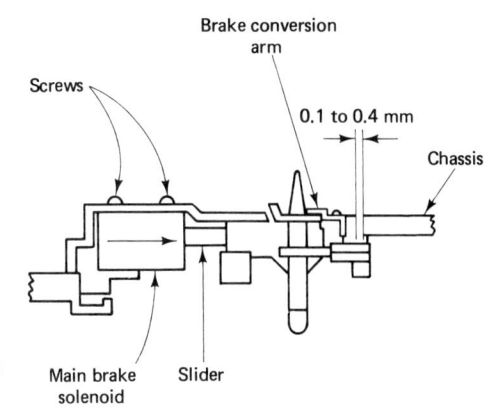

**FIGURE 6-82.** Main brake mounting position (chassis guide adjustment).

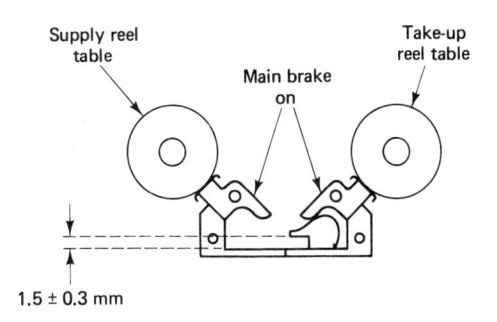

**FIGURE 6-83.** Main brake mounting position (brake conversion and main bracket arm clearance).

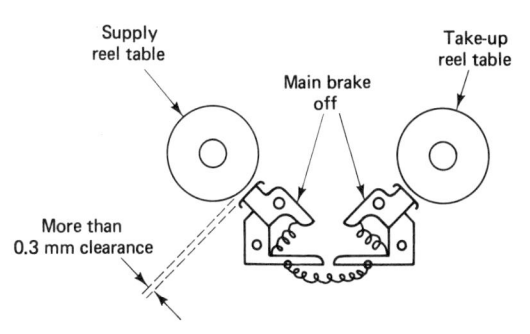

**FIGURE 6-84.** Main brake mounting position (main brake and reel table clearance).

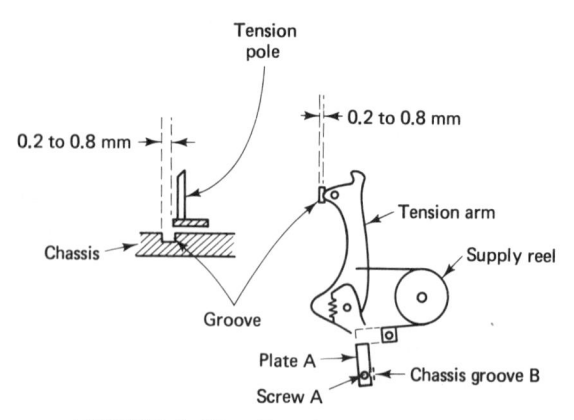

310   Typical Adjustment and Maintenance Procedures

**FIGURE 6-85.**   Tension arm position.

**Tension Arm Pole Position Adjustment.**   Remove the top case and cassette housing. Apply a blind to the end sensor. This can be done by covering the end sensor lamp or the end sensor photo transistors Q81 and Q82 (Sec. 4-6.5, Fig. 4-44). Press the PLAY key and enter the loading mode. After loading is complete, check the pole of the tension arm, viewing from above. Loosen screw A (Fig. 6-85) and adjust plate A so that the distance between the tension arm and the groove of the chassis just beneath the pole is within 0.2 to 0.8 mm. This can be done by loosening screw A so that plate A does not slide smoothly, and then adjusting plate A using a screwdriver, with chassis groove B as a fulcrum. Tighten screw A after adjustment is complete. Recheck the position of the tension arm pole.

**Back-Tension Adjustment.**   Mount the back-tension meter as shown in Fig. 6-86. Loosen screw C and adjust plate B so that the back-tension is 35 to 45 g/cm. This can be done by loosening screw C so that plate B does not slide

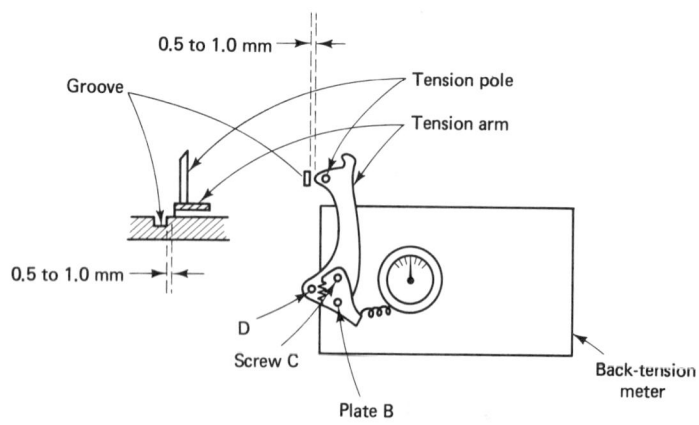

**FIGURE 6-86.**   Back-tension adjustment.

# VHS Mechanical Adjustments 311

smoothly, and then adjusting plate B using a screwdriver, with the convex part D of the tension arm as a fulcrum. Tighten screw C after adjustment is complete. Recheck the back-tension.

### 6-6.4 Supply Sub Brake Position

Check that the clearance between the supply reel table and the supply sub brake is more than 0.3 mm in the PLAY, REC, FF, and REW modes, as shown in Fig. 6-87.

### 6-6.5 Reel Torque and Back-Tension

Using the information in Fig. 6-88, measure the reel torque and back-tension for both the supply and take-up reels.

### 6-6.6 Take-Up Reel Rotation Sensor Position

Loosen the mounting screw to adjust the Hall IC (Sec. 4-6.5, Fig. 4-44) PC board so that the clearance between the Hall IC and the magnetic pulley is 0.5 ± 0.2 mm, as shown in Fig. 6-89.

### 6-6.7 Cylinder Tach Head Position

Loosen the screw to adjust the cylinder tach head (Sec. 1-9.7) bracket so that the clearance between the tach head and the magnet plate is 0.6 ± 0.2 mm, as shown in Fig. 6-90.

### 6-6.8 Microswitch Mounting Position

Turn the worm gear in the loading direction and adjust the installation of the microswitch so that the switch lever is stopped by the click of the plate spring, and is then pushed in farther by 0.5 mm from the position where the microswitch comes ON, as shown in Fig. 6-91.

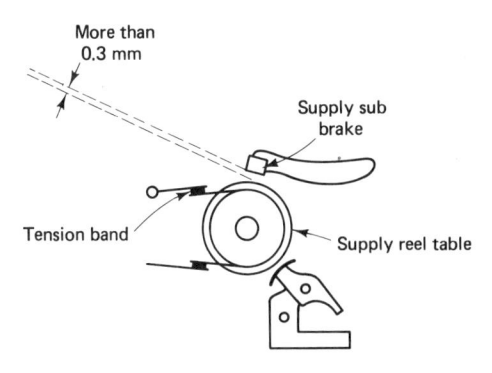

**FIGURE 6-87.** Supply sub brake position.

|  | Mode | Measurement reel | Measurement value | Remark |
|---|---|---|---|---|
| 1 Main brake | Stop | Supply and take-up reel | More than 200 g/cm | Note 1 |
| 2 Slack removal torque | Unloading | Supply reel | 100 to 200 g/cm | Note 1 |
| 3 Rewind torque | Rewind | Supply reel | More than 200 g/cm | Note 1 |
| 4 Take-up torque | Play | Take-up reel | 100 to 200 g/cm | Note 1 |
| 5 FF torque | FF | Take-up reel | More than 200 g/cm | Note 1 |
| 6 Supply back-tension | FF | Supply reel | 4 to 15 g/cm | Note 2 |
| 7 Take-up back-tension | REW | Take-up reel | 4 to 15 g/cm | Note 2 |
| 8 Take-up back-tension<br>1. Rev.<br>2. Loading<br>3. Unloading<br>4. Rec. pause | 1. Rev.<br>2. Loading<br>3. Unloading<br>4. Rec. | Take-up reel | 50 to 120 g/cm | Note 2 |

Note 1: Using the tension measurement reel, pull the tension gauge gradually while the reel stops, and check the value immediately before the reel starts rotation.

Note 2: Wind the cord several turns on the tension measurement reel, pull the tension gauge gradually, and check the value while the reel is rotating.

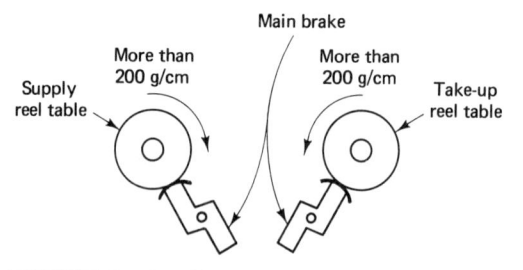

**FIGURE 6-88.** Reel torque and backtension.

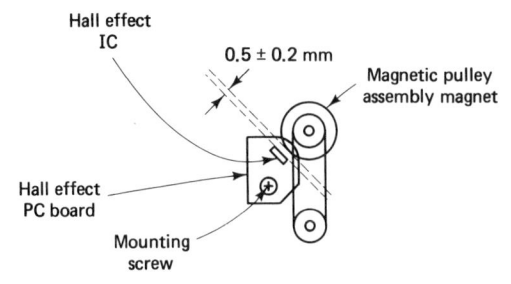

**FIGURE 6-89.** Take-up reel rotation sensor position.

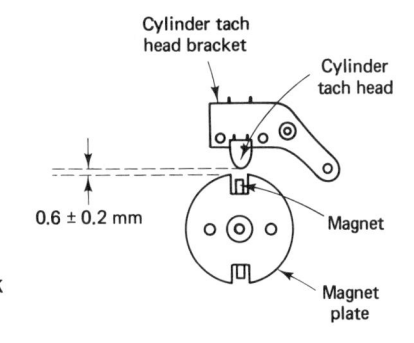

**FIGURE 6-90.** Cylinder tack head position.

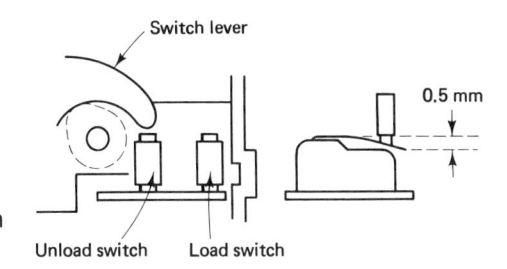

**FIGURE 6-91.** Microswitch mounting position.

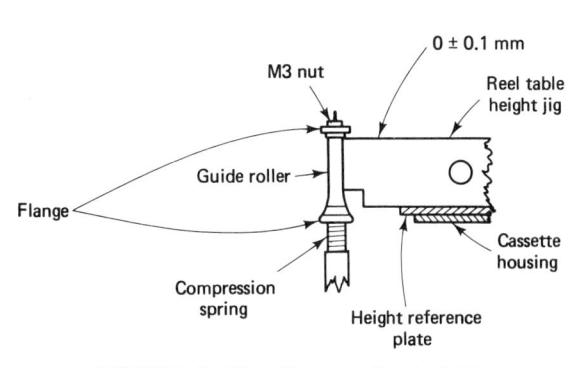

**FIGURE 6-92.** Guide roller height.

*313*

## 6-6.9 Guide Roller Height

Install the Height Reference Plate and place the Reel Table Height Jig, as shown in Fig. 6–92. Adjust the nut so that the clearance is 0 ± 0.1 mm, as shown.

## 6-6.10 Audio/Control Head Mounting Position

The audio/control head must be adjusted for *proper height* (so that the audio and control heads are at their respective tracks on the top and bottom edge of the tape), for *proper azimuth* (so that the audio and control heads are parallel to the direction of tape movement), for *proper relationship to the tracking control* (so that it is possible to obtain proper "lip sync," or coordination between the audio and video display, and for *proper tilt* (so that the head is parallel to the capstan).

**Head Height.** Adjust the head height adjustment nut (Fig. 6–93) so that the audio/control head height is as shown in Fig. 6–94.

**Head Azimuth.** Repeat the procedure of Sec. 6–5.3, using the illustration of Fig. 6–72.

**Proper Relationship to the Tracking Control.** Note that the relationship of the audio/control head and the tracking control is known as the "X-value" on this VCR. The term "X-value" may also be found on some other VCRs. The procedure for X-value adjustment is as follows. Connect an oscilloscope to TP202. Play the color bar signal from the alignment tape. Figure 6–95 shows the variation in the output at TP202 with respect to the positions of the tracking control when the X-value is correctly adjusted. The purpose of the adjustment is so that the output curve matches point b' when the tracking control knob is positioned at the center click or detent.

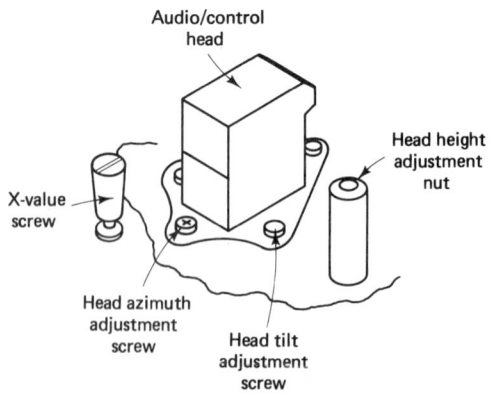

**FIGURE 6-93.** Audio/control head mounting position adjustment points.

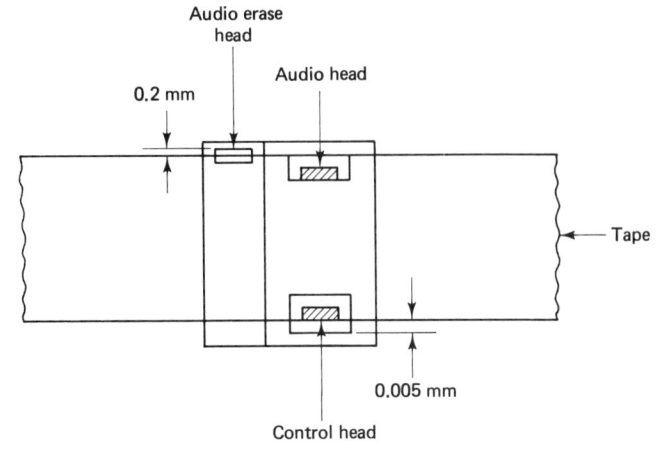

**FIGURE 6-94.** Audio/control head mounting position dimensions.

Adjust the X-value screw (Fig. 6-93) so that the output at TP202 reaches a peak value with tracking control positioned at the center click. Then turn the tracking control 5 to 10° clockwise and counterclockwise and check the positions of the output peaks on the curve. Continue to adjust the X-value screw so that point b' coincides with the output peak when the tracking control is at the center click.

**Tilt.** Adjust the tilt adjustment screw (Fig. 6-93) using the parallel check jig (reel table height gauge) so that the audio/control head is parallel to the capstan.

It is recommended that the electrical adjustments of Sec. 6-5 be rechecked after extensive mechanical adjustment.

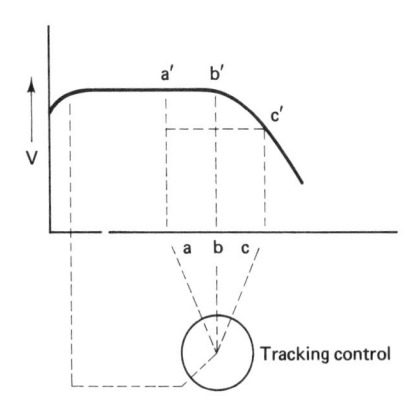

**FIGURE 6-95.** Tracking control relationships.

a = 5 to 10° counterclockwise from center click b
b = center click
c = 5 to 10° clockwise from the center click b

# 7

# Troubleshooting and Service Notes

This chapter describes a series of troubleshooting and service notes for a cross section of VCRs, both Beta and VHS. As discussed in the Preface, it is not practical to provide a specific troubleshooting procedure for each and every VCR. Instead, we describe a universal troubleshooting approach.

It is assumed that you are already familiar with the basics of electronic troubleshooting, including solid-state troubleshooting. If not, and you plan to service VCRs, you are in terrible trouble. Your attention is invited to the author's best-selling *Handbook of Basic Electronic Troubleshooting* (Englewood Cliffs, N.J.: Prentice-Hall, Inc., 1976).

## 7-1 THE BASIC TROUBLESHOOTING FUNCTIONS

Troubleshooting can be considered as a step-by-step logical approach to locate and correct any fault in the operation of equipment. In the case of a VCR, there are seven basic functions required.

First, you must study the VCR using service literature, user instructions, schematic diagrams, and so on, to find out how each circuit works when operating normally. In this way you will know in detail how a given VCR should work. If you do not take the time to learn what is normal, you will never be able to distinguish what is abnormal. For example, some VCRs simply produce a better recording than other VCRs, even in the presence of poor TV broadcast signals.

You could waste hours of precious time (money) trying to make the inferior VCR perform like the quality set if you do not know what is "normal" operation.

Second, you must know the function of, and how to manipulate, all VCR controls and adjustments. It is also assumed that you know how to operate the controls of the TV set used to monitor the VCR playback. An improperly adjusted monitor TV can make a perfectly good VCR appear to be bad. It is difficult, if not impossible, to check out a VCR or TV without knowing how to set the controls. Besides, it will make a bad impression on the customer if you cannot find the EJECT button, especially on the second service call. Also, as a VCR ages, readjustment and realignment of critical circuits are often required.

Third, you must know how to interpret service literature and how to use test equipment. Along with good test equipment that you know how to use, well-written service literature is your best friend. In general, VCR service literature is good as far as procedures and drawings are concerned. Unfortunately, VCR literature is often weak when it comes to descriptions of how circuits operate (theory of operation). The "how it works" portion of most VCR literature is often poorly written, or simply omitted, on the assumption that you and everyone knows the Beta and VHS system theory.

Fourth, you must be able to apply a systematic, logical procedure to locate troubles. Of course, a "logical procedure" for one type of VCR is quite illogical for another. For example, it is quite illogical to check operation of freeze frame or extended play on a VCR not so equipped. However, it is quite logical to check the video, audio, and control tracks on all VCR tapes. For that reason, we discuss logical troubleshooting procedures for various types of VCRs, in addition to basic troubleshooting procedures.

Fifth, you must be able to analyze logically the information of an improperly operating VCR. The information to be analyzed may be in the form of performance, such as the appearance of the picture on a known good monitor TV, or may be indications taken from test equipment, such as waveforms monitored with an oscilloscope. Either way, it is your analysis of the information that makes for logical, efficient troubleshooting.

Sixth, you must be able to perform complete checkout procedures on a VCR that has been repaired. Such checkout may be only simple operation, such as selecting each operating mode in turn and switching through all channels. At the other extreme, the checkout can involve complete readjustment of the VCR, both electrical and mechanical. In any event, some checkout is required after any troubleshooting.

One reason for the checkout is that there may be more than trouble. For example, an aging part may cause high current to flow through a resistor, resulting in burnout of the resistor. Logical troubleshooting may lead you quickly to the burned-out resistor. Replacement of the resistor will restore operation. However, only a thorough checkout will reveal the original high current condition that caused the burnout.

Another reason for after-service checkout is that the repair may have produced a condition that requires readjustment. A classic example of this is where replacement of the video heads requires complete readjustment of both electrical and mechanical components.

Seventh, you must be able to use the proper tools to repair the trouble. As discussed in Chapter 2, VCR services requires all of the common hand tools and test equipment found in TV service, plus many special tools, jigs, and fixtures that are unique to the particular VCR. As a minimum, you must have (and be able to use) tension gauges and various metric tools. These items are generally not familiar to the average TV service technician (unless they also happen to service tape recorders, stereo decks, etc.).

In summary, before starting any troubleshooting job, ask yourself these questions: Have I studied all available service literature to find out how the VCR works (including any special circuits such as remote control, search, freeze frame, etc.)? Can I operate the VCR controls properly? Do I really understand the service literature and can I use all required test equipment or tools properly? Using the service literature and/or previous experience on similar VCRs, can I plan out a logical troubleshooting procedure? Can I analyze logically the results of operating checks, as well as checkout procedures involving test equipment? Using the service literature and/or experience, can I perform complete checkout procedures on the VCR, including realignment, adjustment, and so on, if necessary? Once I have found the trouble, can I use common hand tools to make the repairs? If the answer is "no" to any of these questions, you simply are not ready to start troubleshooting any VCR. Start studying!

## 7-2 THE UNIVERSAL TROUBLESHOOTING APPROACH

The troubleshooting functions discussed in Sec. 7-1 can be summarized into a universal approach with four major steps:

Determine the trouble symptom.

Localize the trouble to a functional unit.

Isolate the trouble to a circuit.

Locate the specific trouble, probably to a specific part.

Let us examine what is being accomplished by each step.

### 7-2.1 Determining Trouble Symptoms

Determining symptoms means that you must know what the equipment is supposed to do when it is operating normally and, more important, that you must be able to recognize when the normal job is not being done. Most people know what a VCR is supposed to do, but no one knows how well each VCR is to perform (and has performed in the past) under all operating conditions (with a given monitor TV, antenna, lead-in, location, etc.).

One problem in determining trouble symptoms of a VCR is that performance must be judged as it appears on a monitor TV. Therefore, as discussed in Sec. 7-4, the condition of the monitor TV must be known and/or checked to determine the trouble symptoms of the VCR. Using the normal and abnormal symptoms produced by the loudspeaker and picture tube of the monitor TV, you must analyze the symptoms to ask the questions: How well is the VCR performing? and Where in the VCR could there be trouble that will produce these symptoms?

The "determining symptoms" step does not mean that you charge into the VCR with screwdriver and soldering tool, nor does it mean that test equipment should be used extensively. Instead, it means that you make a visual check, noting both normal and abnormal performance indications. It also means that you operate the controls to gain further information.

At the end of the "determining symptoms" step, you definitely know that something is wrong and have a fair idea of what is wrong, but you probably do not know just what area of the VCR is faulty. This is established in the next step of troubleshooting.

### 7-2.2 Localizing Trouble to a Functioning Unit

Most electronic equipment can be subdivided into units or areas which have a definite purpose or function. The term "function" is used in VCR troubleshooting to denote an operation performed in a specific area of the unit. For example, in a typical VCR, the functions can be divided into tuner/IF luminance (black and white) video, chroma (color) video, audio, servo, system control, power supply, RF unit (or modulator), and the mechanical section.

To localize the trouble systematically and logically, you must have a knowledge of the functional units of the VCR and must correlate all the symptoms previously determined. Thus, the first consideration in localizing the trouble to a functional unit is a valid estimate of the area in which the trouble might be in order to cause the indicated systems. Initially, several technically accurate possibilities may be considered as the probable trouble area.

As a classic (oversimplified) example, if both picture and sound are poor during playback, the trouble might be in the tuner/IF section, since these functional areas are common to both picture and sound reproduction. On the other hand, if the picture is good but the sound is poor, the trouble is probably in the audio section.

**Use of Diagrams.** VCR troubleshooting involves (or should involve) the extensive use of diagrams. Such diagrams may include a *functional block diagram* and almost always include *schematic* diagrams. *(Practical wiring* *diagrams,* such as found in military-type service literature, are almost never available for VCR service. However, such point-to-point wiring diagrams are usually not necessary since most VCR parts are mounted on PC boards.)

The *overall block diagram* found in some VCR service literature shows the functional relationship of all major sections or units of the VCR. In some literature, each section of the VCR is provided with a block diagram, such as those shown in Chapters 3 and 4. The block diagram is thus the most logical source of information when localizing trouble to a functional unit or section.

The schematic diagram shows the functional relationship of all parts in the VCR. Such parts include all transistors, capacitors, transformers, diodes, and so on. Generally, the schematic presents too much information (not directly related to the specific symptoms noted) to be of maximum value during the localizing step. The decisions being made regarding the probable trouble area may become lost among all the details. However, the schematic is very useful in later stages of the total troubleshooting effort and when a block diagram is not available.

In comparing the block diagram and the schematic during the localizing step, note that each transistor shown on the schematic is usually represented as a block on the block diagram. This relationship is typical in most service literature.

The physical relationship of parts is usually given on PC board component location diagrams (also called parts placement diagrams). These location or placement diagrams rarely show all parts. Instead, they concentrate on major parts, such as transistors, transformers, diodes, and adjustment controls. For this reason, location diagrams are the least useful when localizing trouble. Instead, the location diagrams are most useful when locating specific parts during other phases of troubleshooting.

To sum up, it is logical to use a block diagram instead of a schematic or location diagram when you want to make a valid estimate as to the probable trouble areas (to the faulty functioning unit). The use of a block diagram also permits you to use a troubleshooting technique known as *bracketing* (or good input/bad output). If the block diagram includes major test points, as it may in some well-prepared service literature, the block will also permit you to use test equipment as aids for narrowing down the probable trouble cause. However, test equipment is used more extensively during the isolation step of troubleshooting.

### 7-2.3 Isolating Trouble to a Circuit

After the trouble is localized to a single functional area, the next step is to isolate the trouble to a circuit in the faulty area. To do this, you concentrate on circuits in the area that could cause the trouble and ignore the remaining circuits.

The isolating step involves the use of test equipment, such as meters, oscilloscopes, frequency counters, and signal generators for *signal tracing* and *signal substitution* in the suspected faulty area. By making valid estimates and properly using the applicable diagrams, bracketing techniques, signal tracing, and signal substitution, you can systematically and logically isolate the trouble to a single defective circuit.

Repair techniques or tools to make necessary repairs to the equipment are not used until after the specific trouble is located and verified. That is, you still do not charge into the equipment with soldering tools and screwdrivers at this point. Instead, you are now trying to isolate the trouble to a specific defective circuit so that once the trouble is located, it can be repaired.

### 7-2.4 Locating the Specific Trouble

Although this troubleshooting step mentions only "locate the specific trouble," it includes a final analysis, or review, of the complete procedure, along with using repair techniques to remedy the trouble once it has been located. This final analysis permits you to determine whether some other malfunction caused the part to be faulty or whether the part located is the actual cause of the trouble.

When trying to locate the trouble, inspection using the senses—sight, smell, hearing, and touch—is very important. This inspection is usually performed first, to gather information that may more quickly lead to the defective part. Among the things to look for during the inspection using the senses are burned, charred, or overheated parts, arcing in the high-voltage circuits, and burned-out parts. In VCRs, where it is relatively easy to gain access to the circuitry, a rapid visual inspection should be performed first. In the case of the mechanical parts, a visual inspection is usually the only way to locate a specific problem.

Next in line for locating the specific trouble is the use of an oscilloscope to *observe waveforms,* a meter to *measure voltages,* and a *frequency counter* to *check frequencies* (particularly the very low frequency signals of the servo system). The last test usually is the use of a meter to make *resistance* and *continuity* checks to pinpoint the defective part. After the trouble is located, you make a final analysis of the complete troubleshooting procedure to verify the trouble. Then you repair the trouble and check out the VCR for proper operation.

### 7-2.5 Applying the Troubleshooting Approach to VCR Service

Now that we have established a basic troubleshooting approach, let us discuss how the approach can be applied to the specifics of VCR service. The remainder of this chapter is devoted to examples of how the approach can be used to troubleshoot VCRs.

## 7-3 TROUBLE SYMPTOMS

In the discussion of the localize, isolate, and locate steps, we give examples of how the symptoms can be used as the first step in pinpointing trouble to an area, to a circuit within an area, and finally to a part within the circuit. Before going into these steps, here are some notes regarding symptoms of VCR troubleshooting.

### 7-3.1 Determining Trouble Symptoms

It is not difficult to realize that there definitely is trouble when electronic equipment will not operate. For example, there obviously is trouble when a VCR is plugged in and turned on but fails to either record or playback (and there is no pilot light). A different problem arises when the equipment is operating but is not doing its job properly. Using the same VCR, assume that both record and playback functions are available but that the playback is weak and that there is a buzz in the sound.

Another problem in determining trouble symptoms is improper use of the equipment by the operator. In complex electronic equipment such as radar, operators are usually trained and checked out on the equipment. The opposite is true of home entertainment equipment used by the general public. (The operating instructions supplied with most VCRs are well written and illustrated, but who pays any attention to instructions?)

No matter what equipment is involved, it is always possible for an operator (or customer) to report a "trouble" that is actually a result of improper operation. For these reasons, *you* must first determine the signals of failure, regardless of the extent of malfunction, caused by either the VCR or the customer. This means that you must know how the equipment operates normally, and how to operate the equipment controls.

### 7-3.2 Recognizing Trouble Symptoms

Symptom recognition is the art of identifying the *normal* and *abnormal* signs of operation. A trouble symptom is an undesired *change* in equipment performance or a deviation from the standard. For example, the normal playback from a VCR is a clear, properly contrasted television picture. If noise bands appear on the picture, or faces are blurred, you should recognize this as a trouble symptom because it does not correspond to the normal performance that is expected.

Now assume that the picture is weak, say due to a poor broadcast signal in the area or a defective antenna. If the tuner/IF stages of the VCR do not have sufficient gain to record a good picture under these conditions, you could mistake this for a trouble symptom, unless you really knew the equipment. A poor picture (for this particular mode of VCR operating under these conditions) is not abnormal operation, nor is it an undesired change. Thus, it is not a true trouble symptom and should be so recognized.

### 7-3.3 Evaluation of Symptoms

Symptom evaluation is the process of obtaining more detailed descriptions of the trouble symptoms. The recognition of the original trouble symptom does not in itself provide enough information to decide on the probable cause or causes of the trouble, because many faults produce similar trouble symptoms.

To evaluate a trouble symptom, you generally have to operate controls associated with the symptom and apply your knowledge of electronic circuits, supplemented with information gained from the service literature. Of course, the mere adjustment of operating controls is not the complete story of symptom evaluation. However, the discovery of an incorrect setting can be considered a part of the overall symptom evaluation process.

### 7-3.4 Distinguishing between VCR and TV Trouble Symptoms

One problem in evaluating symptoms in VCR service is the fact that VCRs must be played back through a monitor TV. If the TV is defective or improperly adjusted, the VCR may appear to have troubles. One practical way to confirm troubles in the VCR is to record a few minutes of a broadcast while watching the program carefully. Then play back the recorded material and compare the playback quality to that of the broadcast. Switch between the broadcast and playback. Another practical suggestion for evaluation of trouble symptoms *in the shop* is to have at least one monitor TV of known quality. All VCRs passing through the shop can be compared against the same standard. Of course, the ultimate monitor TV is the industrial receiver/monitor described in Sec. 2-7.2.

## 7-4 LOCALIZING TROUBLE

Localizing trouble means that you must determine which of the major functional areas in a VCR is actually at fault. This is done by systematically checking each area until the actual faulty area is found. If none of the functional areas on your list show improper performance, you must take a return path and recheck the symptom information (and observe more information, if possible). There may be several circuits which could be causing the trouble. The localize step will narrow the list to those in one functional area, as indicated by a particular block of the block diagram.

## 7-4.1 Bracketing Technique

The basic bracketing technique makes use of a block diagram or schematic to localize the trouble to a functional area. Bracketing (sometimes known as the *good input/bad output* technique provides a means of narrowing the trouble down to a circuit group and then to a faulty circuit. Symptom analysis and/or signal-tracing tests are used in conjunction with, or are a part of, bracketing.

Bracketing starts by placing brackets (at the good input and the bad output) on the block diagram or schematic. Bracketing can be mental, or it can be physically marked with a pencil, whichever is most effective for you. No matter what system is used, with the brackets properly positioned, you know that the trouble exists somewhere *between* the two brackets.

The technique is to move the brackets, one at a time (either the good input or the bad output), and then make tests to find if the trouble is within the new bracketed area. The process continues until the brackets localize a circuit group.

The most important factor in bracketing is to find where the brackets should be moved in the elimination process. This is determined from your deductions based on your knowledge of the equipment and on the symptoms. All moves of the brackets should be aimed at localizing the trouble with a minimum of tests.

## 7-4.2 Basic Bracketing in VCR Service

The obvious first move in bracketing for a VCR, once you have determined that there is a problem with the VCR and not the monitor TV or antenna, is to play back an alignment tape or a program tape of known quality. This effectively splits the VCR circuits in half. If the playback is satisfactory, it is reasonable to assume that the heads, playback circuits, and RF unit (or modulator) are good. It is also reasonable to assume that the servo, system control, mechanical section, and power supply are good. If the playback is not satisfactory, all of the foregoing circuits are suspect.

The next logical set of brackets depends on the results of the first tests. If the playback is good, the tuner/IF section and the record circuits are suspect. If the VCR has *video and audio input connectors,* these can be used to localize the trouble to either the tuner/IF or record circuits. You can use *signal injection* and apply video/audio signals (from an NTSC-type generator described in Chapter 2) to the video/audio connectors, record the generator signals, and then play back the recorded material. If the VCR is capable of playing back signals applied at the video/audio input connectors, but not when the signals are applied at the antenna, the problem is in the tuner/IF. If playback is not satisfactory with signals at the video/audio connectors, but a good tape can be played back, the fault is in the record circuits.

If the playback is not good with an alignment or other known good tape, the playback circuits and the RF unit or modulator are suspect. If the VCR has *video and audio output connectors* these can be used to localize the trouble to either the RF unit or playback circuits. You can use *signal tracing* and monitor the audio and video signals (from the good video tape) at the connectors with an oscilloscope (for video) frequency meter or multimeter (for audio). If the VCR is capable of playing back known good recorded material and producing the correct signals at the video/audio output connectors, the problem is in the RF unit. (Typical waveforms at the video/audio output connectors are described in Chapter 6.) If the playback is not satisfactory at the video/audio output connectors, with a known good tape, the trouble is in the playback circuits.

### 7-4.3 Localization with Plug-in Modules

The localization procedure can be modified when the circuits of a VCR are located on plug-in modules, or on PC boards that are easily replaceable. With such VCRs, it is possible to replace each module, in turn, until the trouble is cleared. For example, if replacement of the video board restores normal operation, the defect is in the video section. This can be confirmed by plugging the suspected defective module back into the equipment. Although this confirmation process is not a part of theoretical troubleshooting, it is wise to make the check from a practical standpoint. Often, a trouble symptom of this sort can be caused by the plug-in module making poor contact with the chassis connector or receptacle.

Some service literature recommends that tests be made before arbitrarily replacing all plug-in modules. This is based on the fact that the plug-in modules are not necessarily arranged on a functional area basis. Thus, there is no direct relationship between the trouble symptom and the modules. In such cases, always follow the service literature recommendation. Of course, if plug-in modules are not readily available in the field, you must make tests to localize the trouble to a module (so that you can order the right module, for example).

### 7-4.4 Which Circuit Groups to Test First

When you have localized trouble to more than one circuit group, you must decide which group to test first. Several factors should be considered in making this decision.

As a general rule, if you *make a test that eliminates several circuits,* or circuit groups, make that test first, before making a test that eliminates only one circuit. This requires an examination of the diagrams (block and/or schematic) and knowledge of how the VCR operates. The decision also requires that you apply logic in making the decision.

*Test point accessibility* is the next factor to consider. A test point can be a special jack located at an accessible spot (say at the rear of the housing). The jack (or possibly a terminal) is electrically connected (directly or by a switch) to some important operating voltage or signal path. At the other extreme, a test point can be *any point* where wires join or where parts are connected together.

Another factor (although definitely not the most important) is past experience and history of *repeated VCR failures*. Past experience with identical or similar VCRs and related trouble symptoms, as well as the probability of failure based on records or repeated failures, should obviously have a bearing on the choice of the first test point. However, all circuit groups related to the trouble symptom should be tested, no matter how much experience you may have on the VCR. Of course, the experience factor can help you decide which group to test first.

Anyone who has any practical experience in troubleshooting knows that all of the steps in a localization sequence can rarely proceed in textbook fashion. Just as true is the fact that troubles listed in VCR service literature very often never occur in the VCR you are servicing. These troubles are included in the literature as a guide, not as hard and fast rules. In some cases it may be necessary to modify your troubleshooting procedure as far as localizing the trouble is concerned. The physical arrangement of the VCR may pose special troubleshooting problems. Similarly, experience with similar equipment may provide special knowledge which can simplify localizing the trouble.

## 7-5  ISOLATING TROUBLE TO A CIRCUIT

The first two steps (symptoms and localization) of the troubleshooting procedure give you the initial symptom information about the trouble and describe the method of localizing the trouble to a *probable faulty* circuit group. Both steps involve a minimum of testing. In the isolate step, you will do extensive testing to track the trouble to a specific faulty circuit.

### 7-5.1  *Isolating Trouble in IC and Plug-in Equipment*

It is common to use ICs in many present-day VCRs. For example, in the VCRs described in Chapters 3 and 4, the entire video record and playback circuits (color and black and white) are contained within two or three LSI (large-scale integration) ICs. All parts of the circuit group, except for transformers, coils, and a few miscellaneous parts, are included in the ICs. In such cases, the trouble can be isolated to the IC input and output, but not to the circuits (or individual parts) within the IC. No further isolation is necessary since parts within the IC cannot be replaced on an individual basis. The same condition is true of some VCRs where groups of circuits are mounted on *sealed,* replaceable boards or cards.

(Note that not all plug-in modules or boards are sealed. Some have replaceable parts.)

### 7-5.2 Using Diagrams in the Isolation Process

No matter what physical arrangement is used, the isolation process follows the same reasoning you have used previously; the continuous narrowing down of the trouble area by making logical decisions and performing logical tests. Such a process reduces the number of tests that must be performed, thus saving time and reducing the possibility of error.

A block diagram is a very convenient tool for the isolation process, since a block shows circuits already arranged in circuit groups. That is one major reason that we used block diagrams so extensively in Chapters 3 and 4. As discussed, you may or may not have block diagrams supplied with your service literature. You must work with a schematic (but you can use the blocks of Chapters 3 and 4 for reference, since they are typical of most Beta and VHS circuits).

With either diagram, *if you can recognize* circuit groups as well as individual circuits, the isolation process will be much easier. No matter what diagram you use during isolation, you are looking for three major bits of information: the *signal path* or *flow,* the *waveforms* along the signal paths, and the *operating/adjustment controls* in the various circuits along the signal paths. Typically, signal paths are indicated by solid lines between blocks. Arrows on the lines indicate the direction of signal flow. The operating/adjustment controls are connected to the block representing the circuit affected by the controls. Waveforms are given on the VCR schematic diagrams at several points, usually at the input and output of each circuit group, and possibly at each circuit. (You can compare these waveforms with the "typical" waveforms shown in Chapter 6.) In the simplest form, the isolation step is done by comparing the actual waveforms produced along the signal paths of the VCR circuits against the waveforms shown in the service literature. This is known as *signal tracing.* The isolation step may also involve *injection* or *substitution* of signals normally found along the signal paths. With either technique, you check and compare inputs and outputs of circuit groups and circuits in the signal paths.

### 7-5.3 Signal Tracing versus Signal Substitution or Injection

Both signal tracing and signal substitution (or signal injection) techniques are used frequently in troubleshooting all types of VCR circuits. In effect, you inject signals and monitor signals as described in Chapter 6, but without altering the adjustment controls. The choice between tracing and substitution depends on the test equipment used. As discussed in Chapter 2, some signal generators designed for television service have outputs that simulate signals found in all major signal paths of a VCR (RF, IF, sound IF, audio, video pulses, etc.). If you have such a generator, signal injection or substitution is the logical choice, since you can test all the circuit groups on an individual basis (independent of other circuit groups). However, much effective troubleshooting can be done with signal tracing alone (and this technique is recommended by many technicians).

*Signal tracing* is done by examining the signals at test points with a monitoring device (oscilloscope, meter, frequency meter). In signal tracing, the input probe of the indicating or monitoring device used to trace the signal is moved from point to point, with a signal applied at a fixed point (such as a broadcast signal at the antenna terminals of the VCR, or a known good tape being played back).

*Signal substitution* is done by injecting an artificial signal (from a signal generator, sweep generator, etc.) into a circuit or circuit group to check performance. In signal injection, the injected signal is moved from point to point, with an indicating or monitoring device remaining fixed at one point.

Both signal tracing and substitution are often used simultaneously in troubleshooting VCRs (as is the case for adjustment procedures described in Chapter 6). For example, when troubleshooting the tuner/IF section, it is common to inject a sweep signal at the input and monitor the output to the video circuits with an oscilloscope.

### 7-5.4 Half-Split Technique

The half-split technique is based on the idea of simultaneous elimination of the maximum number of circuit groups or circuits with each test. This saves both time and effort. The half-split technique is used primarily when isolating trouble in a linear signal path but can be used with other types of paths. In using the half-split system, brackets are placed at good input and bad output points in the normal manner, and the symptoms are studied. Unless the symptoms point definitely to one circuit group or circuit that might be the trouble source, the most logical place to make the first test is at a convenient test point *halfway between* the brackets. The bracketing functions described in Sec. 7-4.2 are classic examples of half-splitting.

### 7-5.5 Isolating Trouble to a Circuit within a Circuit Group

Once a trouble is definitely isolated to a faulty circuit group, the next step is to isolate the trouble to the faulty circuit within the group. Bracketing, half-splitting, signal tracing, signal injection, and *knowledge of the signal path* in the circuit group are all important to this step, and are used in essentially the same way as for isolating trouble to the circuit group. This is one reason to study all information available concerning the VCR. If the VCR service literature does not provide suitable information regarding the signal paths, study the typical signal paths described in Chapters 3 and 4.

## 7-6 LOCATING A SPECIFIC TROUBLE

The ability to recognize symptoms and verify them with test equipment will help you to make logical decisions regarding the selection and localization of the faulty group. This ability will also help you to isolate trouble to a faulty circuit. The final step of troubleshooting—locating the specific trouble—requires testing of the various branches of the faulty circuit to find the defective part.

The proper performance of the locate step will enable you to find the cause of trouble, repair it, and return the VCR to normal operation. A follow-up to this step is to record the trouble so that, from the history of the VCR, future troubles may be easier to locate. Also such a history may point out consistent failures which could be caused by a design error.

### 7-6.1 Locating Troubles in Replaceable Modules

Because the trend in modern VCRs (and most other electronic equipment) is toward IC and replaceable-module design, technicians often assume that it is not necessary to locate specific troubles to individual parts. That is, they assume that all troubles can be repaired by replacement of modules and ICs. Some technicians are even trained that way. The assumption is not necessarily true.

While the use of replaceable modules often minimizes the number of steps required in troubleshooting, it is still necessary to check circuit branches to parts outside the module. Front-panel operating controls are a good example of this, since such controls are not located in the replaceable units. Instead, the controls are connected to the terminal of an IC, circuit board, or plug-in module. The system control circuits of a VCR are another example of parts outside of replaceable modules. For example, most of the fail-safe or stop circuits (such as the dew sensor) are located at various points throughout the mechanical section, even though they may all feed to a common circuit board and/or microprocessor.

It is assumed that you are sufficiently advanced in troubleshooting to inspect PC boards and individual parts using the senses (looking for charred resistors, etc.), to test suspected parts (transistors, diodes, etc.) using test equipment or substitution, and to measure waveforms, voltage, resistance, and possibly current. (If not, stay away from any VCR with your soldering tool and screwdriver!) So we will not go into any details concerning these functions which are normally required to locate a specific trouble. However, the remaining paragraphs of this section provide some thoughts that you may not have considered.

### 7-6.2 Duplicating Waveform, Voltage, and Resistance Measurements

If you are responsible for service of one type or model of VCR, it is strongly recommended that you duplicate all the waveform, voltage, and resistance measurements found in the service literature with test equipment of your own. This should be done with a known good VCR that is operating properly. Then when you make measurements during troubleshooting, you can spot even slight variations in voltage, waveform, and so on. Always make the initial measurements with test equipment that will normally be used during troubleshooting. If more than one set of test equipment is used, make the initial measurements with all available test equipment, and record the variations.

## 7-6.3 Adjustments during Troubleshooting

Adjustment of controls (both internal adjustment controls and VCR operating controls) can affect circuit conditions. This may lead to false conclusions during troubleshooting. There are two extremes taken by some technicians during adjustment.

First, the technician may launch into a complete alignment procedure (or whatever internal adjustments are available) once the trouble is isolated to a circuit. No internal control, no matter how inaccessible, is left untouched. The technician reasons that it is easier to make adjustments than to replace parts. While such a procedure will eliminate improper adjustment as a possible fault, the procedure can also create more problems than are repaired. Indiscriminate internal adjustment is the technicians's version of "operator trouble." The problem is particularly bad when mechanical adjustments are performed on a VCR. A slight misadjustment of mechanical parts in the tape path can produce instant disaster and make an otherwise good VCR appear to be beyond repair.

At the other extreme, a technician may replace part after part, where a simple screwdriver adjustment will repair the problem. This usually means that the technician simply does not know how to perform the adjustment procedure or does not know what the control does in the circuit. Either way, a study of the service literature should resolve the situation.

*To take the middle ground,* do not make any internal adjustments during the troubleshooting procedure until trouble has been isolated to a circuit, and then only when the trouble symptom or test results indicate possible maladjustment.

For example, assume that the 3.58-MHz color reference oscillator is provided with an adjustment control that sets the frequency of oscillation (as described in Sec. 6–1.5). If measurements (with a frequency counter) show that the oscillator is off frequency (not 3.579545 MHz ± 5 Hz), it is logical to adjust the oscillator. However, if measurements show only a very low output (but on frequency), adjustment of the oscillator during troubleshooting could be confusing (and could cause further problems).

An exception to this rule is when the service literature recommends alignment or adjustment as part of the troubleshooting procedure. Generally, alignment or adjustment is checked after test and repair have been performed. This assures that the repair (parts replacement) procedure has not upset circuit adjustment.

### 7-6.4 Repairing Troubles

In a strict sense, repairing the trouble is not part of the troubleshooting procedure. However, repair is an important part of that total effort involved in getting the VCR back into operation. Repairs must be made before the VCR can be checked out and made ready for operation.

*Never replace a part* if it fails a second time without making sure that the cause of trouble is eliminated. Preferably, the cause of trouble should be pinpointed before replacing a part for the first time. However, this is not always practical. For example, if a resistor burns out because of an intermittent short, and you have cleared the short, the next step is to replace the resistor. However, the short could happen again, burning out the replacement resistor. If so, you must recheck every element and lead in the circuit.

When replacing a defective part, an *exact replacement* should be used if it is available. If not available, and if the original part is beyond repair, an equivalent *or better* part should be used. Never install a replacement part having characteristics or ratings inferior to those of the original.

Another factor to consider when repairing the trouble is that, if at all possible, the replacement part should be installed in the *same physical location* as the original, with the same lead lengths, and so on. This precaution is generally optional in most low-frequency or d-c circuits, but must be followed for high-frequency applications. In the RF, IF, video, and so on, circuits of the VCR, changing the location of parts may cause the circuit to become detuned or otherwise out of alignment.

### 7-6.5 Operational Checkout

Even after the trouble is found and the faulty part is located and replaced, the troubleshooting effort is not necessarily completed. You should make an operational check to verify that the VCR is free of *all faults* and is performing properly again. Never assume that simply because a defective part is located and replaced the VCR will automatically operate normally again. As a matter of fact, in practical troubleshooting, never assume anything; prove it. Check that the VCR will operate properly in all modes and on all channels normally used. This will ensure that one fault has not caused another. If practical, have the customer recheck all operating modes and channels.

When the operational check is completed and the VCR is "certified" (by you and the customer) to be operating normally, make a brief record of the symptoms, faulty parts, and remedy. This is particularly helpful when you must troubleshoot similar VCRs. Even a simple record of troubleshooting will give you a history of the VCR for future reference.

If the VCR does not perform properly during the operational checkout, you must continue troubleshooting. If the symptoms are the same as (or similar to) the original trouble symptoms, retrace your steps, one at a time. If the symptoms are entirely different, you may have to repeat the entire troubleshooting procedure from the start. However, this is usually not necessary.

## 7-7  TYPICAL VCR INSTALLATION

A VCR must be properly installed if it is to operate satisfactorily. Installation of a VCR involves connecting the VCR to an antenna, a monitor TV, to the power source, and possibly to a stereo (for better sound reproduction), to another VCR to make a print tape, or to a camera and tape recorder (for recording home movies, etc.). Although the following notes apply to specific VCRs (Sony SL-8600 and SL-5800), the connections are typical for many other VCRs.

### 7-7.1  Basic VCR Connections

Figure 7-1 shows the basic connections for a typical VCR and applies to the great majority of VCR installations. As shown, if the antenna cable is a 75-ohm coaxial type, connect the cable directly to the VHF IN terminal with an F-type connector. If the cable is a 300-$\Omega$ ribbon type, connect the cable to an adapter such as the EAC-25 shown, then connect the adapter to the VHF IN terminal.

Connect the VHF OUT connector on the VCR to the VHF antenna terminal on the TV receiver with a 75-$\Omega$ coaxial cable, as shown. If the TV receiver is equipped with 300-$\Omega$ antenna terminals, use the adapter. Connect the UHF OUT terminals on the VCR to the UHF IN terminals on the TV with the 300-$\Omega$ lead in.

Where a combination VHF/UHF antenna is used, separate the VHF and UHF signals using a signal separator and connect the VHF and UHF lead-ins to VHF IN (through the adapter) and UHF IN, respectively, as shown. Most combination antennas have a built-in signal separator, but you may have to provide one for some installations.

**CAUTION:** Connections between the VHF OUT connector of a VCR and the antenna terminals of a TV receiver should be made only as shown in Fig. 7-1, or as specified in the operating instructions. Failure to do so may result in operation that violates the regulations of the FCC regarding the use and operation of RF devices. (You may broadcast TV programs to the entire neighborhood!) Never connect the output of the VCR to an antenna or make simultaneous (parallel) antenna and VCR connections at the antenna terminals of the TV!

### 7-7.2  Connecting a VCR to Multiple Antennas and TV Receivers

Figure 7-2 shows some typical connections where more than one antenna or TV is used. The configuration using the four-channel signal splitter is particularly

Typical VCR Installation 333

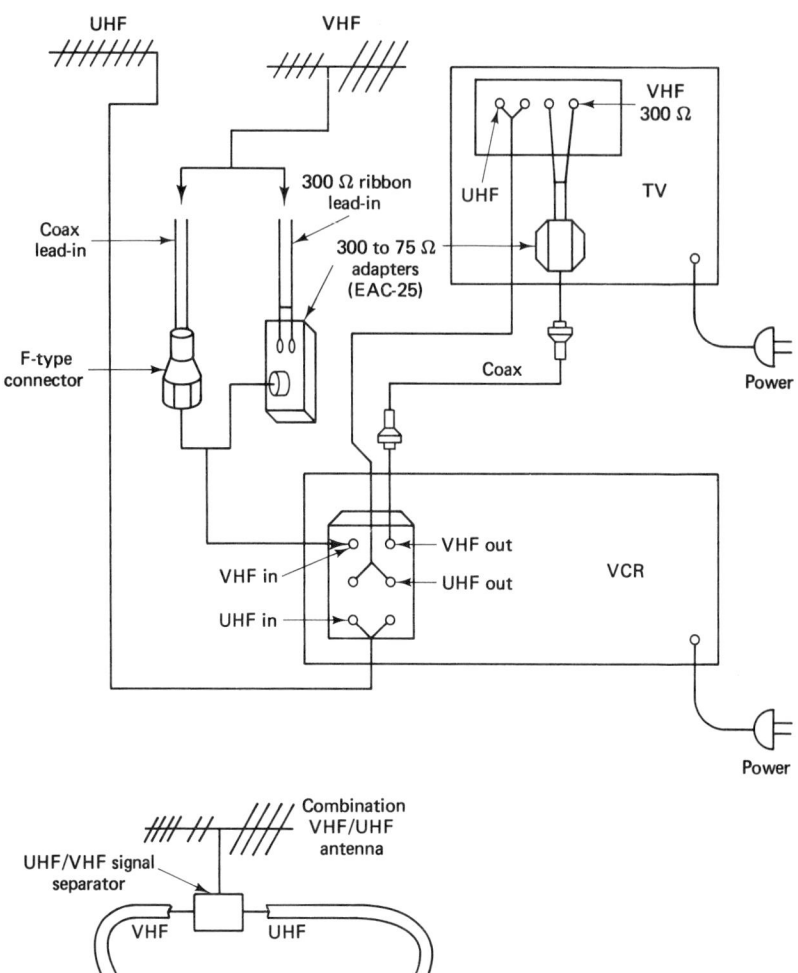

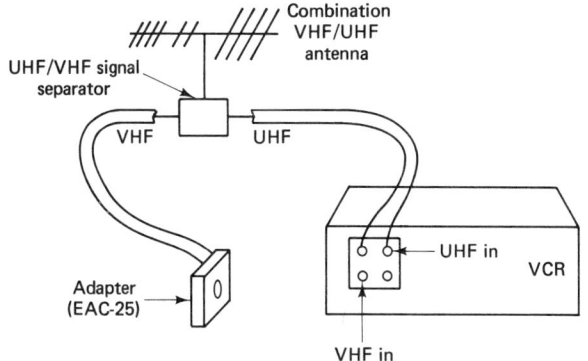

**FIGURE 7-1.** Basic VCR connections.

useful for exhibitions and store displays. Of course, the picture displayed on each TV is the same. The configuration with both signal splitters and VHF/UHF signal separators is handy for those situations where one VCR is used for simultaneous playback through two or more TV sets, or where one TV is used for VCR playback and the other for TV broadcast reception.

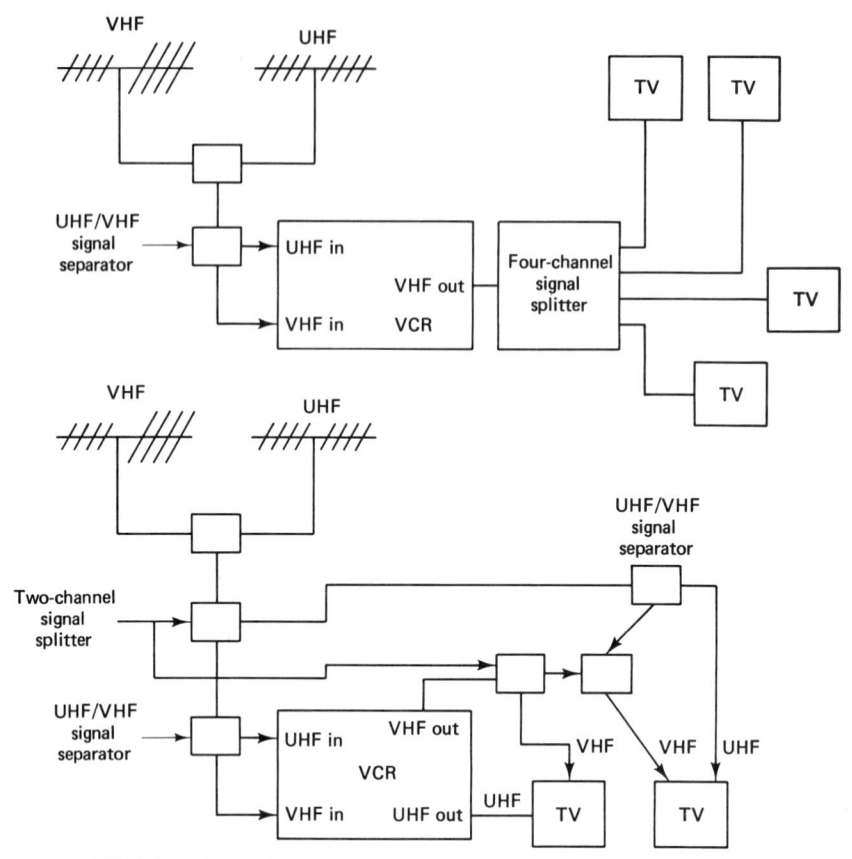

**FIGURE 7-2.** Connecting a VCR to multiple antennas and TV receivers.

### 7-7.3 Connecting a VCR to Stereo Equipment

Figure 7-3 shows how a VCR can be connected to stereo equipment. Most VCRs have similar audio output connections. The use of stereo equipment gives added power and realism to movies and music programs.

### 7-7.4 Copying a Video Tape (Making a Print)

Figure 7-4 shows connections for making copies of video tapes. The process is essentially the same as making a copy of an audio tape. However, keep two points in mind. First, each time a copy is made, the quality of the copy will not be as good as the original. Second, remember the warning with regard to making copies of copyrighted material discussed in Chapter 1.

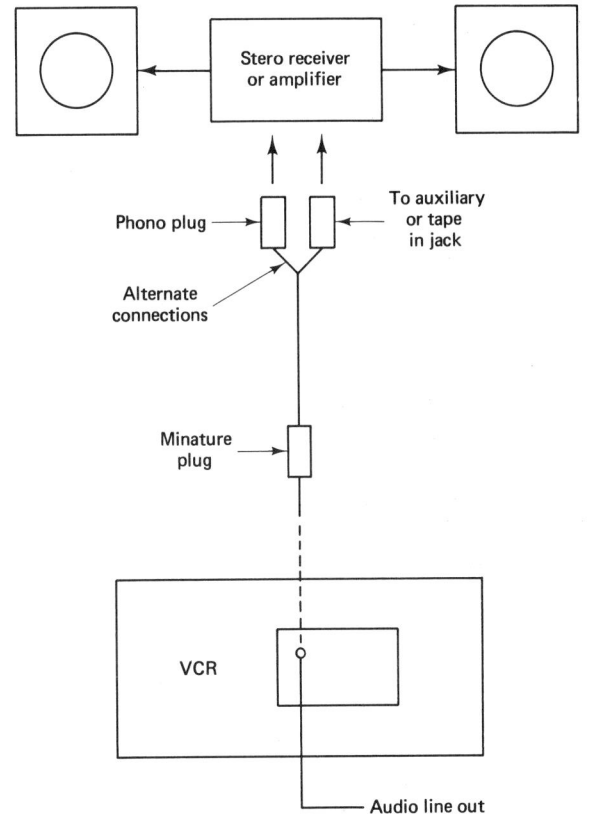

**FIGURE 7-3.** Connecting a VCR to stereo equipment.

### 7-7.5 Connecting a VCR to a Video Camera

Figure 7-5 shows connections between a VCR and a video camera, microphone, or other audio source. In general, cameras must conform to EIA standards. Refer to the service notes regarding cameras in Sec. 7-10. In most VCRs, when an audio source and a microphone are connected simultaneously, the VCR records only the microphone input.

### 7-7.6 Connecting a VCR to a Cable TV (CATV) System

It is recommended that you consult with the cable TV company before installing any VCR. Always follow their recommendations for installation. Also, before operating the VCR with any cable TV system, set the RF unit channel selector on the VCR to channel 3 or 4, whichever is not active in the area. If both channels are

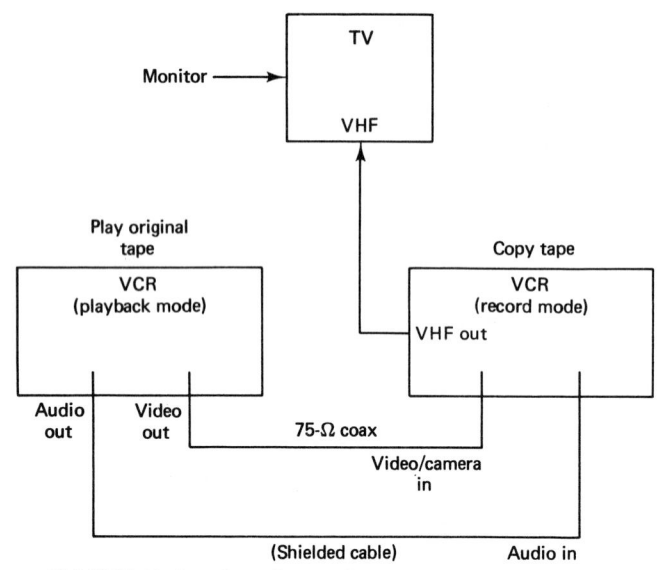

**FIGURE 7-4.** Copying a video tape (making a print).

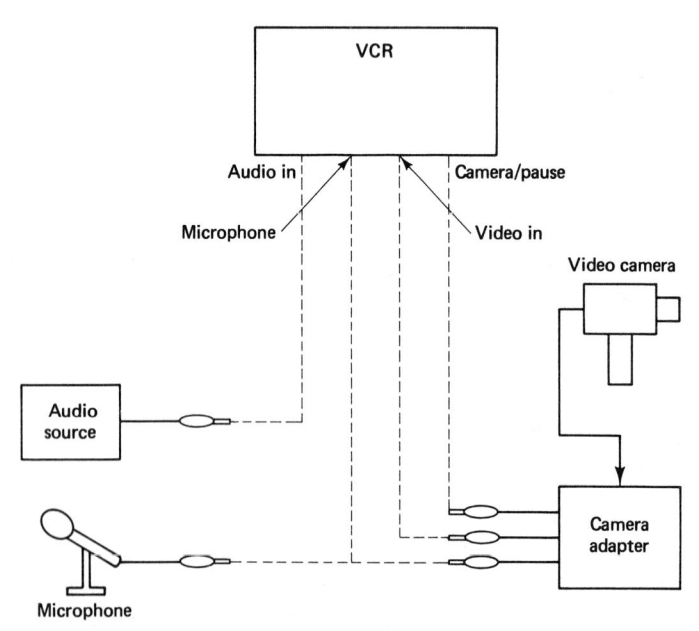

**FIGURE 7-5.** Connecting a VCR to a video camera.

## Typical VCR Installation

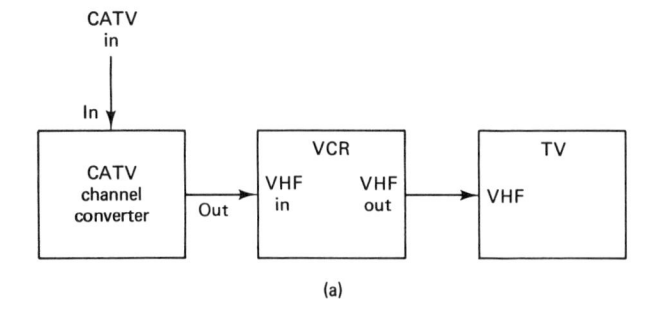

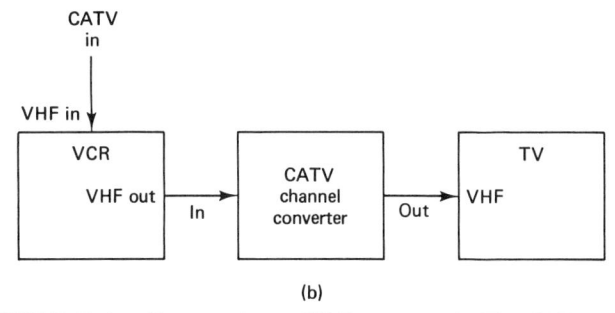

**FIGURE 7-6.** Connecting a VCR to a cable TV (CATV) system.

viewable, check which gives better results by switching between channels 3 and 4.

Figure 7-6 shows two configurations for connecting a VCR to a CATV system. In the configuration of Fig. 7-6a, the CATV cable is connected to the CATV channel converter in the normal manner, and the output of the channel converter is connected to the VCR unit. This configuration is usually recommended. However, it is possible to use the configuration of Fig. 7-6b where the CATV cable is connected directly to the VHF input of the VCR.

With the configuration of Fig. 7-6a, it is possible to record programs from all CATV channels as well as VHF channels 2 through 13. Set the TV channel selector to that of the VCR RF unit channel selector. Set the VCR channel selector to receive the output channel of the converter. Set the VCR program select switch to the VCR position. With this connection, the channel to be viewed or to be recorded is selected on the converter.

With the configuration of Fig. 7-6b, it is possible to view one program from the converter while recording a program on VHF channels 2 through 13 via the VCR. Set the TV channel selector to the output channel of the converter, and the VCR program select switch to the TV position. For playback, set the channel selector on the converter to that of the VCR RF unit channel selector. When the CATV channel converter is not needed, connect the CATV input to the VCR, then connect the VCR and TV receiver in the normal manner.

## 7-7.7 Precautions When Installing a VCR

In addition to any precautions described in the service/operating literature for the VCR, check the following.

The rear of the VCR should be at least 10 cm away from the wall to maintain adequate heat dissipation.

Avoid placing the VCR in areas of high temperature or high humidity.

Make sure to obtain adequate signal level from the antenna. Inside the VCR, the antenna signal is distributed to both the VCR and the TV antenna terminals. Thus, if antenna sensitivity or output is too low, signal loss caused by the distribution may result in deterioration of both the TV picture and video recording quality. Signal loss is particularly a problem for UHF. In extreme cases, a signal booster may be required as described in Sec. 7-7.8.

Check that the fine tuning has been properly adjusted for the *inactive channel* of the TV. The VCR playback output signal uses the inactive channel of the TV but, since this channel is not ordinarily used, the fine tuning may not be precisely adjusted. Play back a tape that you know has good quality and adjust the TV fine tuning to obtain best reception.

Make sure the fine tuning of the VCR is also properly adjusted. Also, recheck all installation wiring and connections.

If you have the job of demonstrating use of the VCR to a customer, go over the operating instructions of the instruction manual with them in detail. Although operation of a VCR is simple to an electronic wizard such as yourself, it may not be so to the general public, since a VCR has many capabilities beyond those of a TV. As a minimum, describe to the customer how to do the following:

Watch the TV.

Record a TV program.

Record one program on the VCR while watching another on the TV set.

Use the automatic recording timer to record while away from home.

Play back a recorded tape.

## 7-7.8 Using a Signal Booster with a VCR

Typically, good TV picture quality requires 65 to 70 dB TV terminal voltage. However, cables and other equipment may result in something less than this. Therefore, when designing a TV reception system to be used with a VCR, it is a good idea to calculate the signal level needed and use a signal booster, if necessary. Generally, a booster requires a minimum input signal of 50 dB. If the input is less than that, S/N ratio deteriorates, noise increases, and picture quality does not improve even with a booster. On the other hand, do not apply an input greater than the booster can handle. If the input is too large, intermodulation distortion can result. The typical input limit in dB equals the maximum output (in dB), less gain (in db). For example, if a booster having a maximum output capability of 95 dB, and a gain of 35 dB, is used with a TV reception system, the input limit should be 95 − 35 = 60 dB.

Figure 7-7 shows the signal paths within a typical VCR, as well as typical connections for a booster and VCR. As shown, a splitter is built into the VCR. This is necessary in order to record a different program from that shown on the

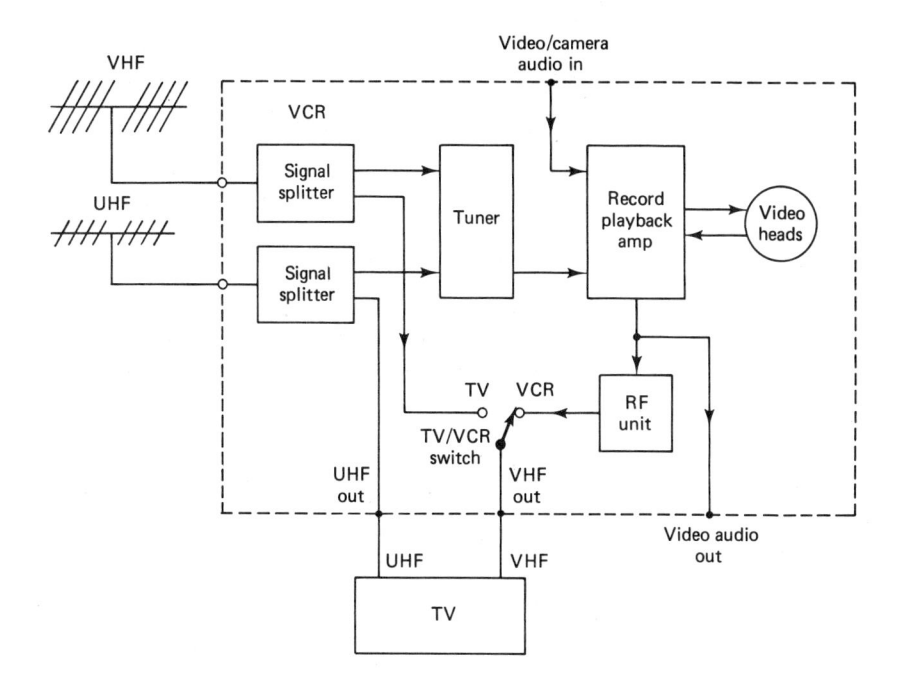

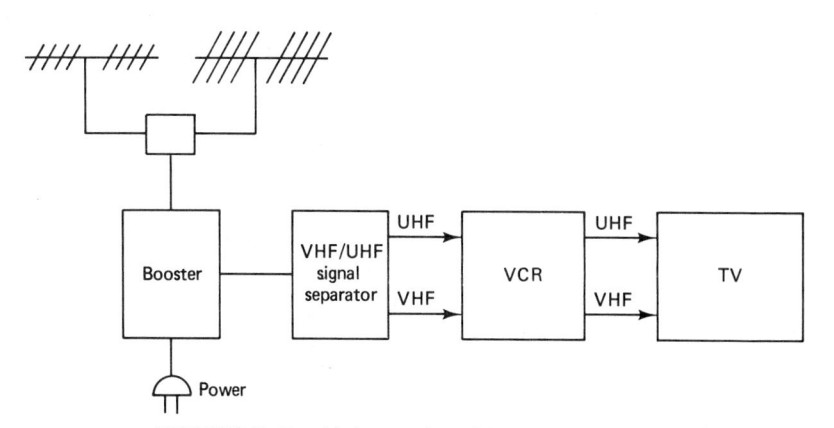

**FIGURE 7-7.** Using a signal booster with a VCR.

TV. The splitters produce about 4 dB attenuation of the antenna signal and can result in picture deterioration, as discussed. Should this be the case, a *splitter booster* connected as shown in Fig. 7-7 will provide the necessary compensation.

## 7-8 TYPICAL VCR OPERATING PROCEDURES

As discussed in Sec. 7-6, many VCR "troubles" are the result of improper operation. For that reason we have included typical operating instructions. Note that the VCR used here as an example (Sony SL-5800) has many advanced operating features not found on all VCRs.

### 7-8.1 TV Adjustment

Before using any TV with a VCR, it is necessary to adjust one channel of the TV to accept the signal from the VCR. The following describes some typical adjustment procedures.

Set the RF unit of the VCR channel selector to channel 3 or 4, whichever is not active in the area. If the TV has a rotary or 10-key tuning-type channel selector, set the VHF channel selector to channel 3 or 4, as required. If the TV has an electronic tuner, use the following procedure:

1. Be sure that the VCR, TV, and antennas are properly connected.
2. Turn on the TV.
3. Turn the power switch of the VCR on.
4. Set the CAMERA/TUNER selector to TUNER. (This connects the input of the VCR recording system to the VCR tuner rather than to the camera input.)
5. Set the program selector to VCR.
6. Select an active channel in the area with a channel select button on the VCR.
7. Select an unused TV channel with a channel select button on the TV.
8. Adjust the selected button so that the TV program selected on the VCR is clearly displayed on the TV screen and the sound is clearly heard.

If a recorded tape of known good quality is available, perform steps 1 through 5, and then tune the TV as follows:

6. Insert a cassette.
7. Press the PLAY button.
8. Select an unused channel button on the TV.

9. Adjust the selected channel button so that the cassette program is clearly displayed on the TV screen and the sound is clearly heard.

### 7-8.2 How to Insert and Remove a Cassette

The procedures for inserting and removing a cassette are described in Sec. 1-10.1 and illustrated in Fig. 1-51.

### 7-8.3 How to Record TV Programs

Figure 7-8 shows the necessary preparations and operating sequence to record TV programs. The following notes apply to typical operating conditions.

**Setting the RECORD MODE Selector.** The VCR used in this example has a RECORD MODE switch that selects the recording speed (and time) for a video cassette. The BIII format provides one-and-a half times the recording length of the BII format, or $4\frac{1}{2}$ hours of recording (instead of 3 hours in the BII mode) with a L-750 cassette. The position of the RECORD MODE selector does not affect the playback speed. Recorded tapes are played back at the correct speed regardless of the position (BII or BIII) of the selector.

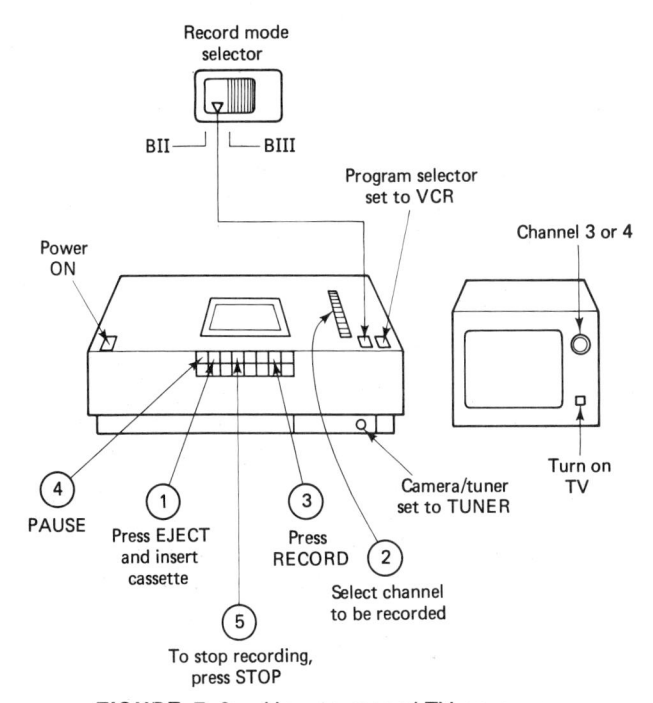

**FIGURE 7-8.** How to record TV programs.

**Stopping the Tape Momentarily during Recording.** Press the PAUSE button. The PAUSE/FREEZE lamp will light. The pause function is electrically activated so the PAUSE button does not lock in the depressed position. During the pause mode, you can observe the picture on the TV screen, but the picture will not be recorded. To resume recording, press the PAUSE button again.

After more than about 5 minutes in the pause mode, the pause function is automatically released and recording begins again. Recordings made with the use of the PAUSE button may show momentary instability in playback at the point where the PAUSE button was pressed. The channel being recorded can be changed in the pause mode.

**Keeping a Recorded Program from Being Accidentally Erased.** If you want to preserve programs on tape, break off the tab on the bottom of the cassette using a screwdriver or similar object, as discussed in Sec. 1-10.2 and shown in Fig. 1-52. With the tab removed, the RECORD button cannot be pressed, and this will prevent accidental recording over a program you want to keep. If you wish to record on a cassette with the tab removed, cover the hole with a piece of cellophane or vinyl tape.

**Full Auto Rewind.** During record, playback, or fast-forward mode, at the end of the tape, the tape will stop automatically; the RECORD, PLAY, or FF button will be released; and after several seconds, the tape will fully rewind. The INDEX position of the SEARCH switch does not function during this rewind, but the MEMORY COUNTER position of this switch functions during the rewind.

### 7-8.4 How to Play Back a Recorded Tape

Figure 7-9 shows the necessary preparations and operating sequence to play back a recorded tape. The following notes apply to typical operating conditions.

**Stopping the Tape Momentarily during Playback (Pause).** Press the PAUSE button. The PAUSE/FREEZE lamp will light. The pause function is electrically activated so that the PAUSE button is not locked in the depressed position. During the pause mode, a *freeze frame* picture will appear on the TV screen.

If a noise band appears in the freeze frame picture, obtain a slow motion picture and adjust the SLOW MOTION TRACKING control until the band disappears. To resume playback, press the PAUSE button again. After more than about 5 minutes in the pause mode of operation, the pause mode will be automatically released, and playback will begin again.

**Moving the Tape Quickly; Rewind, FF (Fast Forward).** To quickly move the tape forward (fast forward), press the FF button. To quickly move the tape

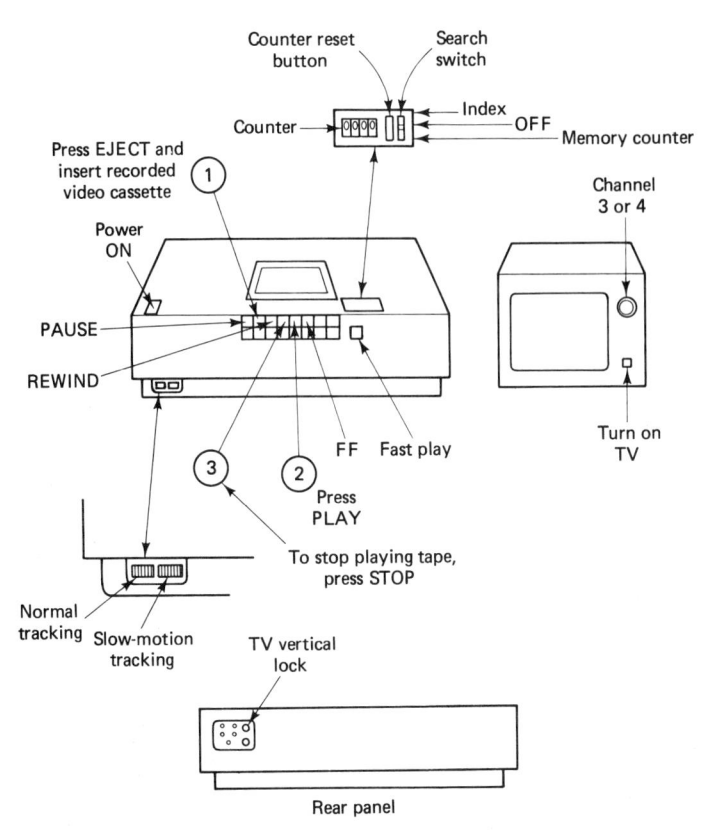

**FIGURE 7-9.** How to play back a recorded tape.

backward (rewind), press the REWIND button. To stop the tape, press the STOP button.

During the fast-forward or rewind mode, the picture from the connected video source, such as a TV station or camera, will appear on the TV screen. In the fast-forward mode, at the end of the tape, the tape will stop automatically, the FF button will be released, and after several seconds, the tape will rewind. In the rewind mode, when the tape is fully rewound, the VCR will stop automatically and the REWIND button will be released.

This particular VCR is designed with logic control. You can change directly from REWIND to FF to PLAY, from FF to REWIND, and from REWIND to FF without pressing STOP in between. Not all VCRs are capable of such operation.

**Getting a Triple-Speed Playback Picture; Fast Play.** Press the PLAY button. Press the FAST PLAY button. The playback speed will be three times that of normal playback. The FAST PLAY lamp will light.

If a noise band appears, turn the NORMAL TRACKING control until the best picture is obtained. To resume normal playback speed, press the FAST PLAY button again.

**Finding a Scene in a Long Video Cassette.** To get a high-speed forward and reverse picture (known as Betascan for this particular VCR), press the PLAY button. Press the FF or REWIND button and hold it down. You will see the forward or reverse picture with high-speed motion on the TV screen. Streaks may appear in the Betascan playback picture, but this is normal. At the point where you want to restart normal playback, release the FF or REWIND button.

To find the point where recorded programs begin, set the SEARCH switch to INDEX. Press the FF or REWIND button. The tape will stop at the nearest beginning of a program, and the depressed button will be automatically released. If the program is not the one you want, press the FF or REWIND button again. The tape will stop at the next beginning of a program.

This operation is possible because a cue signal is recorded on the tape each time the RECORD button is pressed. The VCR will search for the cue signal during the fast forward or rewind mode and stop when the signal is found. During timer recording, the cue signal is recorded at every starting point even though the RECORD button is not depressed at that time. If you want to rewind the whole tape without stopping, set the SEARCH switch to the OFF position and press the REWIND button.

To use the tape counter, make a note of the number corresponding to a particular point on the tape, and use the counter to help you relocate that place. To use the memory function of the counter, stop the tape at the point you want to return to later. Set the SEARCH switch to the MEMORY COUNTER position. Press the tape counter reset button (Fig. 7-9). The counter will indicate 0000. Play or record on the tape. Press the REWIND button when you want to return to that point. The tape will rewind and stop at a point near the tape counter reading 9999 (one count before 0000, to avoid missing the starting point). To rewind the tape farther than 0000, press the REWIND button again. If you want to rewind the whole tape without stopping, set the SEARCH switch to the OFF position and press the REWIND button.

**Eliminating Streaks, Snow, Noise Bands, and Vertical Jitter.** If streaks or snow appear in the playback of tapes or in the fast play picture, turn the NORMAL TRACKING control first in one direction, then in the other, until you get the best picture. Allow a second or two for the new settings to take effect. When playback of this particular tape or the fast play is finished, return the NORMAL TRACKING control to its center detent position.

If a noise band appears in the pause, frame-by-frame, or slow-motion picture, make sure that you have connected the Betascan control unit (a remote control unit for this VCR) and obtain a slow-motion picture. Watching the slow-motion picture, turn the slow-motion picture, turn the SLOW MOTION TRACKING control (Fig. 7-9) until you get the best possible picture. Return the VCR to the previous mode or continue in slow motion, as desired.

*Typical VCR Operating Procedures* 345

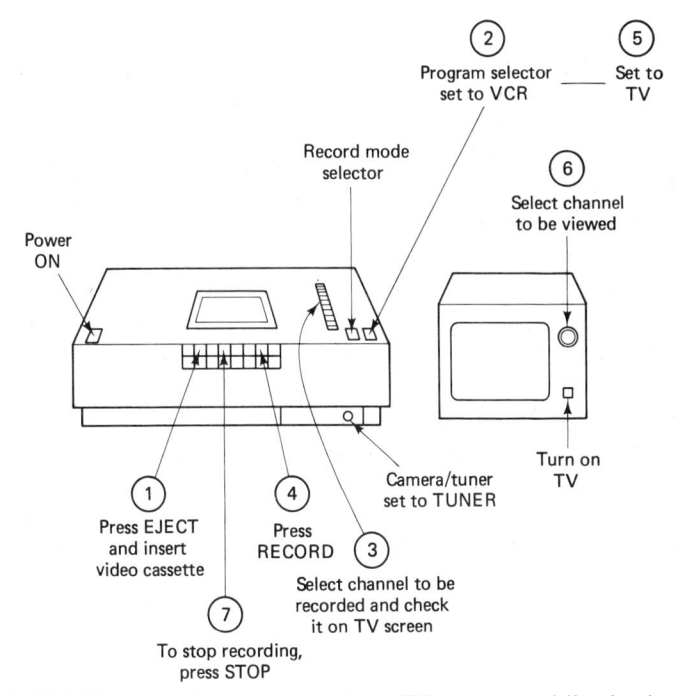

**FIGURE 7-10.** How to record one TV program while viewing another.

If a vertical jitter appears in the pause, frame-by-frame, or slow-motion picture, obtain a freeze picture. Watching the freeze picture, adjust the TV VERT LOCK control for best possible picture. Return the VCR to the previous mode. Once you adjust the TV VERT LOCK control, you do not have to adjust it again unless you change the connected TV receiver. (Note that not all VCRs have a TV vertical lock control.)

Should snow or streaks appear, especially when they appear in all modes of operation, this may indicate the need for cleaning of the video heads (as described in Chapter 6).

## 7-8.5 How to Record One TV Program while Viewing Another

Figure 7-10 shows the necessary preparations and operating sequence to record one TV program while viewing another.

## 7-8.6 How to Use the VCR for Camera Recording

Figure 7-5 shows the connections for camera recording. Figure 7-11 shows the necessary preparations and operating sequence when the VCR is used with a camera. When a compatible camera is used (Sony camera), the recording may be

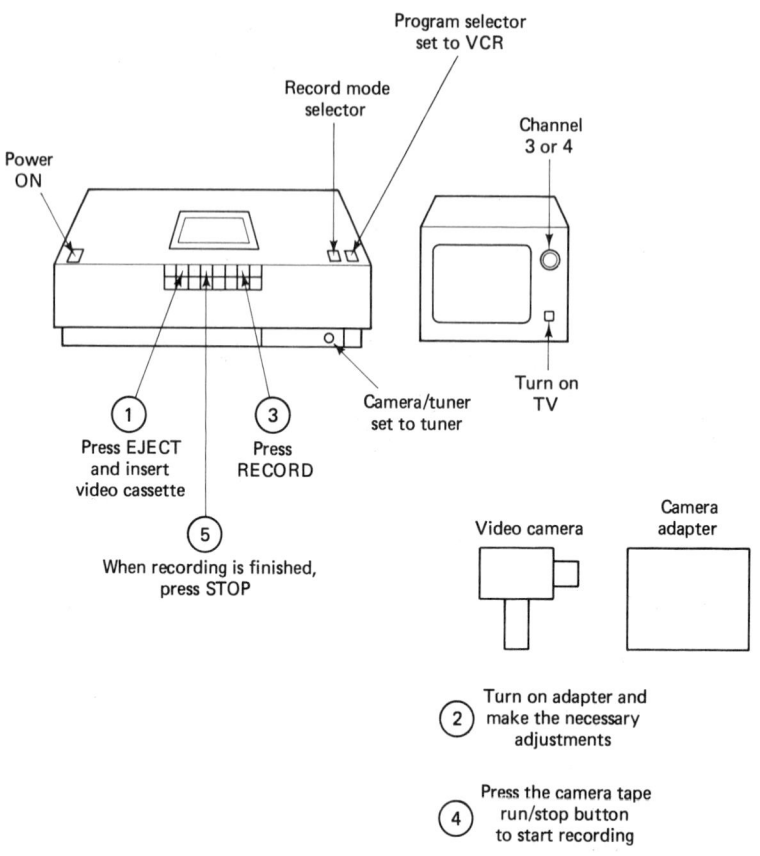

**FIGURE 7-11.** How to use the VCR for camera recording.

started and stopped using the remote control switch on the camera. When other cameras are used, it is recommended that the PAUSE button on the remote control be used. Always follow the operating instructions supplied with the camera.

### 7-8.7 How to Dub Audio onto a Previously Recorded Tape

Figure 7-12 shows the necessary preparations and operating sequence for audio dubbing. If both a microphone and another audio source are connected, the tape will record only the microphone input. TV volume should be turned down to prevent acoustic feedback (whistle-like sounds). When dubbing from the beginning of the tape, press the PLAY button first, then press the AUDIO DUB button.

### 7-8.8 Using the VCR Tuner to Watch TV

With most VCRs, you can view the TV using the VCR tuner. Set the VCR power switch to ON. Set the PROGRAM SELECT switch to VCR. Set the channel

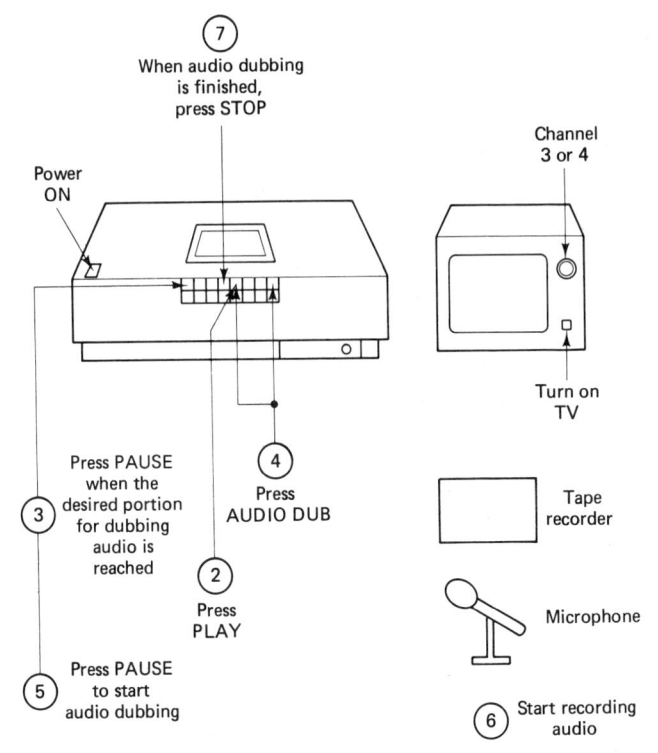

**FIGURE 7-12.** How to dub audio onto a previously recorded tape.

selector of the TV to channel 3 or 4. Select the desired channel using the VCR tuner. With this procedure, you are ready to record a TV program by pressing the RECORD button.

## 7-9 BASIC VCR TROUBLESHOOTING PROCEDURES

Before we get into the detailed service notes described in Sec. 7-10, where we discuss specific troubles related to the major functional sections of a VCR, let us review some simple, obvious steps to be performed before you start troubleshooting. These steps involve such things as checking for proper connections, adjusting the monitor TV, operating the controls in proper sequence, and so on.

### 7-9.1 Simple Diagnostic Procedure

If the video playback or the TV picture is bad, set the program select switch to TV and check picture quality for each TV channel (using the TV channel selector). If the picture quality is still bad, check for defective antenna connections. For example, the antenna may be defective, the VHF and UHF connections may be reversed, the F-type connector plug may be improperly connected, the center wire in the coaxial cable may be broken, or the TV 75 $\Omega$/300 $\Omega$ switch may be in the wrong position. Also check the TV fine tuning.

If the TV picture is good when the program select switch is set to TV but video playback is not good, set the program select switch to VCR, turn the TV to the inactive channel (3 or 4), and check reception on each channel by turning the VCR channel selector. If picture quality is bad or there is no picture on all channels, it is possible that the TV fine tuning is not properly adjusted. If the problem appears only on certain channels, the VCR fine tuning is suspect (as is the VCR tuner).

If picture quality is good when viewing a TV broadcast through the VCR (this is known as E-E operation, as discussed in Chapter 1), try recording and playing back the program.

If noise is apparent (resulting in poor picture quality on playback but not with E-E operation), it is possible that the video heads are dirty (head gaps are slightly clogged). If there is sound but no picture, the video head gaps may be badly clogged. If the playback picture is unstable with a new TV set (never previously used with the VCR), it is possible that the AFC circuits of the TV are not compatible with the VCR. This problem is discussed further in Sec. 7-10. If there is color beat (rainbow-like stripes on the screen) the problem may be interference rather than a failure in the VCR or TV.

### 7-9.2 Operational Checklist

The following checklist describes symptoms and possible causes for some basic VCR troubles.

*Record button cannot be pressed.* Check that there is a cassette installed and that the safety tab has not been removed from the cassette. If necessary, cover the safety tab hole with tape.

*No E-E picture; no picture and sound that you wish to record.* Check that the program select switch is in the correct position. Check the fine tuning on the TV inactive channel.

*No color, or very poor color.* If there is no color on playback, check the fine tuning on the TV inactive channel. Note that if the VCR fine tuning is maladjusted during record, color may appear while recording, but may not appear during playback. Always check fine tuning of both the VCR and TV as a first step when there are color problems.

*Playback picture is unstable.* If you have periodic problems of picture instability, before tearing into the VCR with a pickaxe, check the following points: Has the VCR been operated in an area having a different a-c line frequency? While recording, it is possible that a fringe-area signal was weak (intermittently) so that the sync signal was not properly recorded. During recording, there could have been some interference or large fluctuations in the power supply voltage. The cassette tape could be defective. The tracking control could be improperly adjusted.

*Snow noise appears on the picture during playback only.* Check the tracking control!

*Sound but no picture, and excessive black-and-white snow noise.* Check for very dirty video heads.

*Upper part of picture is twisted or entire picture is unstable.* The time constant of the AFC circuits in the TV is not compatible with the VCR. Refer to Sec. 7-10.

*Tape stops during rewind.* Is the memory counter switch on? If the memory switch is on, the tape stops automatically at 999 during rewind (on most VCRs).

*The rewind and fast-forward buttons cannot be locked or operated.* Is the cassette tape at either end of its travel? If the tape is at the beginning, rewind will not function. Fast forward will not function if the tape is at the end.

*Cassette will not eject.* Is the power on?

*Acoustic feedback (whistle-like sound) when recording with camera and microphone.* Keep the microphone away from the TV. Turn down the TV volume.

*Noise band in the playback picture, picture unstable, with too high or too low pitched sound.* In some VCRs, the tape is automatically locked to the correct speed by the servo. However, most VCRs also require manual switching. For example, the VCR discussed in Sec. 7-8 has a front-panel switch for Beta II and Beta III, as well as a rear-panel switch for Beta I.

## 7-10 VCR SERVICE NOTES

The following notes summarize practical suggestions for troubleshooting all types of VCRs. The notes are arranged by the major function sections of a VCR (video, audio, servo, etc.) in which the troubles are most likely to occur.

### 7-10.1 TV AFC Compatibility Problems

If the AFC circuits of the TV are not compatible with the VCR, *skewing* may result. Generally, the term "skew" or "skewing" in VCRs is applied when the upper part of the reproduced picture is bent or distorted by incorrect backtension on the tape. However, the same effect can be produced when the time constant of the TV AFC circuits cannot follow the VCR playback output. This condition is very rare in newer TV sets (which are designed with VCRs and video disks in mind), and appears in only about 1% of older TV sets (and almost never when the VCR and TV are produced by the same manufacturer). Therefore, do not change the TV AFC circuits unless you are absolutely certain there is a problem. Try the VCR with a different TV. Then try the TV with a different VCR.

Once you are certain that there is a problem of compatibility, you can reduce the time constant of the *integrating circuit* associated with the TV AFC. Figure 7-13 shows the major components of a typical TV AFC integrating circuit. To reduce the time constant, reduce the values of either or both capacitors C1 and C2, reduce the value of R1, and increase the value of R2. It is generally not necessary to change all four values. Be sure to check the stability of the TV horizontal sync after changing any AFC value, since the AFC circuits are usually part of the horizontal sync system in modern TV receivers.

### 7-10.2 Tuner and RF Unit Problems

*Initial setup.* When a VCR is first connected to a TV, it is likely that the unused channel (3 or 4) of the TV is not properly fine tuned. When fine tuning the TV, operate the VCR in the *playback mode* using a known good cassette, preferably with a color program. If you try to fine tune the TV in the record or E-E mode, both the VCR and TV tuners are connected in the circuit, and the picture is affected by either or both tuners. With playback, the picture is dependent only on the TV tuner. Once the normally unused channel of the TV is fine tuned for best picture, the VCR tuner can be fine tuned as necessary.

**Replacing Tuners and RF Units.** In virtually all VCRs, the RF unit (also called the modulator by some manufacturers) must be replaced in the event of failure. No adjustment is possible, and internal parts cannot be replaced on an individual basis. This is because the RF unit is essentially a miniature TV transmitter and must be *type accepted* using very specialized test equipment, as are other transmitters. You must replace the RF unit as a package if you suspect failure. For example, if you have found proper audio and video inputs (and power source) to the RF unit, but there is no output (or low output) at the unused channel, the problem is likely in the RF unit. As a point of reference, a typical RF unit produces 1000 $\mu$V into a 75-$\Omega$ load (or 2000 $\mu$V into a 300-$\Omega$ load) on the selected channel. Signals outside the channel frequency by more than 3.5 MHz are reduced at least 35 dB. Also, there is a 60-dB drop introduced by the change-over switching network between the RF unit and the antenna.

In many VCRs, the tuner is also replaced as a unit in the event of failure.

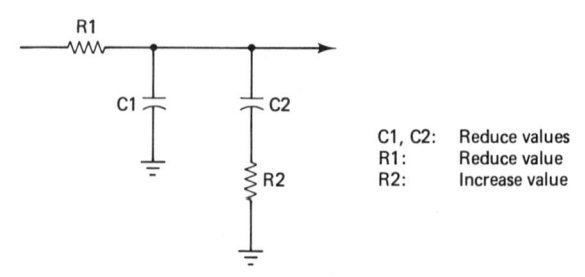

**FIGURE 7-13.** Altering the integrating circuits of a TV AFC.

However, some manufacturers supply replacement parts for their tuners. Also, most manufacturers provide for tuner adjustment as part of service, as discussed in Chapter 6. As a point of reference, a typical VCR tuner (including the IF) produces 1 V peak to peak of video into a 75-$\Omega$ load. Typically, audio output from the tuner is in the -10- to -20-dB range.

### 7-10.3  Copy Problems

It is possible to copy a video cassette using two VCRs. One VCR plays the cassette to be copied, while the other VCR makes the copy. Keep two points in mind when making such copies. First, if the cassette being copied contains any copyrighted material, you may be doing something illegal! Second, a copy is never as good as the original, and copies of copies are usually terrible. Even with professional recording and copying equipment, the quality of a copy (particularly the color) deteriorates with each copying. The quality of a first copy (called the second generation) can be acceptable provided that the original is of very good quality. However, a second copy (third generation) is probably of unacceptable quality. Forget fourth generation (or beyond) copies. So if you are called in to service a VCR that "will not make good copies of other cassettes," explain that the problem probably has no cure.

### 7-10.4  Video Camera Sync and Interlace Problems

If you are to service a VCR that operates properly in all modes, except when used with a known good video camera, the problem may be one of incompatibility. The cameras recommended for use with a VCR (generally of the same manufacturer) should certainly produce good cassette recordings. In general, most cameras designed for use with VCRs, even though of a different manufacturer, are compatible with any VCR. Such cameras are usually designated as having a *2:1 interlace*. Essentially, 2:1 interlace means that both the vertical and horizontal sync circuits of the camera are locked to the same frequency source (possibly the power line) by a definite ratio. When operating with a camera, the sync signal normally supplied by the TV broadcast is obtained from the camera and recorded on the control track of the VCR tape.

Some inexpensive cameras, particularly those used in surveillance work, have a *random interlace* where the horizontal and vertical sync are not locked together. The playback of a recording made with a random interlace camera often has a strong beat pattern (herringbone effect). One way to confirm a random interlace condition is to watch the playback while observing the last horizontal line above the vertical blanking bar. Operate the TV vertical hold control as necessary to roll the picture so that the blanking bar is visible. If the end of the last horizontal line is stationary, the camera has 2:1 interlace and should be compatible. If the end of the last horizontal line is moving on a camera playback, the camera is not providing the necessary sync and probably has random interlace.

### 7-10.5 Wow and Flutter Problems

VCRs are subject to wow and flutter, as are most audio recorders. Wow and flutter are tape transport speed fluctuations that may cause a regularly occurring instability in the picture and a quivering or wavering effect in the sound during record and playback. The longer fluctuations (below about 3 Hz) are called "wow"; shorter fluctuations (typically 3 to 20 Hz) are called "flutter." Wow and flutter can be caused by mechanical problems in the tape transport or by the servo system. Wow and flutter are almost always present in all VCRs, but it is only when they go beyond a certain tolerance that they are objectionable.

If you are to service a VCR where the complaint appears to be excessive wow and flutter, first check the actual amount. This is done using the low-frequency tone recorded on the alignment tape, and a frequency counter connected to the audio line at some convenient point. Typically, the low-frequency tone is in the order of 333 Hz, and an acceptable tolerance would be 0.03%. If necessary, operate the frequency counter in the period mode to increase resolution, as described in Chapter 2.

There are special test instruments used to measure wow and flutter. However, these are not found in most shops (unless they specialize in audio and high-fidelity equipment). Fortunately, any wow and flutter that does not show up using the alignment tape and frequency counter will probably not be objectionable. Keep in mind that wow and flutter can be caused by electrical problems (in the servo) and mechanical problems (tape transport) or a combination of both.

### 7-10.6 System Control Problems

As in the case of describing the theory of operation for system control, it is difficult to generalize about system control service problems. In most modern VCRs, system control is performed by a microprocessor. In effect, you press buttons to initiate the operating mode, and the microprocessor produces the necessary control signals (to operate relays, motors, etc.). Each VCR has its own system control functions. You must learn these functions.

However, system control for all VCRs has certain *automatic stop* functions, such as end-of-tape stop, condensation detector (dew sensor), and so on, which must be accounted for in service. For example, when checking any system control, make certain that all the automatic stop functions are capable of working. Next, make certain that some automatic function has not worked at the wrong time (end-of-tape stop occurs in the middle of the tape). Equally important, make certain that a normal automatic stop is not the cause of your imaginary problem (do not expect the tape to keep moving when the end-of-tape stop has occurred). Also, it is very often necessary to disable the automatic stop function during service. The following notes describe some generalized procedures for checking and testing system control.

*Slack tape sensors* can be checked by visual inspection and by pressing on the tape with your finger to simulate slack tape. If the slack tape sensor includes a microswitch (such as with the Beta), the sensor circuit can be disabled by forcing a match or cardboard against the sensor to keep the microswitch from triggering.

*End-of-tape sensors* can be checked by simulating the end-of-tape condition. For VHS, this involves exposing the phototransistor to light (to simulate the clear plastic tape leader) to trigger automatic stop. For Beta, the end-of-tape foil can be simulated by placing a piece of foil near the surface of the sensing coil. To disable the VHS end-of-tape sensor, it is necessary to cover the phototransistor with opaque tape or a cap. Do not remove the light source for the end-of-tape sensor on a VHS machine! In most cases, removal of the sensor lamp is sensed by a failure circuit which also triggers auto stop. Even with the phototransistor covered, stray light may trigger the auto stop condition.

If the VCR has a switch that is actuated when a cassette is in place, such as the *cassette-in switch* of a Beta VCR, locate the mechanism that actuates the switch and hold the mechanism in place with tape. In many cases, it is possible to operate the VCR through all its modes without a cassette installed if the cassette-in switch can be actuated manually.

*Take-up reel detectors* (VHS) can be checked by holding the take-up reel. This causes the take-up reel clutch to slip (to prevent damage) but the detector will sense that the reel is not turning and produce automatic stop.

Always check that all automatic stop functions work, and that all bypasses and simulations are removed, after any service work.

## 7-10.7 Interchange Problems

When a VCR can play back its own recordings with good quality, but playback of tapes recorded on other machines is poor, the VCR is said to have *interchange* problems, or is unable to interchange tapes. Such problems are almost always located in the mechanical section of the VCR, usually in the tape path. Quite often, interchange problems are the result of improper adjustment. Manufacturers sometimes include interchange adjustments as part of their overall electrical/mechanical adjustments.

The simplest way to make interchange adjustments is to monitor the RF output from the video heads during playback and adjust elements of the tape path to produce a maximum, uniform RF output from a factory alignment tape. Generally, the output is measured at a point after head switching so that both heads are monitored. Always follow the manufacturer's adjustment procedures. Section 6-2.8 describes some typical tape path adjustments involving measurement of the RF output.

## 7-10.8 Servo System Problems

Total failures in the servo system are usually easy to find. If a servo motor fails to operate, check that the power is applied to the motor at the appropriate time. If power is there, but the motor does not operate, the motor is at fault (burned out or open windings, etc.). If the power is absent, track the power line back to the source. (Is the microprocessor delivering the necessary control signal or power to the relay or IC?)

The problem is not so easy to locate when the servo fails to lock on either (or both) record and playback, or locks up at the wrong time (causing the heads to mistrack even slightly). Obviously, if the control signal is not recorded (or is improperly recorded) on the control track during record, the servo cannot lock properly during playback. Therefore, your first step is to see if the servo can play back a properly recorded tape. There are several ways to do this.

There are usually some obvious symptoms when the servo is not locking properly. There will be a horizontal band of noise that moves vertically through the picture if the servo is out-of-sync during playback. The picture may appear normal at times, possibly leading you to think that you have an intermittent condition. However, with a true out-of-sync condition, the noise band *appears regularly,* sometimes covering the entire picture.

Keep in mind that the out-of-sync condition during playback can be the result of servo failure, or the fact that the sync signals (control signals) are not properly recorded on the control track of the tape. To find out if the servo is capable of locking properly, play back a known good tape. If the playback is out-of-sync, you definitely have a servo problem.

The symptoms for failure of the servo to lock during record are about the same as during playback, with one major difference. During record, the head switching point (which appears as a break in the horizontal noise band) appears to move vertically through the picture in a random fashion.

Another way to check if the servo is locking on either record or playback involves looking at some point on the rotating scanner or video head assembly under fluorescent light. When the servo is locked, the fluorescent light produces a blurred pattern on the rotating scanner that appears almost stationary. When the servo is unlocked, the pattern appears to spin. Try observing the scanner of a known good VCR under fluorescent light. Stop and start the VCR in the record mode. Note that the blurred pattern appears to spin when the scanner first starts, but settles down to almost stationary when the servo locks. Repeat this several times until you become familiar with the appearance of a locked and unlocked servo under fluorescent light.

Once you have studied the symptoms and checked the servo playback with a known good tape, you can use the results to localize the trouble in the servo. For example, if the servo remains locked during playback of a known good tape, it is reasonable to assume that the circuits between the control head and servo motors are good. Typically, these circuits include the playback amplifiers, tracking delay network, sample-and-hold circuits, power amplifier, servo motors and brakes, 30PG feedback pulses, and the ramp generators. However, in the more sophisticated VCRs that have both speed and phase control circuits, it is sometimes difficult to localize trouble.

Keep in mind that servo troubles may be mechanical or electrical, and may be the result of either improper adjustment or component failure (or both). As a general guideline, if you suspect a servo problem, start by making the electrical adjustments that apply to the servo. Always follow the manufacturer's adjustment procedures, using the procedures of Chapter 6 as a guide. This may cure the servo problem. If not, following the adjustment procedures will at least tell you if all of the servo control signals are available at the appropriate points in the circuits. (Such control signals include all of the cylinder and capstan control pulses shown in Fig. 1-48.) If one or more of the signals are found to be missing or abnormal during adjustment, you have an excellent starting point for troubleshooting.

There are two points sometimes overlooked when troubleshooting a servo that fails to lock up. First, the free-running speed of the servo may be so far from normal that the servo simply cannot lock up. This problem will usually show up during adjustment. Second, on those VCRs that use rubber belts to drive servo motors, the rubber may have stretched (or be otherwise damaged). If you have replacement belts available, compare the used VCR belts for size and confirmation. Hold a new and used belt on your finger under no strain. If the used belt is larger, or does not conform to the new belt, install the new belt and recheck the servo for proper lockup.

### 7-10.9 Luminance (Black-and-White) Picture Problems

Although the luminance circuits of any VCR are very complex, they are not the major cause of trouble. Mechanical problems are on top of the list, closely followed by servo and system control troubles. Also, although many circuits are involved, all of the circuits are usually found in three or four ICs (in most present-day VCRs). If all else fails, you can replace the few ICs, one at a time, until the problem is solved. (If only mechanical problems were that simple!)

As in the case of other problems, the first step in servicing luminance circuits is to play back a known good tape, or an alignment tape. This will identify the problem as playback or record, or both. Next, run through the electrical adjustments that apply to luminance or picture, using the manufacturer's procedures. Keep the following points in mind when checking performance and making adjustments.

If playback from a known good tape has poor resolution (picture lacks sharpness), look for an improperly adjusted noise canceler, and for bad response in the video head preamps. When making adjustments, study the stairstep or color-bar signals for any transients at the leading edges of the white bars.

If the playback has excessive snow (electrical noise), try adjustment of the tracking control. Mistracking can cause snow noise. Next, try cleaning the video heads, before making any extensive adjustments. (Cleaning the video heads clears up about 50% of all noise or snow problems.) Keep in mind that snow noise can result from mistracking that is caused by a mechanical problem. For example, if there is any misadjustment in the tape path, snow can result. So if you have an excessive noise problem that cannot be corrected by tracking adjustment, head cleaning, or electrical adjustment, try mechanical adjustments, starting with the tape path. Make mechanical adjustments only as a last resort. You may end up with more than noise!

If playback of a known good tape produces smudges on the leading edge of the white parts of a test pattern (from an alignment tape) or a picture, the problem is probably in the preamps, or in adjustments that match the heads to the preamps. The head/preamp combination is not reproducing the high end (5 MHz) of the video signals. The adjustment procedures will usually show the head/preamp response characteristics.

If you observe a herringbone (beat) pattern in the playback of a known good tape, look for carrier leak. There is probably some unbalance condition in the FM demodulators or limiters, allowing the original carrier to pass through the demodulation process. If very excessive carrier passes through the demodulator, you may get a negative picture. Recheck all carrier leak adjustments.

Most adjustment procedures include a check of the video output level (typically 1 V peak to peak). If the VCR produces the correct output level when playing back an alignment tape but not from a tape recorded on the VCR, you probably have a problem in the record circuits. As an example, the record current may be low. (One symptom of low record current is snow or excessive noise.) Another common problem in the record circuits is white clip adjustment. Most adjustment procedures include a white clip adjustment and a record current check.

To sum up luminance troubleshooting, if you play back an alignment tape, or at least a known good tape, and follow this with head cleaning and a check of the recommended alignment procedures, you will probably have no difficulty in locating most black-and-white picture problems.

### 7-10.10 Color Problems

As in the case of luminance circuits, the color circuits of a VCR are very complex, but not necessarily difficult to troubleshoot (nor do they fail as frequently as the mechanical section). Again, the first step in color troubleshooting is to play back an alignment tape, followed by a check of all adjustments pertaining to color. As in the luminance circuits, when performing the adjustment procedures, you will be tracing the signal through the color circuits.

There are two points to remember when making the checks. First, most color circuits are contained within ICs, possibly the same ICs as the luminance circuits. Similarly, the color and luminance circuits are interrelated. If you find correct inputs and power to an IC, but an absent or abnormal output, you must replace the IC. A possible exception in the color circuits are the various filters and traps located outside the IC.

Second, in most VCRs, the CW input to the color converters comes from the same source for both record and playback (from crystal-controlled oscillators). If you get good color on playback but not on record, the problem is definitely in the record circuits. However, if you get no color on playback of a known good tape, the problem could be in the color playback circuits or in the common CW signal source. Therefore, a good place to start color circuit signal tracing is to check any common source CW signals. Then check any AFC circuit signals (629 or 688 kHz) and any APC signals (3.57 or 3.58 MHz). If any of these signals are missing (or abnormal), the color will be absent or abnormal.

The following notes describe some typical VCR color circuit failure symptoms, together with some possible causes.

If the hue control of the TV must be reset when playing back a tape that has just been recorded, check the color subcarrier frequency. Use a frequency counter.

If you get a "barber pole effect," indicating a loss of color lock, the AFC circuits are probably at fault. Check that the AFC circuit is receiving the H-sync pulses and that the VCO is nearly on-frequency, even without the correction circuit. For example, Sec. 6–5.2 describes a procedure for checking the VCO, with the correction input voltage shorted.

If you get bands of color several lines wide on saturated colors (such as alternate blue and magenta bands on the magenta bar of a color-bar signal), check the APC circuits, as well as the 3.58-MHz oscillator frequency.

If you get the herringbone (beat) pattern during a color playback, try turning the color control of the TV down to produce a black-and-white picture. If the herringbone pattern is removed on black and white, but reappears when the color control is turned back up, look for leakage in both the color and luminance circuits. For example, there could be a carrier leak from the FM luminance section beating with the color signals, or there could be leakage of the 4.27-MHz signal into the output video.

If you get flickering of the color during playback, look for failure of the ACC system. It is also possible that one video head is bad (or that the preamps are not balanced), but such conditions will show up as a problem in black-and-white operation.

If you have what appears to be very severe color flicker on a Beta VCR, you may be losing color on every other field. This can occur if the phase of the 4.27-MHz reference signal is not shifted 180° at the H-sync rate when one head is making its pass. The opposite head works normally, making the picture appear at a 30-Hz rate.

If you lose color after a noticeable dropout, look for problems in the burst

ID circuit. It is possible that the phase reversal circuits have locked up on the wrong mode after a dropout. In that case, the color signals have the wrong phase relation from line to line, and the comb filter is canceling all color signals. Check both inputs (3.58-MHz input from the reference oscillator, and the video input signal) applied to the burst ID circuit. If the two signals are present, check that the burst ID pulses are applied to the switchover FF.

Again, keep in mind that all the color circuit functions discussed here may be contained within one or two ICs, and cannot be checked individually. Therefore, you must check inputs, outputs, and power sources to the IC, and then end up replacing the IC. Good luck! You will need it!

# Index

## A

ACC (automatic color control), 189, 190
Adjustments, 245
AFC (automatic frequency control), 62
   loop, 35
   problems, 349
AFT (automatic fine tune) adjustments, 248
AGC (automatic gain control), 103
Alignment tapes, 98
Analyst generator, 84
APC (automatic phase control), 60
   loop, 36
Audio:
   adjustments, 250, 263, 301-3
   Beta, 156
   dubbing, 346
   dub mode, 233
   recording, 40
Auto-loading (tape), 38
Automatic rewind, 241
Azimuth recording, 32, 52

## B

Backporch, 14
Baseband signals, 96
Beta, 31
   circuits, 100
Black and white:
   troubleshooting, 355
   TV, 6
Black clip adjustments, 260
Booster, signal, 338
Burst:
   amplification, 114
   deemphasis, 118

## C

Cable TV, 335
Camera, 335
  interface, 351
  pause, 154
  recording, 345
Capstan phase control, 196
Capstan servo, 29, 129
Carrier phase inverter, 36
Cassette:
  inserting and removing, 341
  video, 1, 70
CATV (cable television), 335
Chroma, 13, 17, 111
Chrominance, 30, 40
Cleaners, 276
Cleaning, 245, 277, 286
Color:
  adjustments, 293–301
  burst, 14
  generators, 87
  troubleshooting, 356
  TV broadcast, 13
  under, 30
  VHS, 188
Comb filter, 35
  adjustments, 256
Control system:
  Beta, 144
  troubleshooting, 352
  VHS, 206
Copying tapes, 334
Copy problems, 351
Copyright problems, 78
Counter:
  cue/memory, 214
  frequency, 93
  tape, 241
Crosstalk, 32
Cue/memory counter, 214
Cylinder:
  phase control, 192
  VHS, 24

## D

Dark clip, 175
  adjustments, 297
Delay, 1H, 191
Delay line, 1H, 35
Detector adjustments, 248
Dew sensor, 147, 212
  adjustments, 250
Direct injection, 82
DOC (dropout control), 44, 103, 180
  adjustments, 262
Down-converted color, 30
Drum (Beta), 24
  servo, 29, 120
Dubbing, 346

## E

Edge-noise canceler, 182
E-E, 42
  adjust, 259, 264
Eject mode, 242
Electrical adjustments:
  Beta, 245
  VHS, 287
End sensor, tape, 212
Erase, 159
Error signal, servo, 28

## F

Fast forward mode, 234
Fast search mode, 237
Field (TV screen), 26
Fixtures, test and troubleshooting, 96
Flutter problems, 352
FM:
  demodulator, 182
  detector adjustments, 248

Index 363

deviation adjustments, 260
  recording, 23
Forward sensor, 148
Frame (TV screen), 26
Frequency counters, 93
F-search, 153

**G**

Guardband, 31, 52

**H**

Heads:
  recording, 40
  resonance adjustments, 261
  switching adjustments, 252, 289
  video, 24
Helical scan, 24
High density recording, 31

**I**

IF adjustments, 304-7
Industrial video receiver/monitor, 95
Installation, VCR, 332-40
Interchange problems, 353
Interface, camera, 351
Interlaced scanning, 10
Introduction, VCR 1

**J**

Jitter, 344

**K**

Keyed rainbow generator, 87

**L**

Lapping cassette, 99
Leakage currents, 76
Linear staircase, 89
Line period (1H), 34
Loading:
  motor, 216
  tape, 68, 225
Lockout, record, 72
Lubrication, 245, 277-86
  oils, 276
Luminance, 13, 17, 30, 40, 53, 55, 102, 172
  adjustments, 259, 293-301
  troubleshooting, 355

**M**

Magnetic recording, 18
Maintenance, 245
  timetable, 276
Malerase, 70
Mechanical adjustments:
  Beta, 266-76
  VHS, 307-15
Mechanical operation, 222
Memory, counter, 214, 241
Meters, 92
Metric standards, 98
Microprocessor, system control, 144
Moisture condensation, 77, 147, 212
Monitor, video, 97
Motor drive, 203
Multiple antennas, 332
Muting, 160

**N**

Negative picture, 55
Noise bands (in picture display), 344

NTSC color generators, 88
NTSC color signals, 15

## O

Operation, VCR, 340
  checklist, 348
Oscilloscopes, 90

## P

Pattern generator, 84
Pause mode, 232
Period, line (1H), 34
Phase-locked-loops, 35
PI color recording, 33, 52
Playback, 43, 55, 116, 178, 342
  amplifier, 103
  color, 59, 189
Play mode, 229
Power supply adjustments, 246
Precautions, 338
Probes, 94

## R

Rainbow generator, 87
Record, 53, 172, 341
  amplifier, 103
  Beta, 39
  chroma, 113
  color, 57, 188
  lockout, 70
  mode, 231
Recording:
  azimuth, 32, 52
  current adjustments, 263
  FM, 23
  high density, 31
  magnetic, 18

PI color, 33
  video, 22
Reel base sensor, 148
  lock sensor, 212
  motor adjustments, 252
  motor drive, 152
Reverse phenomenon, 55
Reverse search, 240
Rewind mode, 238
Rewind sensor, 148
RF:
  AGC adjustments, 249
  switching pulse, 50
  troubleshooting, 350
Rotary transformer, 49
Rotational drawings, 243
R-search, 153

## S

Safety precautions, 74
Scanner, 25
Scanning, interlaced, 10
Search, 344
Sensors, 146, 212
  troubleshooting, 353
Service, 316-32
  notes, 349
Servo, 45, 65
  adjustments, 252, 287-93
  I (Beta), 120
  II (Beta), 136
  systems, 27
  VHS, 192
Signal booster, 338
Signal generators, 78
Snow (in picture display), 344
Speed control, cylinder, 200
Staircase pattern, 89
Stereo, 334
Still mode, 231
Streaks (in picture display), 344

Sweep/marker generator, 78
Switching pulse, 50
System control:
  Beta, 144
  troubleshooting, 352
  VHS, 206

## T

Tape, 18
  counter, 241
  end sensors, 146, 212
  loading, 38
  loading (VHS), 68
  protection, 219
  run system, 223
  sensor adjustments, 250
Tapes:
  alignment, 98
  copying, 334
Test equipment, 73
Test patterns, 86
Timer (Beta), 160
Timing:
  charts, 243
  phase, 150
  phase adjustments, 251
Tools, service, 73
Tracking adjustments, 254
Troubleshooting, VCR, 316–32, 347
Tuner:
  adjustments, 246, 304–7
  Beta, 165
  troubleshooting, 350
TV:
  adjustments, 340
  AFC problems, 349
  broadcast system, 6
TVH and TVV, 91

## V

Vertical jitter, 344
VHS, 52
  circuits, 171
Video:
  adjustments, 249, 256
  camera, 335
  cassette, 70
  cassette, inserting and removing, 341
  circuits, 101
  IF adjustments, 247
  heads, 24
  monitor, 95
  recording, 22
VTR (video tape recorder), 2
V-V, 42

## W

White clip, 175
  adjustments, 260, 297
Wow problems, 352

## Y

Y-FM demodulator, 109
Y-FM modulator, 103

## Z

Zero guard bands, 31, 52